Comparative Physiology of the Vertebrate Kidney

Published on behalf of The American Physiological Society by Springer

This book is published on behalf of the American Physiological Society (APS) by Springer. Access to APS books published with Springer is free to APS members.

APS publishes three book series in partnership with Springer: *Physiology in Health and Disease (formerly Clinical Physiology), Methods in Physiology, and Perspectives in Physiology* (formerly *People and Ideas*), as well as general titles.

William H. Dantzler

Comparative Physiology of the Vertebrate Kidney

Second Edition

William H. Dantzler
Department of Physiology
University of Arizona, College of Medicine
Tucson, Arizona
USA

The first edition was published as volume of the book series "Zoophysiology".

ISBN 978-1-4939-8123-6   ISBN 978-1-4939-3734-9   (eBook)
DOI 10.1007/978-1-4939-3734-9

1st edition: © Springer-Verlag Berlin Heidelberg 1989
2nd edition: © The American Physiological Society 2016
Softcover reprint of the hardcover 2nd edition 2016

Printed on acid-free paper

This Springer imprint is published by Springer Nature
The registered company is Springer Science+Business Media LLC New York

Preface

Like the first edition, this new edition emphasizes the comparative approach to understanding vertebrate renal function. I remain convinced that this approach is of particular value in understanding both the details of renal function at the cellular and molecular levels and the renal role in regulating vertebrate fluid volumes and solute concentrations. My exposure to this approach first occurred during a student research experience in the laboratory of the late Wilbur H. Sawyer, who also provided an introduction to the works of Homer W. Smith and August Krogh. The importance of this approach was reinforced by doctoral and postdoctoral research in the laboratory of the late Bodil Schmidt-Nielsen. It has been confirmed through years of personal experience since then.

My research and my understanding of renal function have been aided through the years by collaboration and discussion with numerous students and postdoctoral associates. Of particular importance in developing my views on comparative renal function, on the relationship of renal structure to function, and on transport processes have been my long-term associations with my colleagues and friends, Eldon J. Braun, Thomas L. Pannabecker, and Stephen H. Wright at the University of Arizona; Stefan Silbernagl at the University of Würzburg, Würzburg, Germany; and Varanuj Chatsudthipong at Mahidol University, Bangkok, Thailand. Much of my personal research over the years has been supported by grants from the United States National Science Foundation and National Institutes of Health.

I thank my fellow scientists for sometimes allowing me to note their unpublished observations. I also thank them and their publishers for permitting me to use many published figures. Finally, and most importantly, I thank my wife, Barbara, not only for enduring my absorption with my work but also for providing the constant encouragement and support that has so greatly aided my progress.

Tucson, AZ   William H. Dantzler
Spring, 2016

Contents

Contents

Chapter 1
Introduction

Abstract This chapter points out the general importance of the kidney in regulating the internal environment of all vertebrates. It then discusses the importance of the comparative approach in understanding renal physiology. Next, it briefly indicates the environmental requirements and the possible roles of the kidney in regulating the water and solute composition of the internal environment. Finally, it suggests some possible useful reviews on comparative renal physiology.

Keywords Kidney importance • Comparative approach • Internal environment • Water • Solutes • Regulation • Published reviews

Kidneys play a role—usually a major role—in regulating water and solute composition of the internal environment of all vertebrates. In this book, I examine the physiological functions of the kidneys from a comparative viewpoint with emphasis on nonmammalian vertebrates. I consider mammalian renal function for comparison with nonmammalian renal function and as a frame of reference for some discussions. However, I make no attempt to consider the vast literature on mammalian renal function in detail. In fact, I do not give the studies on nonmammalian vertebrates in complete detail. Instead, I summarize primarily the major findings and the important unanswered questions raised by those findings, for I do not intend this volume as an all-inclusive reference. I intend it as a reasonably comprehensive and integrated picture of comparative renal function for the biological scientist or advanced student of biology who has some knowledge of physiology and a desire to know more about renal function in vertebrates and for the mammalian renal physiologist who wishes to obtain a broader view of renal function. I hope that some of these readers will be sufficiently intrigued by the many unsolved problems of renal function in nonmammalian vertebrates that they will try to solve them with modern techniques and new experimental designs.

1.1 The Comparative Approach

The comparative approach to understanding renal function has a number of advantages. First, a consideration of those physiological processes that make it possible for different species to survive under diverse environmental conditions not only

W.H. Dantzler, *Comparative Physiology of the Vertebrate Kidney*,
DOI 10.1007/978-1-4939-3734-9_1

1

increases our understanding of environmental adaptations but also may suggest those scientific studies most likely to reveal important general physiological principles. Second, the comparison of renal function in a variety of species often permits the exploration of a basic physiological mechanism in an exaggerated form in a single species (August Krogh principle) (Krogh 1929) or through the use of a simple preparation.

The comparative approach to physiological studies has often been suggested as a means of shedding light on the evolution of physiological processes. However, the information available on renal function in nonmammalian vertebrates is so fragmented and so frequently limited to a single species or a few species and so seldom involves genetic details that attempts to reach major conclusions about evolution of mechanisms result only in unwarranted speculation. Therefore, although I make comparisons of renal mechanisms and adaptations among the major vertebrate groups throughout the book, particularly in Chap. 8, and I occasionally note the apparent independent evolution of comparable mechanisms in separate vertebrate groups (e.g., the mechanism for concentrating the urine in birds and mammals), I avoid general conclusions about the phylogenetic evolution of renal mechanisms.

1.2   Environmental Requirements in Regulating the Water and Solute Composition of the Internal Environment

Vertebrates, both mammalian and nonmammalian, survive and, indeed, thrive under a very wide range of environmental conditions. Some live in arid environments in which conservation of water is of great importance. Others live in completely aqueous freshwater environments in which excretion of excess water and retention of important inorganic solutes are essential. Still others live in completely aqueous marine environments. Here, the high osmolality of the surrounding seawater either requires osmotic conformity with the medium or the retention of water and the excretion of excess solutes, including not only sodium and chloride but also large quantities of ingested divalent ions, such as calcium, magnesium, phosphate, and sulfate. Even some of those species that conform to a marine environment, however, must adjust the solute composition of their internal environment. Many species, particularly terrestrial ones, tolerate variations in their solute and water requirements; other species, particularly aquatic ones, even move from one environmental extreme to another, for example, from seawater to freshwater.

1.3   Possible Roles of Kidneys in Regulating Water and Solute Composition of the Internal Environment

As noted above, kidneys generally play a role in regulating the solute and water composition of the internal environment. Some information on this role is available for species from the major vertebrate classes: cyclostomes (Agnatha), both the myxini (hagfishes) and petromyzones (lampreys); elasmobranchs (Chondricthyes); bony fishes (Osteichthyes), primarily teleosts, but also lungfish; amphibians (Amphibia), both urodeles and anurans; reptiles (Reptilia), primarily Testudinea, Squamata, and Crocodilia, but also Rhynchocephalia; Birds (Aves); and mammals (Mammalia). However, only among the mammals is the kidney essentially of sole importance in regulating the solute and water composition of the internal environment. Among all other vertebrate groups, extrarenal routes for the regulation of solute and water movements, or postrenal modification of ureteral urine, or both are also important. For example, among the cyclostomes, the role of the kidney in the regulation of solute and water balance appears to be substantially greater in the lampreys than in the hagfish, but in both groups regulation of solute and water movement across the gills and, possibly, the integument or within the gastrointestinal tract may be significant. Renal function is clearly important in solute and water balance in marine elasmobranchs, but a specialized extrarenal route (rectal gland) for sodium chloride excretion also exists and ion and water movements across the gills may be regulated. Among the euryhaline teleosts, the kidneys are particularly important in adaptation of the animals to freshwater or seawater, but regulation of renal function is clearly coordinated with regulation of ion and water movement across the gills. Although renal function is very important in regulating solute and water excretion in amphibians, it is also coordinated with postrenal ion and water movements across the bladder or cloaca and with ion and water movements across the integument. Among reptiles and birds, renal function, although highly significant in regulating the composition of the internal environment, again must be coordinated with the regulation of postrenal transport of ions and water across the bladder, cloaca, or colon and, in some species, with the regulation of ion excretion by extrarenal salt glands. In some reptilian species, there may be regulation of ion and water movements across the integument. Unfortunately, the exact quantitative relationships among the various routes of solute and water excretion have not been delineated for any species of nonmammalian vertebrate. Nevertheless, regulation of renal function must, in some way, be carefully integrated and coordinated with regulation of these other routes of water and solute movement, as well as with the behavior of the animals, in their total adaptation to their environment. These coordinated relationships certainly deserve careful quantitative study.

Even for renal function alone, many variations exist among vertebrates. Present knowledge is insufficient to permit a complete description of these variations or the mechanisms underlying them. Nevertheless, in this volume I attempt to describe and analyze all those renal functions in nonmammalian vertebrates for which there is some information available. For this purpose, I begin with a brief description of both gross and fine structure of the kidneys. I then move to a consideration of the initial process in urine formation, primarily ultrafiltration of the plasma at the renal glomerulus but also secretion of fluid by the renal tubules. I then cover transport of inorganic ions, fluid, and organic substances by the renal tubules and regulation of these transport processes. I consider renal aspects of acid–base balance only in terms of tubular transport. Following these discussions, I cover the processes involved in producing urine hypoosmotic or hyperosmotic to plasma, their relative importance, and their regulation. Throughout the book, I compare and contrast renal mechanisms among the vertebrates. However, I devote a final chapter (Chap. 8) to an integrated (although, by necessity, somewhat simplified) summary of renal function for each vertebrate group and summary comparisons of several major renal functions for which adequate information is available among vertebrate groups.

1.4   Useful Reviews on Comparative Renal Physiology

The present volume covers all major aspects of vertebrate renal physiology from a comparative viewpoint. Other reviews consider specific renal functions from a comparative viewpoint, renal function in specific animal groups only, or comparative renal function in nonmammalian vertebrates only. Unfortunately, some of these are quite dated. Nevertheless, because a number of these provide excellent coverage of some of these topics, I have a provided a selection here. References to many other shorter reviews on specific aspects of renal function are provided throughout the following chapters.

The most recent reviews cover renal function in individual vertebrate classes only. The most recent of these are found in the sections of the chapter "Osmoregulation and Excretion" (Larsen et al. 2014) in *Comprehensive Physiology* (online: comprehensivephysiology.com). Another series of recent chapters covering renal function in individual vertebrate classes are found in the volume *Osmotic and Ionic Regulation: Cells and Animals* (Evans 2009).

All other reviews are much older. The most recent of these on integrative physiology of the vertebrate kidney is the chapter "Vertebrate Renal System" (Braun and Dantzler 1997) in *Handbook of Physiology. Comparative Physiology* (now online: comprehensivephysiology.com). Other similar reviews were written close to the same time as the first edition of this book. These include the chapters "Comparative Aspects of Renal Function" (Dantzler 1992a) in *The Kidney: Physiology and Pathophysiology*, 2nd edition and "Comparative Physiology of the Kidney" (Dantzler 1992b) in *Handbook of Physiology. Renal Physiology* (now online: comprehensivephysiology.com).

Finally, a series of even older review chapters or monographs are available on single major groups of vertebrates. These include reviews on fishes (Hickman and Trump 1969), reptiles (Dantzler 1976) and birds (Skadhauge 1973, 1981).

References

Braun EJ, Dantzler WH (1997) Vertebrate renal system. In: Dantzler WH (ed) Handbook of physiology: Comparative physiology. Oxford University Press, New York, pp 481–576

Dantzler WH (1976) Renal function (with special emphasis on nitrogen excretion). In: Gans CG, Dawson WR (eds) Biology of Reptilia, vol 5, Physiology A. Academic Press, London, pp 447–503

Dantzler WH (1992a) Comparative aspects of renal function. In: Seldin DW, Giebisch G (eds) The kidney: Physiology and pathophysiology, 2nd edn. Raven, New York, pp 885–942

Dantzler WH (1992b) Comparative physiology of the kidney. In: Windhager EE (ed) Handbook of physiology-renal physiology. Oxford Press, New York, pp 415–474

Evans DH (ed) (2009) Osmotic and ionic regulation. Cells and animals. CRC Press, Boca Raton, FL

Hickman CP Jr, Trump BF (1969) The kidney. In: Hoar WS, Randall DJ (eds) Fish physiology, vol I, Excretion, ion regulation, and metabolism. Academic, New York, pp 91–239

Krogh A (1929) The progress of physiology. Am J Physiol 90:243–251

Larsen EH, Deaton LE, Onken H, O'Donnell M, Grosell M, Dantzler WH, Weihrauch D (eds) (2014) Osmoregulation and excretion. Compr Physiol 4:405–573

Skadhauge E (1973) Renal and cloacal salt and water transport in the fowl (*Gallus domesticus*). Dan Med Bull 20:1–82

Skadhauge E (1981) Osmoregulation in Birds. Springer, Berlin

Chapter 2
Renal Morphology

Abstract This chapter first covers the gross internal and external renal morphology of fishes, amphibians, reptiles, birds, and mammals. This discussion begins by considering the segments of the prototypical vertebrate nephron. It then examines the variations in nephron segments from this prototype and the arrangement of the nephrons within the kidney in each vertebrate class. In particular, it emphasizes certain specific variations in gross nephron structure such as the aglomerular nephrons in some fishes and a few reptiles, the extremely complex nephrons of elasmobranch fishes, and the presence and arrangement of reptilian-type (loopless) and mammalian-type (looped) nephrons in birds. The discussion of gross morphology also includes a brief consideration of the blood vessels, emphasizing the variation in the complexity of glomerular capillaries in different vertebrate classes and the presence of a renal venous portal system in the kidneys of all amphibians, reptiles, birds, marine teleost fishes, and probably euryhaline teleost fishes. The chapter then briefly explores the fine structure of the glomerulus, especially the filtration barrier, and of the nephron segments found in each vertebrate class. It ends with a short discussion of the juxtaglomerular apparatus found in mammals and portions of it found in nonmammalian vertebrate species.

Keywords Comparative renal morphology • Gross structure • Fine structure • External morphology • Internal morphology • Glomerulus structure • Tubule structure • Vasculature structure

2.1 Introduction

Similarities and differences exist among vertebrate classes in gross external morphology, internal organization, and cellular structure of the kidneys. Since these similarities and differences are related to similarities and differences in renal function, some knowledge of structure is required for understanding function. However, this volume is not a treatise on renal morphology and only those known morphological features that appear most important for an understanding of comparative renal function are discussed in this chapter. Some additional structural details, not covered here, are discussed in the context of function in later chapters.

© The American Physiological Society 2016
W.H. Dantzler, *Comparative Physiology of the Vertebrate Kidney*,
DOI 10.1007/978-1-4939-3734-9_2

2.2   Gross External and Internal Morphology

2.2.1   General Considerations

The prototypical vertebrate nephron consists of a glomerulus followed by a neck segment, a proximal tubule, an intermediate segment, a distal tubule, and, finally, a collecting tubule and duct system (Figs. 2.1 and 2.2). However, variations in the occurrence and structure of these specific nephron components (Fig. 2.2), in the relationship of different nephron segments to their blood supply and to each other, and in the relationship of individual nephrons to each other are found within and between vertebrate classes. These relationships also help to define the gross external morphology of the kidney. The most significant variations in these structures and their relationships are considered below and later will be considered in more detail in relation to function.

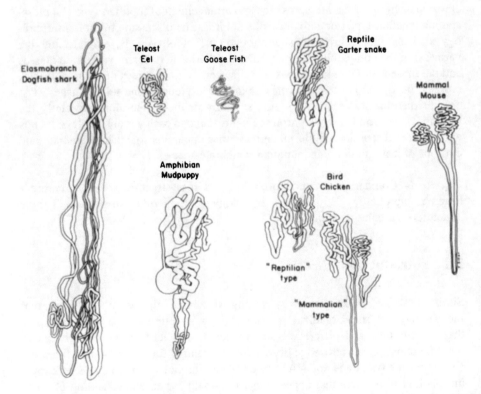

Fig. 2.1 Representations of fish (elasmobranch, *Squalus acanthias*; glomerular teleost, *Anguilla rostrata*; aglomerular teleost, *Lophius piscatorius*), amphibian (*Necturus maculosus*), reptilian (*Thamnophis sirtalis*), avian (domestic fowl, *Gallus gallus*), and mammalian (*Mus flavicolis*) nephrons drawn to a single scale (After Long and Giebisch 1979; Marshall and Grafflin 1928; reproduced from Dantzler 1992, with permission)

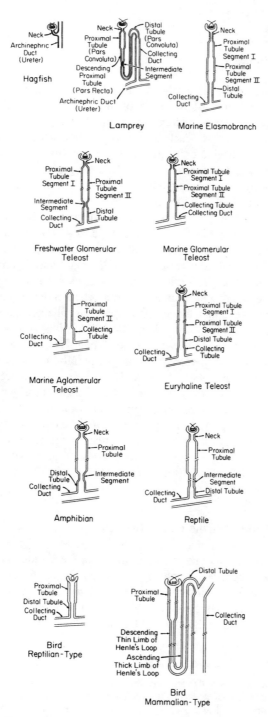

Fig. 2.2 Diagrammatic representations of nephrons from the major classes of nonmammalian vertebrates to show the major nephron segments. No attempt was made to draw nephrons to an exact scale, although some attempt was made to indicate relative sizes for fish and bird nephrons.

2.2.2 Fishes

2.2.2.1 Cyclostomes

Among the cyclostomes, the adult Myxini (hagfishes), which conform to their marine environment, at least in terms of the osmolality of their body fluids, retain paired apparently nonfunctional pronephric kidneys as well as paired functional mesonephric kidneys. Each of the paired mesonephric kidneys contains 30–35 very large, oval glomeruli arranged segmentally on the medial side of a primitive archinephric duct (ureter) (Fig. 2.2). Each glomerulus is connected to the archinephric duct by a short, non-ciliated neck segment (Fig. 2.2) (Fels et al. 1998; Hickman and Trump 1969). There are no other nephron segments.

The adult Petromyzones (lampreys), which apparently do not conform to their environment, thus maintaining body fluids hypoosmotic to a marine environment and hyperosmotic to a freshwater environment, have nephrons similar in gross structure to more advanced vertebrates (Hickman and Trump 1969; Logan et al. 1980). Each has a glomerulus, a ciliated neck segment, a proximal tubule, an intermediate segment, a distal tubule, and a collecting duct (Fig. 2.2). Although early studies suggested that the urinary space of adjacent glomeruli might be continuous in some species (Regaud and Policard 1902; Youson 1975), the most recent studies indicate that this is not so (Logan et al. 1980). Each nephron apparently has a major loop segment arranged parallel to its own collecting duct (Fig. 2.2) (Logan et al. 1980), but the functional significance of this loop is unknown. The nephrons do not appear to be arranged in a manner that permits them to function in concert to produce urine hyperosmotic to the plasma (vide infra; Chap. 7). In the sea lamprey, *Petromyzon marinus*, the entire loop consists of distal tubule (Youson and McMillan 1970a), whereas in the river lamprey, *Lampetra fluviatilis*, the descending limb of the loop consists of proximal tubule and the ascending limb of distal tubule (Fig. 2.2) (Morris 1972).

2.2.2.2 Stenohaline Marine Elasmobranchs

Marine elasmobranchs, like hagfishes, conform to the osmolality of their environment, but, unlike hagfishes, much of the osmolality of the extracellular fluid is determined by the concentrations of urea and trimethylamine oxide (TMAO). Nephrons in these animals contain all the standard vertebrate components noted above (Figs. 2.1 and 2.2) (Hickman and Trump 1969), but the arrangement of the

---

Fig. 2.2 (continued) Breaks in the nephrons indicate that the lengths of those segments may be much greater, relative to other segments, than actually shown. Except for those nephrons in which a loop structure was parallel to the collecting ducts (lamprey and avian mammalian-type), no attempt was made to show the shape of the nephron segments (Dantzler 1992, with permission)

nephrons is highly complex, arguably the most complex of any vertebrate kidney. Although the elasmobranch kidney is not organized into discrete cortical and medullary regions like the mammalian kidney, the nephrons are arranged to permit countercurrent flow within the dorsolateral region of the kidney (see Fig. 2.3, and its legend for a detailed description of this arrangement) (Boylan 1972; Deetjen and Antkowiak 1970; Evans and Claiborne 2009; Lacy et al. 1975). Moreover, the five nephron segments so arranged and the peritubular capillaries arranged in countercurrent fashion among them are encapsulated in a peritubular sheath (Fig. 2.3), which may serve to create a microenvironment in which some form of countercurrent exchange or, possibly, countercurrent multiplication can operate. Such a process may be very important in the retention of urea by the kidneys of marine elasmobranchs (vide infra; Chap. 6).

2.2.2.3 Euryhaline and Stenohaline Freshwater Elasmobranchs

Euryhaline elasmobranchs when in freshwater and stenohaline freshwater elasmobranchs maintain their extracellular osmolality above that of freshwater but reduce it below that of seawater, primarily by reducing the urea concentration (Evans and Claiborne 2009). The nephron structure and arrangement appear to vary depending on whether the species is a euryhaline one or a stenohaline freshwater one. For example, the euryhaline Atlantic stingray (*Dasyatis Sabina*) has a renal structure essentially the same as that described above for marine elasmobranchs (Fig. 2.3) (Lacy and Reale 1999). However, the kidney of the stenohaline freshwater stingray (*Potamotrygon* sp.) has the dorsolateral bundle zone (described in Fig. 2.3) replaced by a peripheral complex zone containing the glomeruli and lacking the peritubular sheath surrounding the tubule segments (Lacy and Reale 1999). The ventromedial sinus zone (described in Fig. 2.3) is replaced with a central sinus zone (Lacy and Reale 1999). The *Potamotrygon* kidney also lacks loops III and IV of the tubules shown for the marine elasmobranch kidney in Fig. 2.3 (Lacy and Reale 1999). It is assumed that the kidneys of other stenohaline freshwater species would also have this modified structure and that the absence of nephron loops III and IV is related to the inability of the nephrons in these species to reabsorb filtered urea (vide infra; Chap. 6), but there are no physiological studies on this possible function.

2.2.2.4 Stenohaline Marine Teleosts

Stenohaline marine teleosts maintain the osmolality of their body fluids well below that of their environment. Grossly, their kidneys are divided into an anterior head kidney containing lymphoid, hematopoietic, and glandular tissue and a posterior trunk kidney containing the renal tissue, but in many species the two kidneys are partially or completely fused and cannot be distinguished by external examination (Hickman and Trump 1969). The nephrons of glomerular stenohaline marine

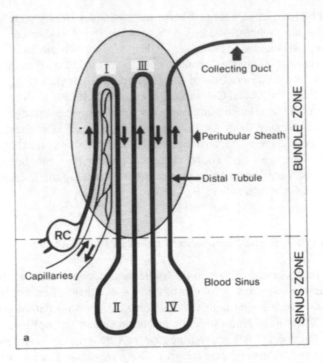

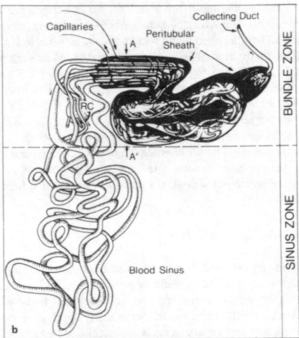

Fig. 2.3 Diagram of elasmobranch nephron in bundle zone (dorsolateral region) and sinus zone (ventromedial region). The dorsal surface is parallel to the top of the page. (*Part a*) Simplified diagram showing renal corpuscle (RC) (Glomerulus) and four highly stylized nephron loops (I–IV).

teleosts generally contain, in addition to a glomerulus, a neck segment, two or three proximal segments that constitute the major portion of the nephron, sometimes an intermediate segment between the first and second proximal segments, and a collecting tubule emptying into the collecting duct system (Fig. 2.2).

As would be expected in marine animals that do not have to dilute the urine to excrete excess water (vide infra; Chap. 7), the distal tubule is absent in almost all species (Hickman and Trump 1969) and, where present, may actually indicate some degree of euryhalinity. Although the entire proximal tubule has a brush border, only the first segment is ultrastructurally similar to the proximal convoluted tubule of mammals (Hickman and Trump 1969).

As would also be appropriate for marine fish that need to minimize their urine volume, some thirty species have nephrons lacking a glomerulus (Fig. 2.2) (Evans and Claiborne 2009; Hickman and Trump 1969). These aglomerular nephrons typically have only a single proximal segment with a brush border similar to that of the second proximal segment of the nephrons of glomerular teleosts and a collecting tubule (Fig. 2.2).

2.2.2.5   Stenohaline Freshwater Teleosts

Stenohaline freshwater teleosts, which maintain the osmolality of their body fluids well above that of their environment, have nephrons that differ from those of stenohaline marine teleosts. Grossly, the head and trunk kidneys are fused to a large extent (Hickman and Trump 1969). The nephrons of glomerular species typically contain, in addition to a glomerulus, a ciliated neck segment, an initial proximal segment with a prominent brush border, a second proximal segment with

---

Fig. 2.3 (continued) A peritubular sheath surrounds the countercurrent system of nephron segments (loops I, III, and the distal tubule) and anastomosing capillary loops in the bundle zone (*gray* color shows area enclosed by peritubular sheath). *Small arrows* indicate the direction of tubular fluid and blood flow. (*Part b*) Schematic drawing of the pathway of the skate nephron in the bundle zone (dorsolateral region) and in the sinus zone (ventromedial region) showing some of the nephron complexity. As shown diagrammatically in *Part a*, the entering limbs of nephron loops I and III and the distal tubule pierce the peritubular sheath near the renal corpuscle and extend to the opposite end of the sheath. Close to the renal corpuscle, the five tubule segments (loops I, III, and the distal tubule) located in the bundle zone are covered by the peritubular sheath and run parallel to each other (to emphasize this distinctive course, they have been drawn side by side in one plane and not assembled into a bundle as they actually are). The tubule bundle and surrounding peritubular sheath then become convoluted, and the parallel course of the tubules is lost as the loops wrap around each other (shown schematically in *Part b*). For simplicity, the opposite end of the peritubular sheath, where the distal tubule emerges, has been drawn away from the renal corpuscle on the far right side of the diagram and the schematic. The distal tubule pierces the sheath at this point to join the collecting duct, whereas the other two nephron segments loop back and retrace their path, finally exiting the sheath where they entered it. Capillaries also enter and exit the peritubular sheath at its renal corpuscle terminus and form an anastomotic network around and within the bundle (Lacy et al. 1985, with permission)

a less prominent brush border, an intermediate segment, and a distal tubule emptying into the collecting duct system (Fig. 2.2). The kidneys of most stenohaline freshwater teleosts, as might be expected in animals that need to excrete much water, tend to have more nephrons with larger glomeruli than those of stenohaline marine teleosts. Interestingly, the glomeruli of some apparently stenohaline freshwater teleosts that survive adaptation to seawater atrophy and disappear (Elger and Hentschel 1981). The few aglomerular freshwater teleosts (apparently species that evolved in seawater and later invaded freshwater) seem to have nephrons entirely like those of marine aglomerular species (Fig. 2.2) (Hickman and Trump 1969).

2.2.2.6 Euryhaline Teleosts

These animals, which can maintain the osmolality of their body fluids above that of the environment when adapted to freshwater and below that of the environment when adapted to seawater, have nephrons most similar to those of stenohaline freshwater teleosts. These typically have a glomerulus (often smaller and less vascular than in stenohaline freshwater species), a first proximal segment, a second proximal segment, a variably present short intermediate segment, a distal tubule, and a collecting tubule emptying into the collecting duct system (Fig. 2.2) (Hickman and Trump 1969). Interestingly, some euryhaline species are actually aglomerular (e.g., the toadfish, *Opsanus tau*) (Baustian et al. 1997; Lahlou et al. 1969). Moreover, in glomerular euryhaline species, the number of glomeruli filtering can vary from sea water to freshwater (Brown et al. 1980) (vide infra; Chap. 2).

2.2.3 *Amphibians*

The external shape of the mesonephric kidneys of amphibians, whose habitats range from completely aqueous to arid terrestrial, varies substantially among species. They tend to be elongated with some evidence of segmentation in urodeles and relatively short and compact in anurans. However, basic internal renal organization is rather similar for all amphibians and can be illustrated by that of the frog kidney shown in Fig. 2.4. All nephrons contain a glomerulus, ciliated neck segment, proximal tubule, ciliated intermediate segment, and distal tubule emptying into the collecting duct system (Figs. 2.1, 2.2, and 2.4) (vide infra; Sect. 2.3 for cellular segmental subdivisions). Of particular note, as in most other nonmammalian vertebrates, the nephrons empty at right angles into the collecting ducts. There are no discrete cortical and medullary regions and no lengthened intermediate nephron segments arranged parallel to collecting ducts as in avian and mammalian kidneys (2.4).

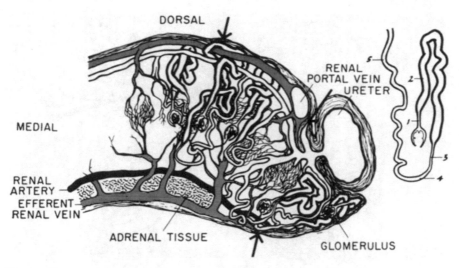

Fig. 2.4 Cross-section of frog kidney, adapted from von Mollendorf (1930). *1* neck segment; *2* proximal tubule; *3* intermediate segment; *4* distal tubule; *5* collecting duct. *Arrows* designate typical micropuncture sites (Long 1973, with permission)

2.2.4 *Reptiles*

Like the mesonephric kidneys of adult amphibians, the metanephric kidneys of adult reptiles, whose habitats also range from completely aqueous to arid terrestrial, show marked variation in their external morphology. This is undoubtedly due to the extreme variation in body form within the class Reptilia, as exemplified by the ophidians on one hand and chelonians on the other. As examples, the kidneys of saurians are compact, somewhat triangular-shaped structures joined at the posterior end, those of ophidians are long and thin, and those of chelonians are constrained by the shape of the carapace (Dantzler and Bradshaw 2009) (Dantzler, personal observations). These varying kidney shapes also determine the gross arrangement of the nephrons. For example, ophidian nephrons lie side by side in neat parallel rows and attach at roughly right angles to major collecting ducts (Fig. 2.5) (Dantzler, personal observations), whereas saurian nephrons branch off the collecting ducts more obliquely and are arranged in compact bunches (O'Shea et al. 1993). Again, however, although the external shape of the kidneys varies among the Reptilia, the internal organization of these organs is reasonably uniform. This basic organization is illustrated diagrammatically for the snake kidney in Fig. 2.5. The nephrons are generally composed of all standard components—glomerulus, ciliated neck segment, proximal tubule, ciliated intermediate segment, and distal tubule (Fig. 2.2)—and they empty at right angles into collecting ducts (Fig. 2.5). However, a few nephrons without glomeruli have been described in some ophidians (Regaud and Policard 1903) and in one saurian species (O'Shea et al. 1993). Again, as in amphibian kidneys, in reptilian kidneys there are no

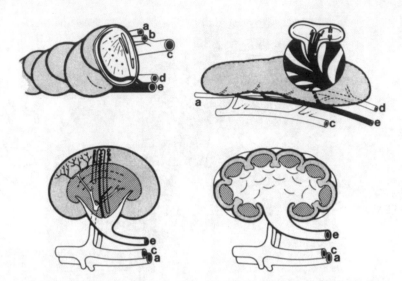

Fig. 2.5 Diagram of reptilian (snake) kidney (*upper left*), avian kidney (*upper right*), unipapillary mammalian kidney (*lower left*), and multirenculated mammalian kidney (*lower right*). Letter notations are consistent for all four figures and are as follows: *a* renal vein; *b* renal artery (reptilian kidney only); *c* aorta; *d* renal portal vein (reptilian and avian kidneys); *e* ureter. Note that nephrons and collecting ducts (*solid black*) in reptilian and avian kidneys are continuous with ureters (also shown in *solid black* in these kidneys) (Dantzler and Braun 1980, with permission)

discrete cortical and medullary regions and no nephrons with elongated intermediate loops arranged parallel to collecting ducts as in avian and mammalian kidneys (Fig. 2.5).

2.2.5 Birds

The kidneys of birds vary less from species to species in gross external appearance than those of amphibians, reptiles, or even mammals. They are prominent, flattened, retroperitoneal organs (Figs. 2.5 and 2.6), deeply recessed into the bony synsacrum (fused lumbar, sacral, and caudal vertebrae). The kidneys are crossed by major nerve trunks and blood vessels securing them tightly in place. Each kidney usually consists of three divisions—cranial, medial, and caudal—which may or may not be apparent superficially, depending on the species (Johnson 1968).

The avian kidney can be divided into cortical and medullary regions, but the boundary between these regions is not as sharp as it is in most mammalian kidneys. Moreover, the avian kidney contains nephrons resembling those found in the kidneys of most other nonmammalian vertebrates and nephrons resembling those found in the kidneys of mammals (Figs. 2.1, 2.2, 2.5, and 2.6). Most nephrons (e.g., about 90 % in Gambel's quail, *Callipepla gambelii*; about 70 % in European

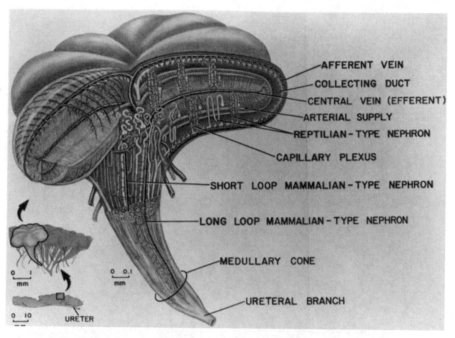

AFFERENT VEIN

COLLECTING DUCT

CENTRAL VEIN (EFFERENT)

ARTERIAL SUPPLY

REPTILIAN - TYPE NEPHRON

CAPILLARY PLEXUS

SHORT LOOP MAMMALIAN - TYPE NEPHRON

LONG LOOP MAMMALIAN - TYPE NEPHRON

MEDULLARY CONE

URETERAL BRANCH

URETER

Fig. 2.6 Illustration of detailed organization of avian kidney. The whole kidney is shown at the lower left with two successive enlargements of sections of the kidney showing increasing internal detail. The largest section shows detail of nephrons forming a single medullary cone. Near the surface of the kidney are the small loopless reptilian-type nephrons arranged in a radiating pattern around a central vein to form a single lobule. In the deeper regions of the cortex are the larger mammalian-type nephrons with highly convoluted proximal tubules, loops of Henle, and distal tubules. A gradual transition occurs from loopless reptilian-type nephrons to the longest-looped, mammalian-type nephrons. The loops of Henle in parallel with collecting ducts and vasa recta are bound together within a medullary cone in an arrangement that allows the production of urine hyperosmotic to the plasma (vide infra; Chap. 7) (Braun and Dantzler 1972, with permission)

starlings, *Sturnus vulgaris*) are of the type found in other nonmammalian verte-brates (Braun 1978; Braun and Dantzler 1972). These nephrons (referred to as "reptilian-type" or "loopless" nephrons) are located in the superficial or cortical region of the kidney (Figs. 2.5 and 2.6). They are very simple, even simpler than most true reptilian nephrons. Each one is comprised of a small glomerulus, no neck segment, a proximal tubule with a few simple folds, no intermediate segment, and a distal tubule with only one main loop before it empties at right angles into a collecting duct (Figs. 2.1, 2.2, 2.6, and 2.7) (Braun and Dantzler 1972; Wideman et al. 1981). In some, but not all nephrons, there is a short connecting tubule between the end of the distal tubule and the collecting duct (Wideman et al. 1981). Like the nephrons in teleostean, amphibian, and reptilian kidneys, these reptilian-type nephrons are not arranged in a manner that would be expected to permit them to contribute directly to the production of urine hyperosmotic to the plasma. Instead, these nephrons are arranged in a radiating pattern about the long

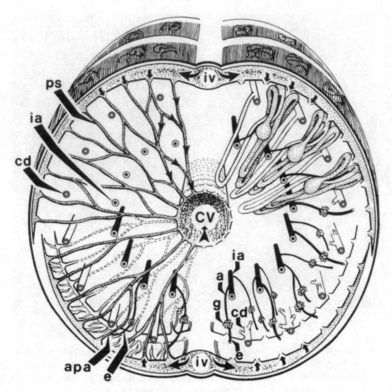

Fig. 2.7 Schematic cross-section of a single cortical lobule in avian kidney. Renal portal blood (*arrows*) flows through the interlobular veins (*iv*) and empties directly into the freely anastomosing spaces between tubules (peritubular sinuses) (*ps*) to be recollected by the central vein (*cv*). Other abbreviations: *a* afferent arteriole; *apa* arterial-portal anastomosis; *cd* collecting duct; *e* efferent arteriole; *g* glomerular capillaries; *ia* intralobular artery (Wideman et al. 1981, with permission)

axis of a vein (central efferent vein) to form cylindrical cortical units (cortical lobules) (Figs. 2.6 and 2.7). These cortical lobules are themselves grouped in radiating patterns from central points over the entire surface of the kidney (Fig. 2.6).

Deep to these cortical lobules containing the small reptilian-type nephrons are larger nephrons in which the intermediate segment is lengthened into a loop of Henle like that in mammalian nephrons (Fig. 2.6). These nephrons (referred to as "mammalian-type" or "looped" nephrons) make up about 10 % of the nephrons in Gambel's quail and about 30 % of the nephrons in starlings (Braun 1978; Braun and Dantzler 1972). Like true mammalian nephrons, these nephrons have, in addition to a larger glomerulus than the reptilian-type nephrons, a convoluted proximal tubule, a straight proximal tubule, a thin descending limb of Henle's loop, a thick ascending limb of Henle's loop, which always begins before the bend, a distal convoluted tubule, and possibly a terminal collecting duct or connecting tubule (E. J. Braun, personal communication) (Figs. 2.1 and 2.2). However, the transition from the reptilian-type to the mammalian-type nephrons is gradual, not abrupt, so that there are transitional nephrons of intermediate size and relatively short loops of Henle in the area between the superficial cortical lobules and the deeper medullary region (Fig. 2.6).

The loops of Henle of the mammalian-type nephrons, the vasa recta, which arise from the efferent arterioles of the glomeruli of mammalian-type nephrons, and the collecting ducts, which drain both the reptilian-type and mammalian-type nephrons from each radial group of cortical lobules, are arranged in parallel and bound together by a connective tissue sheath into a tapering structure called a medullary cone (Figs. 2.5 and 2.6). This arrangement, as in the mammalian kidney, would be expected to permit the avian kidney to produce a urine hyperosmotic to the plasma (vide infra; Chap. 7). As in the reptilian and amphibian kidneys and in contrast to the mammalian kidney, however, there is no renal pelvis. The nephrons, collecting ducts, and ureter are continuous, and, therefore, each medullary cone terminates as a branch of the ureter (Figs. 2.5 and 2.6). The number of medullary cones per kidney is relatively constant for a given species but varies markedly from species to species (Johnson 1968).

The general gross organizational pattern of many cortical areas with their associated medullary cones, the real lobes of the avian kidney, resembles the pattern of the extreme multirenculated kidney found in cetacean and some bovine mammals (Fig. 2.5). In both birds and cetaceans, there are many lobes at varying depths throughout the kidney, but in the cetaceans these lobes are separated by connective tissue whereas in birds they tend to be fused into a single mass (Johnson 1968). In addition to this general similarity of the organization of the avian kidney to the cetacean kidney, the individual lobes of the avian kidney resemble the unipapillary kidneys of small mammals (Fig. 2.5).

2.2.6   Mammals

The general external appearance of the paired, retroperitoneal kidneys of mammals is strikingly different from the external appearance of the kidneys of non-mammalian vertebrates (Fig. 2.5). However, as noted above, the internal organization of the mammalian kidney does bear a distinct resemblance to that of the avian kidney (Fig. 2.5). Mammalian nephrons have glomeruli, proximal convoluted and straight tubules, loops of Henle with thick and thin limbs, distal convoluted tubules, connecting tubules, and collecting ducts (Figs. 2.1 and 2.8). Although the cortical organization of the mammalian kidney differs from that of the avian kidney, the very simple cortical nephrons (nephrons with all their segments in the cortex) of multirenculated mammalian kidneys (Sperber 1944) do resemble the larger reptilian-type or transitional nephrons of the avian kidney. Moreover, the mammalian and avian kidneys resemble each other in their possession of loops of Henle arranged in parallel with vasa recta and collecting ducts in renal papillae and medullary cones (Figs. 2.5, 2.6, and 2.8). Indeed, neonatal rat kidneys have only short-looped nephrons resembling the largest mammalian-type avian nephrons

Fig. 2.8 Schematic
representation of three
sections of rat kidney in
which three components
are illustrated separately.
A The arterial vessels and
capillaries; *B* the venous
vessels; *C* collecting duct
with a nephron with a short
loop of Henle and a nephron
with a long loop of Henle
attached. Labels for regions
of the kidney are as follows:
C cortex; *SZ* subcortical
zone; *OM* outer medulla; *IM*
inner medulla (Kriz 1970,
with permission)

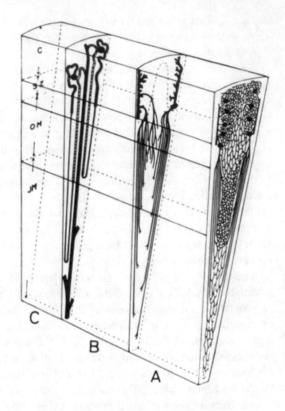

(Liu et al. 2001). This similarity appears to result from parallel evolution as
birds and mammals adapted to life on dry land and suggests that there may be a
unique general way for the kidneys to eliminate solutes efficiently in a solution
more concentrated than the plasma (see Chap. 7 for additional details on the three-
dimensional structural arrangement of loops of Henle, vasa recta, and collecting
ducts in relation to the urine concentrating mechanism).

Two basic structural types of mammalian kidneys exist: the unipapillary
type (one renculus) and the compound or multirenculated type in which this
basic unipapillary unit is repeated within the kidney (Fig. 2.5) (Oliver 1968). The
repeating unipapillary units in compound or multirenculated kidneys can be discrete or combined into a single compound mass enclosed by a continuous cortex
(Oliver 1968). Modifications of the unipapillary kidney, such as the crest kidney
(single cortex with multiple medullary and papillary regions) are also observed.

Which one of these general structural types is found in a given mammalian
species is related not to phylogeny but to body mass (Oliver 1968; Sperber 1944).
Small mammals have unipapillary kidneys. In mammals larger than rabbits, the
unipapillary kidneys are superseded by crest kidneys. The crest kidney, in turn,
tends to disappear in animals larger than kangaroos, although animals as large as
giraffes, camels, and horses still have crest kidneys. The multirenculated kidney

first begins to appear in animals the size of otter and beaver and is present in a number of very large animals, including hippopotamus, elephant, and whale.

The change in form from unipapillary to multirenculated, i.e., many small, unipapillary units, may be related to the physical force required to move fluid along the nephrons and the need to reabsorb filtered solutes and water (Calder and Braun 1983; Dantzler and Braun 1980; Sperber 1944). The hydrostatic pressure of the arterial blood in the glomerular capillaries provides the driving force for the flow along the tubules (vide infra; Chap. 3), but mean arterial blood pressure does increase proportionately with body size (Calder 1981). Because resistance to flow along a nephron will increase as nephron length increases, nephron length would not be expected to increase proportionately with body size (Calder and Braun 1983; Dantzler and Braun 1980). Indeed, proximal tubule length does not increase continuously with increasing body and kidney mass; rather, it tends to become constant at about the same body and kidney mass at which the unipapillary kidney is replaced by the crest kidney (Oliver 1968; Sperber 1944). In very large mammals with multirenculated kidneys, the proximal tubule length is even shorter than in small mammals with unipapillary kidneys. Although an increase in tubule diameter would reduce resistance to flow, it would interfere with reabsorption of filtered solutes and water by increasing diffusion distances (Calder and Braun 1983). And, indeed, tubule diameter does not increase continuously (Calder and Braun 1983). Apparently, to accommodate the absolute increase in metabolism and corresponding increase in excretory end products with an increase in body size, the number of nephrons increases. However, there also appears to be an optimal number of nephrons (the number contained in the largest unipapillary kidney) that can function together appropriately in the concentrating and diluting process. As the number of required nephrons exceeds this optimum, the unipapillary kidney gives way to the multirenculated kidney.

2.2.7 General Blood Vessel Patterns

Kidneys of all vertebrate species with glomerular nephrons receive an arterial blood supply. However, the number of renal arteries supplying the kidneys varies substantially among classes, species, and even individual animals. The arterial divisions within the kidney also vary substantially among vertebrate classes and species depending on the gross internal architecture discussed above. In all cases, however, each glomerulus is supplied by an afferent arteriole which breaks up into a glomerular capillary network. In the case of the primitive lampreys, each afferent arteriole apparently supplies more than one glomerulus, and each glomerulus apparently contains capillaries arising from more than one afferent arteriole (Hentschel and Elger 1987), but in all other vertebrate species examined, each glomerulus is supplied by only one afferent arteriole.

The complexity of the glomerular network varies markedly among the vertebrates, as demonstrated by Bowman (Bowman 1842) nearly 175 years ago. In

Fig. 2.9 Scanning electron micrograph of a cast of a mammalian (rat) glomerulus with its many capillary loops (*CL*) and adjacent renal vessels. The afferent arteriole (*A*) takes origin from an interlobular artery at *lower left*. The efferent arteriole (*E*) branches to form the peritubular capillary plexus (*upper left*) (Tisher and Madsen 1986, with permission)

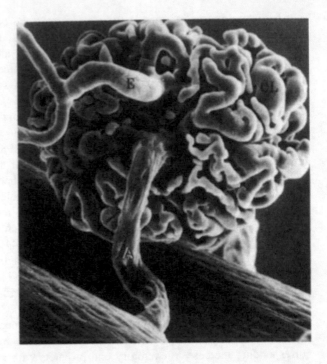

general, the capillary network of most fishes, amphibians, reptiles, and mammals consists of thoroughfare channels that anastomose freely (Fig. 2.9). However, the extent of the anastomoses and the resulting complexity of the capillary network are much greater in the glomeruli of mammals than in the glomeruli of nonmammalian vertebrates. Interestingly, the renal glomeruli of birds, homeotherms like mammals with a high sustained mean arterial blood pressure, have the simplest glomerular network (Dantzler and Braun 1980). The glomerular tuft of the smallest, reptilian-type nephrons may consist of a single capillary loop with no cross-branching or anastomoses (Fig. 2.10) (Braun 2009; Dantzler and Braun 1980) (E. J. Braun personal communication). The glomerular tuft of the larger, mammalian-type nephrons may consist of a single unbranched capillary coiled around the periphery of the renal capsule (Fig. 2.10), although occasional anastomoses or cross-branching also may occur (Braun 2009; Dantzler and Braun 1980) (E. J. Braun personal communication).

In all glomerular species, the glomerular capillaries unite into a single efferent arteriole that leaves the glomerular capsule (Figs. 2.9 and 2.10). The efferent arterioles draining the glomerular capillaries of nephrons in glomerular fishes, amphibians, reptiles, reptilian-type nephrons of birds, and superficial and midcortical nephrons of mammals break up into a network of sinusoids or capillaries surrounding the tubules. The pattern of capillaries or sinusoids, which varies among the classes and species, is particularly complex in mammals (Beeuwkes and Bonventre 1975). Of special importance for tubule function, the kidneys of

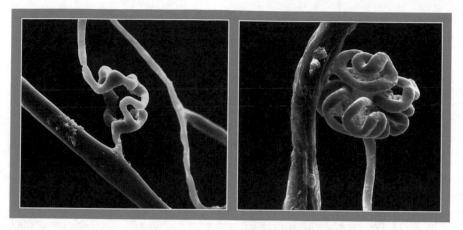

Fig. 2.10 Scanning electron micrographs of negative casts of avian glomeruli with their simple capillary loop structure. At left is a cast of capillary from a reptilian-type or loopless nephron. At right is a cast of capillary from a mammalian-type or looped nephron. The casts of avian glomerular capillaries appear to show tufts consisting of one unbranched capillary (Courtesy of E. J. Braun, University of Arizona)

all amphibians, reptiles, birds, marine teleost fishes, and, probably, euryhaline teleost fishes have a renal venous portal system that contributes blood to the peritubular capillaries or sinusoids. As alluded to above, the efferent arterioles from the glomeruli of the mammalian-type nephrons of birds and from the glomeruli of the inner cortical or juxtamedullary nephrons of mammals give rise to vasa recta that supply the avian medullary cones (Fig. 2.6) or mammalian medulla (Fig. 2.8) (vide infra; Chap. 7 for more details on three-dimensional vasa recta arrangement and role in urine concentrating mechanism). The medullary cones of the avian kidney receive no venous portal blood (Wideman et al. 1981). Finally, the aglomerular nephrons of some marine teleosts receive only a renal portal blood supply (Beyenbach 1985).

2.3 Fine Internal Structure

2.3.1 General Considerations

The fine structure of kidneys, particularly ultrastructure, although extensively studied in mammals, has received only sporadic and inconsistent study in nonmammalian vertebrates. Studies of kidney ultrastructure in nonmammalian vertebrates have been performed primarily on species that have been studied most extensively physiologically, but few descriptions are complete. Nevertheless, a few general descriptions for nonmammalian vertebrates, primarily in comparison to those for mammals, are given here. More detailed descriptions, where available and where significant, are mentioned later in conjunction with descriptions of specific physiological processes.

2.3.2   Glomerulus

The renal glomerulus, the site of production of an ultrafiltrate of the plasma in all glomerular nephrons, is composed of (*1*) the capillary network described above, (*2*) a central region of mesangial cells with their matrix material, (*3*) the parietal layer of Bowman's capsule composed of (*a*) the basement membrane that covers the outside of the capillaries and (*b*) the podocytes, and (*4*) the visceral layer of Bowman's capsule, which is continuous with the epithelium of the proximal tubule (Fig. 2.11). The general ultrastructure of the total filtration barrier—the fenestrated capillary endothelium, the basement membrane, and the visceral epithelial cells or podocytes with filtration slits covered by a slit diaphragm between them—is similar for all glomerular species as illustrated for a mammalian kidney in Fig. 2.12. However, there are some qualitative and quantitative differences among classes and species in the detailed structure of the major components of this barrier.

The fenestrae of the capillary endothelium, the innermost or proximal component of the glomerular filtration barrier, vary in both number and size. These

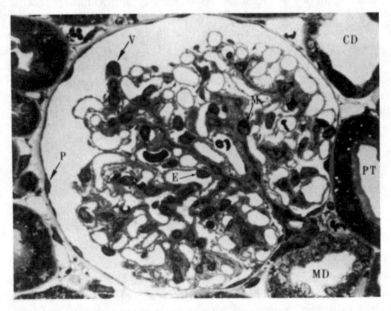

Fig. 2.11 Light micrograph of a normal glomerulus from rat, demonstrating the four major cellular components: mesangial cell (*M*), endothelial cell (*E*), visceral epithelial cell (*V*), and parietal cell (*P*). Other abbreviations: *CD* collecting duct; *PT* proximal tubule (×750) (Tisher and Madsen 1986, with permission)

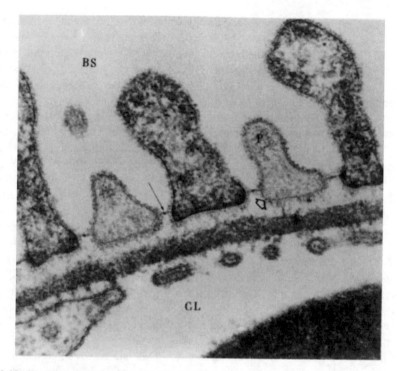

Fig. 2.12 Electron micrograph of normal rat glomerulus fixed in a 1 % glutaraldehyde solution containing tannic acid. Note the relationship between the apparent three layers of the glomerular basement membrane and the presence of podocytes (*P*) (foot processes) imbedded in the lamina rara externa of the basement membrane (*thick arrow*). The filtration slit diaphragm with the central filament (*thin arrow*) is especially evident between the individual foot processes. The fenestrated endothelial lining of the capillary loop is shown below the basement membrane. Other abbreviations: *CL* capillary lumen; *BS* Bowman's space (×120,000) (Tisher and Madsen 1986, with permission

fenestrae are few or absent in hagfish (Heath-Eves and McMillan 1974), lampreys (Youson and McMillan 1970b), elasmobranchs (Bargman and von Hehn 1971), and teleosts (Bulger and Trump 1968), considerably more numerous in amphibians (Schaffner and Rodewald 1978) and reptiles (Anderson 1960; Peek and McMillan 1979a), and very common in birds (Casotti and Braun 1996) and mammals (Tisher and Madsen 1986) (see also Fig. 2.12). In those mammals studied (rats and humans), the fenestrae are approximately circular, varying in diameter from about 50 to 100 nm, depending on the species and study (Casotti and Braun 1996; Nielsen et al. 2012; Tisher and Madsen 1986), and are uniformly distributed over the entire endothelium. In birds (domestic fowl, *Gallus gallus*), the fenestrae are apparently uniformly distributed and are elliptical with mean dimensions of about 148 and 110 nm for the major and minor axes, respectively (Casotti and Braun 1996). In amphibians (bullfrog, *Rana catesbiana*), the fenestrae are apparently circular, are not so uniformly spaced as in mammals, and vary widely in diameter from about 100 to over 300 nm (Schaffner and Rodewald 1978). In mammals and birds, at least, the endothelial cells are covered with a negatively charged glycocalyx consisting of polyanionic glycosaminoglycans and glycoproteins (Casotti and Braun 1996; Nielsen et al. 2012). These details have yet to be studied in representatives of other vertebrate classes.

The basement membrane, the middle component of the filtration barrier, has generally been shown in electron micrographs to consist of a central electron-dense layer, the lamina densa, with two much thinner electron-lucent layers on either side of it, the lamina rara interna and the lamina rara externa (Fig. 2.12) (Nielsen et al. 2012). Although the lamina densa clearly exists and appears to reflect a true basement membrane, a number of studies on mammals have indicated that the two laminae rarae are artifacts of fixation and that the three-layer structure of the basement membrane does not exist (Chan et al. 1993; Chan and Inoue 1994; Reale and Luciano 1990). A single electron-dense layer also appears to reflect the true basement membrane in the glomeruli of amphibians (mudpuppy, *Necturus maculosus*; bullfrog, *Rana catesbeiana*) (Miner 2009; Schaffner and Rodewald 1978) and birds (domestic fowl, *Gallus gallus*) (Casotti and Braun 1996). Although no such detailed studies are available on species of other vertebrate classes and there is still some uncertainty about this situation even in those species studied, at present it appears most appropriate to consider the electron-dense layer (Fig. 2.12) to be the true glomerular basement membrane. This lamina densa varies in thickness among different vertebrate classes and even within a single species. For example, it appears to be about 50–80 nm in rats (Latta 1970; Latta et al. 1975) and about 30 to over 400 nm in bullfrogs (*Rana catesbeiana*) (Schaffner and Rodewald 1978). Electron micrographs also show strong anionic staining of the glomerular basement membrane in rats (Kanwar and Farquhar 1979) but very little such staining in birds (Casotti and Braun 1996). In all species, this electron-dense layer appears fibrous and in mammals is composed of collagen IV, laminins, fibronectins, and sulfated proteoglycans (Nielsen et al. 2012; Pollak et al. 2014). The composition has yet to be studied in nonmammalian vertebrates.

The outermost or distal layer of the filtration barrier, consisting of epithelial cells with foot processes and filtration slits with slit diaphragms between the processes, is covered with a negatively charged glycocalyx in mammals, birds, and amphibians just as the endothelial cells are (Casotti and Braun 1996; Kanwar and Venkatachalam 1992; Schaffner and Rodewald 1978). The foot processes rest on the basement membrane just as the endothelial cells do and, in mammals, are anchored to it by integrins. The filtration slit diaphragm covering the filtration slits (Fig. 2.12) consists of a continuous central filament attached to the foot processes by regularly spaced cross-bridges in mammals and amphibians (Nielsen et al. 2012; Schaffner and Rodewald 1978). In mammals the slit diaphragm is now known to be a complex of proteins, primarily nephrins, but also involving podocin and cadherins (Nielsen et al. 2012; Pollak et al. 2014). No such detail about structure of the filtration barrier has yet been determined for nonmammalian vertebrates, although studies on zebra fish may soon reveal such information for teleosts and indicate the similarities and dissimilarities to mammals. In addition, the role of the various components of the filtration barrier in permitting or preventing filtration of macro-molecules in any species is the subject of active study and considerable controversy (vide infra; Chap. 3).

On one side of the capillaries, the endothelial cells actually rest on the mesangial cells rather than on the basement membrane and covering podocytes described above, but it is not clear that any filtration occurs across this region. However, in all species examined (Kriz and Kaissling 1992; Peek and McMillan 1979a; Tisher and Madsen 1986), myofibrils are found in the mesangial cells, which may permit them to play a role in regulating the area available for filtration (vide infra; Chap. 3). The number of such myofibrils may vary among the vertebrate species, which may account for some variation in the regulation of filtration. The amount of extraglomerular mesangial tissue also varies among species, and this too may have important physiological consequences in the regulation of the glomerular filtration rate (vide infra; Chap. 3).

2.3.3 Proximal Tubule

Ultrastructural studies have demonstrated three cell types (designated P1, P2, and P3) along the mammalian proximal tubule (Fig. 2.13 (Kriz and Kaissling 1992; Tisher and Madsen 1986). These are not intermixed but change from P1 to P2 to P3 serially along the length of the tubule. They are differentiated primarily by height of the brush border, amount of interdigitation of the basolateral membrane, and number and arrangement of the mitochondria among the interdigitations (Fig. 2.13). Although all true proximal tubules of nonmammalian vertebrates have a brush border and numerous mitochondria, there are no clear consistent changes in cell type along the tubule even where functional differences exist (W. H. Dantzler and R. B. Nagle unpublished observations) (Dantzler 1974; Hickman and Trump 1969; Maunsbach and Boulpaep 1984). The deep interdigitations of the basolateral membrane with the linear arrangement of the mitochondria between them are absent from the proximal tubule cells of nonmammalian vertebrates (Figs. 2.14 and 2.15) (W. H. Dantzler and R. B. Nagle unpublished observations) (Dantzler et al. 1986; Hickman and Trump 1969; Maunsbach and Boulpaep 1984; Roberts and Schmidt-Nielsen 1966). However, there is often significant amplification of the basal cell membrane alone, the degree of which can vary substantially among species in the same class (Fig. 2.14 and 2.15) (W. H. Dantzler and R. B. Nagle unpublished observations) (Dantzler et al. 1986; Maunsbach and Boulpaep 1984; Roberts and Schmidt-Nielsen 1966). Finally, tight junctions between cells are generally very short; complex lateral intercellular spaces are apparent in the proximal tubules of all vertebrates.

Fig. 2.13 Drawings of the proximal tubule ultrastructure in mammalian kidneys. To demonstrate the cellular interdigitations, neighboring cells and their processes are covered by a stippled texture. (*a*) P1 segment. The cellular interdigitations are most extensive and extend to the apical ends of the cells. The brush border is dense and high. (*b*) P2 segment. The cellular interdigitations are decreased; apically, the cells have a smooth outline. The brush border is less dense and reduced in height. Peroxysomes (*cross-hatched profiles*) are numerous. (*c, d*) P3 segments. The cellular interdigitations are drastically reduced. Regarding the brush border, considerable interspecies differences occur. In rabbit (*c*), as in most species, microvilli are scanty and short. In contrast, in rat (*d*) the brush border of P3 is the highest among all proximal tubule segments (Kriz and Kaissling 1992, with permission)

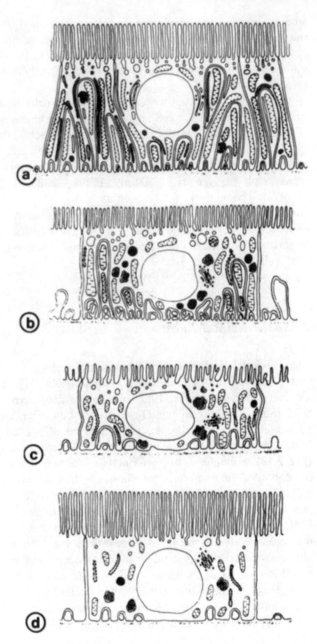

2.3.4 Early Distal Tubule

Cells of the first portion of the distal tubule ("early distal tubule" in many renal papers and texts), beginning after the macula densa in mammals and after the end of the thin intermediate segment in most nonmammalian vertebrates, have common

AMBYSTOMA   NECTURUS

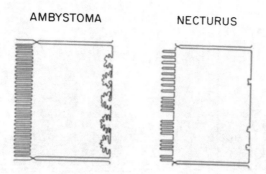

Fig. 2.14 Schematic drawings comparing proximal tubule cells in the urodele amphibians, *Ambystoma tigrinum* (tiger salamander) and *Necturus maculosus* (mudpuppy). Note the higher, more uniform, and more dense brush border of the apical membrane and greater amplification of the basal membrane in *A. tigrinum* than in *N. maculosus*. Neither species has the large amplification and intercellular interdigitations of the lateral cell membranes observed in mammals (see Fig. 2.13) (After Maunsbach and Boulpaep 1984, with modifications)

ultrastructural features in freshwater teleosts (Anderson and Loewen 1975; Hickman and Trump 1969), amphibians (Hinton et al. 1982; Stanton et al. 1984), birds (Nicholson 1982), and mammals (Kaissling and Kriz 1979). The primary common features are deep basolateral infoldings with elongated mitochondria occupying the cytoplasm within these infoldings (Figs. 2.15 and 2.16). These features are also characteristic of cells of intermediate segment IV of nephrons in the bundle zone of elasmobranch nephrons (Lacy and Reale 1985).

Although the distal tubule cells of a few reptiles (e.g., gecko, *Hemidactylus* sp.) also have these common features (Fig. 2.15), those of most reptiles (e.g., horned lizard, *Phrynosoma cornutum*; Galapagos lizard, *Tropidurus* sp.; blue spiny lizard, *Sceloporus cyanogenus*; crocodile, *Crocodylus acutus*; garter snake, *Thamnophis sirtalis*) do not (Davis et al. 1976; Davis and Schmidt-Nielsen 1967; Peek and McMillan 1979b; Roberts and Schmidt-Nielsen 1966). Instead, these cells have extensive lateral interdigitations, a large, often irregular nucleus, and ovoid or spherical mitochondria (Fig. 2.15). The type of distal cell found in reptiles does not appear to be related to the ability of a given species to produce a dilute urine (vide infra; Chap. 7).

2.3.5 Late Distal and Collecting Tubules

There is considerable confusion about the structural distinctions between the distal portions of the distal tubule ("late distal tubule" in many renal papers and textbooks) and the collecting tubule or collecting duct into which it empties in nonmammalian vertebrates (Stoner 1985). In fact, except for fishes (Hentschel and

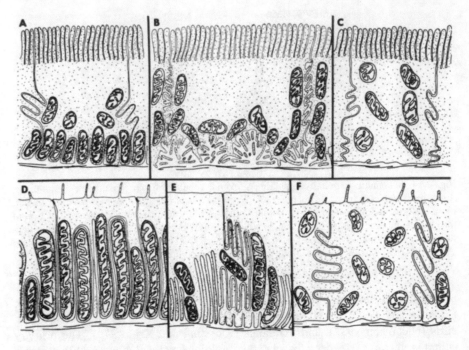

Fig. 2.15 Schematic drawings showing characteristics of surfaces of proximal and distal tubule cells of mammals and reptiles. (*A*) Mammalian proximal cell type (less detailed than in Fig. 2.13). (*B*) Gecko (*Hemidactylus* sp.) proximal cell type. Note amplification of basal surface area. (*C*) Horned lizard (*Phrynosoma cornutum*) and Galapagos lizard (*Tropidurus* sp.) proximal cell type. Note lack of amplification of basal surface area. (*D*) Mammalian thick ascending limb or early distal convoluted tubule cell type. (*E*) Gecko (*Hemidactylus* sp.) distal cell type. Note deep basolateral infoldings with elongated mitochondria within them similar to those in mammalian cells. (*F*) Horned lizard (*Phrynosoma cornutum*) and Galapagos lizard (*Tropidurus* sp.) distal cell type. Note lack of deep basal infoldings but presence of extensive lateral interdigitations (Roberts and Schmidt-Nielsen 1966, with permission)

Elger 1987) and amphibians (Stanton et al. 1984), a clear distinction may not exist. Among most species studied, the cells throughout the distal tubule ("early" and "late" portions) are of the type discussed above. However, the cells of the late distal tubule of amphibians are much taller than those of the early distal tubule and have very large nuclei and deep basal infoldings (Fig. 2.16) (Stanton et al. 1984). The cells of the late distal tubule of some fishes (e.g., *Polypterus sengalus*) and of some saurian reptiles (e.g., *S. cyanogenys*) are also taller than those of the early distal tubule and have less marked lateral interdigitations (Davis et al. 1976; Hentschel and Elger 1987). The amphibian collecting tubule, like the mammalian cortical collecting tubule or duct, consists primarily of light principal cells with some dark, mitochondria-rich intercalated cells interspersed among them (Fig. 2.16) (Stanton et al. 1984). The transition from the distal tubule into the collecting tubule is

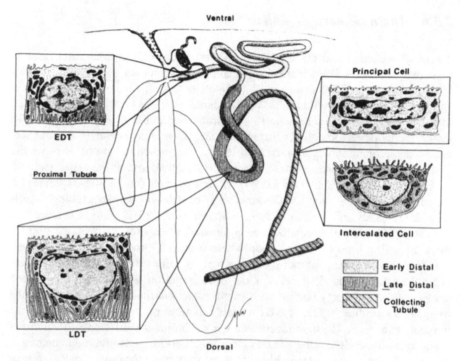

Fig. 2.16 Nephron of amphibian *Amphiuma means* (Congo eel). Glomerulus and proximal tubule drawn for reference. *Shading* indicates distal nephron segments. *Lines* from schematics to tubule segments indicate cellular composition of each segment. Abbreviations: *EDT*, early distal tubule or "diluting segment"; *LDT*, late distal tubule (Stanton et al. 1984, with permission)

gradual with the appearance of increasing numbers of intercalated and principal cells near the end of the distal tubule as it becomes the collecting tubule (Fig. 2.16) (Stanton et al. 1984). Such a gradual transition from the distal tubule through the connecting tubule also occurs in the mammalian kidney. Among fishes, intercalated cells are found only in the collecting tubules of primitive lampreys, Polypteri, and lung fish (Hentschel and Elger 1987). The late distal tubule and connecting tubule of some reptiles (e.g., *S. cyanogenys*) also contain light and dark cells that may correspond to the principal cells and intercalated cells described in other species (Davis et al. 1976). However, the light cells appear to be mucus-secreting (Davis et al. 1976). Dark or intercalated cells also begin to appear at the end of the distal tubule of birds, apparently marking the transition into the collecting tubule or duct (Nicholson 1982). However, the avian collecting tubule or duct contains mucin-secreting cells as well as these dark cells with the mucin-secreting cells becoming the predominant type in the more distal portions of the collecting-duct system (Nicholson 1982).

2.3.6 Juxtaglomerular Apparatus

The distal nephron is closely associated with its corresponding glomerular vascular pole, particularly the afferent arteriole, in amphibians, reptiles, birds, and mammals (S. D. Yokota, R. A. Wideman, and W. H. Dantzler unpublished observations) (Kriz and Kaissling 1992; Morild et al. 1985; Stanton et al. 1984), an arrangement that may be of functional significance in glomerular regulation (vide infra; Chap. 3). However, the development of a juxtaglomerular apparatus appears to be complete only among mammals and, possibly, birds. The principal components of the juxtaglomerular apparatus of mammals are the macula densa of the distal nephron, the renin-producing granular cells of the afferent arteriole (and in some species also the efferent arteriole), and the Goormaghtigh or extraglomerular mesangial cells (Fig. 2.17) (Kriz and Kaissling 1992; Nielsen et al. 2012). The macula densa consists of an area of specialized cells situated within the wall of the cortical thick ascending limb of Henle's loop where it passes through the angle between the afferent and efferent arterioles of its own glomerulus (Fig. 2.17). These cells are taller than the surrounding cells of the distal nephron with an apically placed nucleus and are densely packed with small mitochondria not associated with the basolateral membrane. They have a variable number of slender microvilli on the luminal membrane. The tight junctions are not significantly different from those of the surrounding cells, but there are no interdigitations of the lateral membranes and the intercellular spaces are variably dilated, depending on the physiological state of the animal (Kriz and Kaissling 1992; Nielsen et al. 2012). The basal surface of these macula densa cells abuts the Goormaghtigh cells and variable portions of the afferent and efferent arterioles (Fig. 2.17).

Among nonmammalian vertebrates, renin-producing granular cells are present in the glomerular arterioles of teleost, dipnoan (lungfish), amphibian, reptilian, and avian kidneys (Nishimura and Bailey 1982), but macula densa cells are found only in avian kidneys (Morild et al. 1985; Sokabe and Ogawa 1974). Even for avian kidneys, the presence of a true macula densa has been questioned

Fig. 2.17 Schematic of juxtaglomerular apparatus in mammals. Smooth muscle cell walls of afferent and efferent arterioles are shown in *solid black*. Goormaghtigh cells and mesangial cells are shown by their nuclei in *black* (Kriz and Kaissling 1992, with modifications)

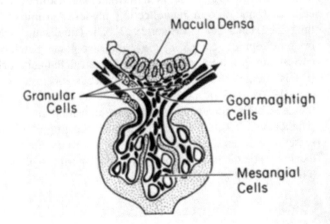

(Wideman et al. 1981), but the most recent anatomical study provides good evidence of a macular densa in both reptilian-type and mammalian-type nephrons (Morild et al. 1985). Moreover, all studies indicate the presence of extensive extraglomerular mesangial (Goormaghtigh) cells in the avian kidney (Morild et al. 1985; Wideman et al. 1981).

References

Anderson BG, Loewen RD (1975) Renal morphology of freshwater trout. Am J Anat 143:93–114
Anderson E (1960) The ultramicroscopic structure of a reptilian kidney. J Morphol 106:205–240
Bargman W, von Hehn G (1971) Über das Nephron der Elasmobranchier. Z Zellforsch Mikrosk Anat 114:1–21
Baustian MD, Wang SQ, Beyenbach KW (1997) Adaptive responses of aglomerular toadfish to dilute sea water. J Comp Physiol B 167:61–70
Beeuwkes R III, Bonventre JV (1975) Tubular organization and vascular-tubular relations in the dog kidney. Am J Physiol 229:695–713
Beyenbach KW (1985) Comparative physiology of the proximal tubule. Renal Physiol 8:222–236
Bowman W (1842) On the structure and use of the malpighian bodies of the kidney, with observations on the circulation through that gland. Roy Soc Lond Philos Trans 132:57–80
Boylan JW (1972) A model for passive urea reabsorption in the elasmobranch kidney. Comp Biochem Physiol [A] 42:27–30
Braun EJ (1978) Renal response of the starling (Sturnus vulgaris) to an intravenous salt load. Am J Physiol Renal Physiol 234:F270–F278
Braun EJ (2009) Osmotic and ionic regulation in birds. In: Evans DH (ed) Osmotic and ionic regulation: cells and animals. CRC Press, Boca Raton, FL, pp 505–524
Braun EJ, Dantzler WH (1972) Function of mammalian-type and reptilian-type nephrons in kidney of desert quail. Am J Physiol 222(3):617–629
Brown JA, Oliver JA, Henderson IW, Jackson BA (1980) Angiotensin and the single nephron glomerular function in the trout Salmo gairdneri. Am J Physiol Regul Integr Comp Physiol 239:R509–R514
Bulger RE, Trump BF (1968) Renal morphology of the English sole (Parophrys vetulus). Am J Anat 123:195–226
Calder WA (1981) Scaling of physiological processes in homeothermic animals. Ann Rev Physiol 43:301–322
Calder WA III, Braun EJ (1983) Scaling of osmotic regulation in mammals and birds. Am J Physiol Regul Integr Comp Physiol 244:R601–R606
Casotti G, Braun EJ (1996) Functional morphology of the glomerular filtration barrier of Gallus gallus. J Morphol 228:327–334
Chan FL, Inoue S (1994) Lamina lucida of basal lamina: an artefact. Microsc Res Tech 28:48–59
Chan FL, Inoue S, Leblond CP (1993) The basal laminas of cryofixed or aldehyde-fixed, freeze-substituted tissues are composed of a lamina densa and do not contain a lamina lucida. Cell Tissue Res 273:41–52
Dantzler WH (1974) PAH transport by snake proximal renal tubules: differences from urate transport. Am J Physiol 226(3):634–641
Dantzler WH (1992) Comparative aspects of renal function. In: Seldin DW, Giebisch G (eds) The kidney: physiology and pathophysiology, 2nd edn. Raven, New York, pp 885–942
Dantzler WH, Bradshaw SD (2009) Osmotic and ionic regulation in reptiles. In: Evans DH (ed) Osmotic and ionic regulation: cells and animals. CRC Press, Boca Raton, FL, pp 443–503
Dantzler WH, Braun EJ (1980) Comparative nephron function in reptiles, birds, and mammals. Am J Physiol Regul Integr Comp Physiol 239:R197–R213

Dantzler WH, Brokl OH, Nagle RB, Welling DJ, Welling LW (1986) Morphological changes with Na+-free fluid absorption in isolated perfused snake tubules. Am J Physiol Renal Physiol 251: F150–F155

Davis LE, Schmidt-Nielsen B (1967) Ultrastructure of the crocodile kidney (Crocodylus porosus) with special reference to electrolyte and fluid transport. J Morphol 121:255–276

Davis LE, Schmidt-Nielsen B, Stolte H (1976) Anatomy and ultrastructure of the excretory system of the lizard, Sceloporus cyanogenys. J Morphol 149:279–326

Deetjen P, Antkowiak D (1970) The nephron of the skate. Bull Mt Desert Biol Lab 10:5–7

Elger B, Hentschel H (1981) The glomerulus of a stenohaline fresh-water teleost, Carassius auratus gibelio, adapted to saline water. Cell Tissue Res 220:73–85

Evans DH, Claiborne JB (2009) Osmotic and ionic regulation in fishes. In: Evans DH (ed) Osmotic and ionic regulation: cells and animals. CRC Press, Boca Raton, FL, pp 295–366

Fels LM, Kastner S, Stolte H (1998) The hagfish kidney as a model to study renal physiology and toxicology. In: Jorgensen JM, Lonholt JP, Weber RE, Molte H (eds) The biology of hagfishes. Chapman and Hall, London, pp 347–363

Heath-Eves MJ, McMillan DB (1974) The morphology of the kidney of the Atlantic hagfish, Myxine glutinosa (l). Am J Anat 139:309–334

Hentschel H, Elger M (1987) The distal nephron in the kidney of fishes. Adv Anat Embryol Cell Biol 108:1–151

Hickman CP Jr, Trump BF (1969) The kidney. In: Hoar WS, Randall DJ (eds) Fish physiology, vol I, Excretion, ion regulation, and metabolism. Academic, New York, pp 91–239

Hinton DE, Stoner LC, Burg M, Trump BF (1982) Heterogeneity in the distal nephron of the salamander (Ambystoma tigrinum): a correlated structure function study of isolated tubule segments. Anat Rec 204:21–32

Johnson OW (1968) Some morphological features of the avian kidney. The Auk 85:216–228

Kaissling B, Kriz W (1979) Structural analysis of the rabbit kidney. Adv Anat Embryol Cell Biol 56:1–23

Kanwar YS, Farquhar MG (1979) Anionic sites in the glomerular basal lamina. In vivo and in vitro localization to the laminae rarae by cationic probes. J Cell Biol 81:137–153

Kanwar YS, Venkatachalam M (1992) Ultrastructure of the glomerulus and juxtaglomerular apparatus. In: Windhager EE (ed) Handbook of Physiology. Renal Physiology. American Physiological Society (Oxford University Press), New York, pp 3–40

Kriz W (1970) Organization of structures within the renal medulla. In: Schmidt-Nielsen B, Kerr DWS (eds) Urea and the kidney. Excerpta Medica, Amsterdam, pp 342–357

Kriz W, Kaissling B (1992) Structural organization of the mammalian kidney. In: Seldin DW, Giebisch G (eds) The kidney: physiology and pathophysiology, 2nd edn. Raven, New York, pp 707–777

Lacy ER, Reale E (1985) The elasmobranch kidney. II. Sequence and structure of the nephrons. Anat Embryol 173:163–186

Lacy ER, Reale E (1999) Urinary system. In: Hamlett WC (ed) Sharks, Skates, and Rays. Johns Hopkins University Press, Baltimore, MD, USA, p. 353-397.

Lacy ER, Reale E, Schlusselberg DS, Smith WK, Woodward DJ (1985) A renal countercurrent system in marine elasmobranch fish: a computer assisted reconstruction. Science 227:1351–1354

Lacy ER, Schmidt-Nielsen B, Galaske RG, Stolte H (1975) Configuration of the skate (Raja erinacea) nephron and ultrastructure of two segments of the proximal tubule. Bull Mt Desert Biol Lab 15:54–56

Lahlou B, Henderson IW, Sawyer WH (1969) Renal adaptations by Opsanus tau, a euryhaline aglomerular teleost, to dilute media. Am J Physiol 216:1266–1272

Latta H (1970) The glomerular capillary wall. J Ultrastruct Res 32:526–544

Latta H, Johnston WH, Stanley TM (1975) Sialoglycoproteins and filtration barriers in the glomerular capillary wall. J Ultrastruct Res 51:354–376

Liu W, Morimoto T, Kondo Y, Iinuma K, Uchida S, Imai M (2001) "Avian-type" renal medullary tubule organization causes immaturity of urine-concentrating ability in neonates. Kidney Int 60:680–693

Logan AG, Moriarty RJ, Morris R, Rankin JC (1980) The anatomy and blood system of the kidney in the river lamprey, *Lampetra fluviatilis*. Anat Embryol 158:245–252

Long WS (1973) Renal handling of urea in *Rana catesbeiana*. Am J Physiol 224:482–490

Long WS, Giebisch G (1979) Comparative physiology of renal tubular transport mechanisms. Yale J Biol Med 52:525–544

Marshall EK Jr, Grafflin AL (1928) The structure and function of the kidney of *Lophius piscatorius*. Bull Johns Hopkins Hosp 43:205–230

Maunsbach AB, Boulpaep EL (1984) Quantitative ultrastructure and functional correlates in proximal tubule of *Ambystoma* and *Necturus*. Am J Physiol Renal Physiol 246:F710–F724

Miner JH (2009) The amphibian kidney's filtration barrier: where is the basement membrane? Am J Physiol Renal Physiol 297:F549–F550

Morild I, Mowinckel R, Bohle A, Christensen JA (1985) The juxtaglomerular apparatus in the avian kidney. Cell Tissue Res 240:209–214

Morris R (1972) Osmoregulation. In: Hardesty MW, Potter IC (eds) The biology of lampreys, vol II. Academic, London, pp 193–239

Nicholson JK (1982) The microanatomy of the distal tubules, collecting tubules and collecting ducts of the starling kidney. J Anat 134:11–23

Nielsen S, Kwon T-H, Fenton RA, Praetorius J (2012) Anatomy of the kidney. In: Taal MW, Chertow GM, Marsden PA, Skorecki K, Yu ASL, Brenner BM (eds) Brenner and Rector's the kidney, 9th edn. Elsevier (Saunders), Philadelphia, pp 31–93

Nishimura H, Bailey JR (1982) Intrarenal renin-angiotensin system in primitive vertebrates. Kidney Int 22:S185–S192

O'Shea JE, Bradshaw SD, Stewart T (1993) Renal vasculature and excretory system of the agamid lizard, *Ctenophorus ornatus*. J Morphol 217:287–299

Oliver J (1968) Nephrons and kidneys. Hoeber, New York

Peek WD, McMillan DB (1979a) Ultrastructure of the renal corpuscle of the garter snake *Thamnophis sirtalis*. Am J Anat 155:83–102

Peek WD, McMillan DB (1979b) Ultrastructure of the tubular nephron of the garter snake *Thamnophis sirtalis*. Am J Anat 154:103–128

Pollak MR, Quaggan SE, Hoenig MP, Dworkin LD (2014) The glomerulus: the sphere of influence. Clin J Am Soc Nephrol 9:1461–1469

Reale E, Luciano L (1990) The lamina rarae of the glomerular basement membrane. Their manifestation depends on the histochemical and histological techniques. Contrib Nephrol 80:32–40

Regaud C, Policard A (1902) Notes histologiques sur la secretion renale. C R Soc Biol (Paris) 54:91–93

Regaud C, Policard A (1903) Sur l'existence de diverticules du tube urinpare sans relations avec les corpuscles de Malpighi, chez les serpents, et sur l'indépendence relative des fonctions glomérulaires et glandulaires dur rein en general. C R Soc Biol (Paris) 55:1028–1029

Roberts JS, Schmidt-Nielsen B (1966) Renal ultrastructure and excretion of salt and water by three terrestrial lizards. Am J Physiol 211:476–486

Schaffner A, Rodewald R (1978) Glomerular permeability in the bullfrog *Rana catesbeiana*. J Cell Biol 79:314–328

Sokabe H, Ogawa M (1974) Comparative studies of the juxtaglomerular apparatus. Int Rev Cytol 37:271–327

Sperber I (1944) Studies on the mammalian kidney. Zoologiska Bidrag Fran Uppsala 22:249–435

Stanton B, Biemesderfer D, Stetson D, Kashgarian M, Giebisch G (1984) Cellular ultrastructure of *Amphiuma* distal nephron: effects of exposure to potassium. Am J Physiol Cell Physiol 247: C204–C216

Stoner LC (1985) The movement of solutes and water across the vertebrate distal nephron. Renal Physiol 8:237–248

Tisher CC, Madsen K (1986) Anatomy of the kidney. In: Brenner BM, Rector FC Jr (eds) The kidney, 3rd edn. Saunders, Philadelphia, pp 3–60

von Mollendorf W (1930) Harn-und Geschlechtsapparat. In: Handbuch der mikroskopischen Anatomie des Menschen, sect I, vol VII, Springer, Berlin

Wideman RF Jr, Braun EJ, Anderson GL (1981) Microanatomy of the renal cortex in the domestic fowl. J Morphol 168:249–267

Youson JH (1975) Absorption and transport of ferritin and exogenous horseradish peroxidase in the opisthonephric kidney of the lea lamprey. I Renal corpuscle. Can J Zool 53:571–581

Youson JH, McMillan DB (1970a) The opisthonephric kidney of the sea lamprey of the great lakes, *Petromyzon marinus* L. I. the renal corpuscle. Am J Anat 127:207–232

Youson JH, McMillan DB (1970b) The opisthonephric kidney of the sea lamprey of the great lakes, *Petromyzon marinus* L. II. Neck and proximal segments of the tubular nephron. Am J Anat 127:233–258

Chapter 3
Initial Process in Urine Formation

Abstract This chapter primarily concerns filtration of fluid by the glomeruli. It begins with a detailed discussion of the forces involved in the production of an ultrafiltrate of the plasma at each glomerulus, including the variations in these forces and the filtration coefficient examined to date in a few species of fishes, amphibians, reptiles, and mammals. It then examines the values for single-nephron filtration rates measured to date under control circumstances in representatives of all five vertebrate classes and some possible reasons for observed differences. Next, the chapter considers quantitative changes in whole-kidney glomerular filtration rates, single-nephron glomerular filtration rates, and the number of filtering nephrons under various conditions of hydration, dehydration, and salt and water loads and then evaluates the various hormonal, neural, and other factors, including autoregulation, that help to regulate single-nephron glomerular filtration rates. As a further consideration with regard to glomerular function, the chapter discusses the filtration of protein, its measurements, and its variation among vertebrates. Finally, the chapter considers the secretion of fluid by the renal tubules both as the primary process in urine formation in aglomerular nephrons and a contributor to urine formation in glomerular nephrons in a number of vertebrate species, even including some mammals.

Keywords Glomerular filtration • Ultrafiltrate • Filtration forces • Whole-kidney glomerular filtration rate (GFR) • Single-nephron glomerular filtration rate (SNGFR) • Intermittent filtration • Filtration regulation • Protein filtration

3.1 Introduction

The initial process in urine formation consists of the delivery of water and plasma solutes into the lumen of the proximal tubule. In glomerular nephrons, this process primarily involves the production of an ultrafiltrate of the plasma. The rate of production of such an ultrafiltrate at the individual glomerulus is the primary, if not always the sole, determinant of the rate at which fluid is delivered to the corresponding proximal tubule and therefore is the initial determinant of the volume and composition of the final urine. The filtration process, which relies on arterial hydrostatic pressure maintained for other functions, is well suited to the

© The American Physiological Society 2016
W.H. Dantzler, *Comparative Physiology of the Vertebrate Kidney*,
DOI 10.1007/978-1-4939-3734-9_3

rapid elimination of large volumes of fluid without high energy costs. Since it also can be altered quickly in some species, it can play an important role in the regulation of excretory water losses.

In aglomerular nephrons of teleost fishes and, most likely, those of ophidian reptiles, secretion of fluid by the proximal tubule is the initial process in urine formation and is thus the initial determinant of the volume and composition of the final urine (Berglund and Forster 1958; Beyenbach 1982; Forster 1953; Hickman and Trump 1969). Since this process is not well suited to the rapid elimination of large volumes of fluid, it is not surprising that aglomerular nephrons evolved initially in marine teleosts. However, secretion of fluid by proximal tubules apparently can contribute to urine formation in glomerular as well as aglomerular nephrons under a number of circumstances (vide infra) (Beyenbach 1982, 1986, 2004; Beyenbach and Fromter 1985; Forster 1953; Hickman 1968; Schmidt-Nielsen and Renfro 1975; Yokota et al. 1985a).

3.2   Filtration of Fluid by Glomeruli

3.2.1   Process of Ultrafiltrate Formation at Renal Glomeruli

The glomerular ultrafilter produces a nearly protein-free (vide infra) filtrate of the plasma with which it is in Donnan equilibrium (Navar 1978; Renkin and Gilmore 1973). If the filtration process by the individual glomerulus is modeled by considering the total capillary network as a cylindrical tube of equivalent surface area, the single nephron glomerular filtration rate (SNGFR) may be described by the following equation containing the major factors involved in the process (Yokota et al. 1985a):

$$\text{SNGFR} = L_p A \left[(P_{GC} - P_{BS}) - \pi_{GC} \right] \tag{3.1}$$

In Eq. (3.1), P_{GC} is the outwardly directed hydrostatic pressure in the glomerular capillaries, which favors filtration. Although P_{GC} must decrease along the capillary network for blood to flow from the afferent to the efferent arteriole, this decrease appears to be very small (vide infra) and, therefore, in the usual treatments of the above equation P_{GC} is considered to be constant along the capillaries. P_{BS}, the inwardly directed hydrostatic pressure in Bowman's space, which opposes filtration, is also considered to be constant. The hydrostatic pressure gradient across the capillary wall is then the difference between P_{GC} and P_{BS}. Because the protein that is filtered into Bowman's space is generally negligible (vide infra) and can be ignored, the colloid osmotic pressure of the capillary plasma, π_{GC}, is equivalent to the gradient for colloid osmotic pressure across the capillary wall that opposes the hydrostatic pressure driving filtration. The difference between these two opposing pressure gradients is the net ultrafiltration pressure (P_{UF}).

As filtration progresses along the length of the capillaries, the protein concentration in the capillaries increases reciprocally with the fraction of fluid remaining in them, and the π_{GC} increases as an exponential function of the protein concentration (Landis and Pappenheimer 1963). Filtration will continue only so long as the hydrostatic pressure gradient exceeds the opposing osmotic pressure gradient (i.e., as long as there is a positive net P_{UF}).

The hydraulic conductivity, L_p, or the aqueous permeability of the capillary wall determines the rate of filtration per unit area of filtration surface for any given P_{UF}. The rate of filtration is directly proportional to both L_p and the area available for filtration, A, as indicated in Eq. (3.1). Under steady-state conditions, if filtration equilibrium is reached (i.e., if the opposing transmural hydrostatic and osmotic pressure gradients become equal), filtration will cease and, if other variables remain constant, the fraction of the plasma flow filtered (filtration fraction) will remain constant. As long as filtration equilibrium is reached somewhere along the tube (the glomerular capillaries), the rate of filtration is a linear function of the plasma flow. However, if the rate of plasma flow is great enough that filtration equilibrium is not achieved somewhere along the tube (the glomerular capillaries), then the rate of filtration will be dependent on hydraulic conductivity (L_p), area of filtering surface (A), and magnitude of the net ultrafiltration pressure (P_{UF}).

An accurate measurement of either L_p or A would permit determination of the other from that value, the net P_{UF}, and the SNGFR using Eq. (3.1). However, there is no direct way to measure L_p in the glomerular capillaries. As far as A is concerned, the glomerular capillaries, with the apparent exception of the simplest avian ones (Fig. 2.10), are not simple tubes but complex branching networks (vide supra; Chap. 2). The total area available for filtration is a function of the diameter, length, and number of capillary branches. In addition, the specific morphology of the capillary network, including the branching pattern and the capillary dimensions, and the microrheological properties of the blood determine the distribution of blood flow and, thus, the area actually used for filtration. Because neither the surface area (A) used for filtration nor the hydraulic conductivity (L_p) can be measured directly, they are usually treated together as their product, the ultrafiltration coefficient (K_f), which can be determined from the measured SNGFR and net P_{UF} as indicated in the simplified equation:

$$\text{SNGFR} = K_f[(P_{GC} - P_{BS}) - \pi_{GC}] \qquad (3.2)$$

The factors discussed above determine the SNGFR for any given glomerulus, but there are many glomeruli in each kidney. The whole-kidney glomerular filtration rate (GFR) is, of course, the sum of all single nephron glomerular filtration rates at any given time. Although the whole-kidney GFR is readily measured by clearance methods (vide infra), it is not readily determined by summing the SNGFRs. There are several reasons for this. First, the number of nephrons varies among species and even from kidney to kidney and is not easily determined accurately. Second, the morphological characteristics of individual nephrons and their corresponding functional characteristics vary among species and within individual kidneys.

Therefore, the SNGFR for a given glomerulus cannot be considered representative of all other glomeruli even in that kidney. Third, the micropuncture method of measuring SNGFR only applies to those nephrons with glomeruli or tubule loops available on some renal surface. And, although it is theoretically possible to measure the SNGFRs in all nephrons in one kidney at a given time by the constant infusion sodium ferrocyanide technique of deRouffignac et al. (1970), in fact this is not practical technically. Nevertheless, despite the heterogeneity of glomerular structure and function, the glomeruli within a kidney of a given species are often grouped into general functional categories based on their size, morphological complexity, or location within the kidney, and the SNGFRs of a few nephrons in each group are used as an average for the entire group (vide infra).

In addition to the morphological differences among glomeruli and the resulting functional differences, individual glomerular filtration rates may change over time. Not only may glomeruli have different rates of continuous filtration, but filtration within a single glomerulus may vary with time and may even cease transiently. Therefore, in some species, the percent of nephrons filtering at any given time may vary depending upon the physiological state of the animal (vide infra).

The production of a true ultrafiltrate of the plasma and the roles described above for the glomerular capillary hydrostatic and colloid osmotic pressure and intracapsular hydrostatic pressure in this process were first demonstrated by micropuncture studies on amphibians (frogs, *Rana pipiens*; mudpuppies, *Necturus maculosus*) (Hayman 1927; Wearn and Richards 1924; White 1929). Since those initial studies, ultrafiltration has been demonstrated by micropuncture measurements in primitive cyclostome fishes (petromyzonta, river lampreys, *Lampetra fluviatilis*) (McVicar and Rankin 1985), reptiles (snakes, *Thamnophis sirtalis, Storeria occipitomaculata*, and *S. dekayi*) (Bordley and Richards 1933), and mammals (rats) (Brenner et al. 1971). Curiously, in one group of cyclostome fishes (Myxini, Pacific hagfish, *Epitatretus stoutii*), direct micropuncture measurements indicate that an apparent ultrafiltrate, at least a colloid-free fluid, is produced at the glomerulus in the absence of a positive net ultrafiltration pressure (when capillary colloid osmotic pressure apparently equals or exceeds the capillary hydrostatic pressure) (Riegel 1978, 1986a, b). This apparent paradox has yet to be resolved and certainly demands further study, especially since the rates of apparent filtration at single glomeruli in hagfish are very high, approaching those of mammals (vide infra).

No direct collections from Bowman's capsule or measurements of the forces involved in ultrafiltration have been made in other fishes or birds. Nevertheless, the excretion by the kidneys of teleost and elasmobranch fishes and birds of molecules that are only filtered by mammalian and amphibian nephrons has led to the general acceptance of the process of ultrafiltration and of the forces involved in that process for the glomerular nephrons of all vertebrates.

Over the years, the general pressure relationships involved in glomerular ultrafiltration, first demonstrated in studies on amphibians, have been explored in more detail, not only in amphibians but also in primitive fishes, reptiles, and mammals. The relevant measurements are summarized in Table 3.1. Micropuncture measurements on one species of urodele amphibian (Congo eel, *Amphiuma means*) (Persson

Table 3.1 Examples of pressures involved in glomerular ultrafiltration

	$P\bar{A}$ (mmHg)	$P\overline{GC}$ (mmHg)	PBS (mmHg)	π_{aff} (mmHg)	π_{eff} (mmHg)	π_{GC} (mmHg)	PUF_{aff} (mmHg)	PUF_{eff} (mmHg)	$\overline{PUF}$ (mmHg)	K_f (nl min^{-1} mmHg^{-1})	References
Fishes											
River Lamprey											
Lampetra fluviatilis											
Freshwater	21.6	16.1	3.9	8.6	9.0	8.8	5.4	4.9	3.4	1.68	McVicar and Rankin (1985)
20 % Seawater	18.7	12.3	3.0	8.6	9.3	9.0	0.7	0	0.35	–	
Amphibians											
Congo eel											
Amphiuma means	17.3	12.6	4.8	5.3	7.8	9.1	2.5	0	0.65	7.64	Persson (1981)
Reptiles											
Garter snake											
Thamnophis spp.	38	22	2	17	–	–	3	–	–	0.2–1.7	Dantzler and Bradshaw (2009), S. D. Yokota (personal communication)
Mammals											
Laboratory rat											
Rattus norvegicus (Munich-Wistar strain)	95	45.0	10	18	35	26	17	0	9	4.8	Brenner et al. (1971), Deen et al. (1973)

Values are approximate means estimated from the references. $P\bar{A}$ indicates mean arterial pressure; $P\overline{GC}$, mean hydrostatic pressure within the glomerular capillaries; PBS, pressure in Bowman's space; π_{aff}, colloid osmotic pressure at afferent end of glomerular capillaries (generally within afferent arteriole); π_{eff}, colloid osmotic pressure at efferent end of glomerular capillaries (generally within efferent arteriole of immediately attached peritubular capillaries); π_{GC}, mean colloid osmotic pressure within the glomerular capillaries (arithmetic mean of π_{aff} and π_{eff}); PUF_{aff}, net ultrafiltration pressure at afferent end of glomerular capillaries; PUF_{eff}, net ultrafiltration pressure at efferent end of glomerular capillaries; $\overline{PUF}$, mean net ultrafiltration pressure; K_f, ultrafiltration coefficient

1978a, b, 1981) and on a widely studied strain of rat (Brenner et al. 1971, 1972) show that under normovolemic conditions, filtration equilibrium is reached by the efferent end of the capillaries (i.e., net ultrafiltration pressure at this point is zero) (Table 3.1). As noted above, this observation indicates that the filtration rate is influenced by the plasma flow through the capillaries (Brenner et al. 1972). In both amphibians and mammals, the sum of the colloid osmotic pressure at the efferent end of the capillaries and the hydrostatic pressure in Bowman's space essentially equals the randomly measured hydrostatic pressure in the glomerular capillaries, suggesting that the latter pressure does, in fact, decrease little along the length of the capillary network (Brenner et al. 1972; Persson 1981). This suggestion is further supported in amphibians by the identity of the hydrostatic pressure in the afferent arteriole and the randomly measured hydrostatic pressure in the glomerular capillaries (Persson 1981). These observations support the concept that, in both amphibians and mammals, filtration equilibrium is reached because of an increase in the colloid osmotic pressure along the length of the capillaries as filtration occurs. However, no exact measurements of the pressure profile along the capillary network from the afferent to the efferent end or of the site of filtration equilibrium have been made for either amphibians or mammals.

Because filtration equilibrium is reached before the end of the glomerular capillaries in these amphibians and mammals, a unique value for the ultrafiltration coefficient cannot be obtained. However, the ultrafiltration coefficient for *A. means* estimated from the pressure data in these studies (Table 3.1) (Persson 1981) is similar to that calculated even less directly for frog and *N. maculosus* (Renkin and Gilmore 1973). And this value for *A. means* is nearly twice that determined by micropuncture measurements on hypervolemic rats during filtration disequilibrium (Table 3.1) (Deen et al. 1973). Because the ultrafiltration coefficient is the product of the capillary area available for filtration and the capillary hydraulic conductivity, the difference between the values for amphibians and mammals may result primarily from an apparently greater area available for filtration in amphibian than in mammalian glomeruli (Renkin and Gilmore 1973) rather than from a difference in hydraulic conductivities. Unfortunately, as noted above, there are no really accurate measurements of the glomerular capillary area available for filtration, much less the area actually used, in any species.

In the one species of fish, the primitive, anadromous river lamprey (*L. fluviatilis*), in which filtration has been evaluated by micropuncture, the attainment of filtration equilibrium depends on whether the animals are adapted to freshwater or brackish water (McVicar and Rankin 1985). In freshwater-adapted animals, filtration equilibrium is not achieved along the glomerular capillaries, i.e., there is a significant positive net ultrafiltration pressure remaining at the efferent end of the capillaries (Table 3.1) (McVicar and Rankin 1985). Under these circumstances, filtration is particularly sensitive to hydrostatic pressure. In addition, a unique value for the ultrafiltration coefficient can be calculated (Table 3.1) (also McVicar and Rankin 1985). This value is similar to the lower estimates in rats (Deen et al. 1973) and dogs (Navar et al. 1977). In lampreys adapted to brackish water (20 % seawater), however, filtration equilibrium is achieved (Table 3.1), preventing determination of a unique value for the ultrafiltration coefficient (McVicar and Rankin 1985). In addition, as noted above, under these circumstances, filtration should be influenced by plasma flow along the capillaries.

The hydrostatic pressure in the glomerular capillaries of the brackish-water-adapted lampreys is about three-fourths of that in the glomerular capillaries of freshwater-adapted animals (Table 3.1) (also McVicar and Rankin 1985). This decrease in glomerular capillary pressure in the brackish-water-adapted animals appears to result both from a small decrease in mean arterial pressure and an increase in the pressure drop across the afferent arterioles (Table 3.1). Since the colloid osmotic pressure of the systemic plasma is unchanged when the animals are adapted to brackish water, net positive ultrafiltration pressure at the afferent end of the glomerular capillaries is markedly reduced (Table 3.1), resulting in a reduced SNGFR (vide infra; Table 3.2). As in mammals (rats) and amphibians (Congo eels), filtration equilibrium in these brackish-water-adapted lampreys appears to result primarily from a rise in the colloid osmotic pressure along the glomerular capillaries (Table 3.1) rather than a decrease in capillary hydrostatic pressure. However, as for other animals, no direct measurements of the pressure profile along the capillary network have been made for this species.

The only direct micropuncture measurements of the pressures involved in glomerular ultrafiltration yet made in reptiles (garter snakes; *Thamnophis* spp.) are less complete than those in fishes, amphibians, and mammals (Table 3.1) (S. D. Yokota personal communication) (Dantzler and Bradshaw 2009). Random measurements of the hydrostatic pressure in the glomerular capillaries suggest that, as in fishes, amphibians, and mammals, this pressure remains relatively constant along the capillary network (S. S. Yokota personal communication). However, because plasma colloid osmotic pressure has only been measured at the afferent end of the capillaries, net ultrafiltration pressure is only available at this point in the capillary network (Table 3.1). Therefore, it is not possible to know with certainty whether filtration equilibrium is reached along the capillary network. However, given the low ultrafiltration pressure at the afferent end of the capillaries, it seems likely that this is the case. If this is true, then, as noted above, the SNGFR will be particularly sensitive to changes in renal plasma flow and a unique value for the ultrafiltration coefficient cannot be determined.

Nevertheless, it is possible to estimate ultrafiltration coefficients from the net ultrafiltration pressure at the afferent end of the capillaries (Table 3.1) and the measured SNGFRs for this snake species (Table 3.2) (Dantzler and Bradshaw 2009). These estimates yield ultrafiltration coefficient values ranging from 0.2 to 1.7 nl min^{-1} mmHg^{-1} (Table 3.1) (Dantzler and Bradshaw 2009). The largest of these values is about the same as that in the freshwater-adapted river lamprey but less than one-fourth that in the Congo eel and only about one-third that in Munich-Wistar rats (Table 3.1). However, given the paucity of data, the measurements of ultrafiltration coefficients in garter snakes are probably less reliable than those in the other species.

Table 3.2 Examples of single nephron glomerular filtration rates (SNGFR)

	Conditions	SNGFR nl min^{-1}	References
Fishes			
Myxinoidea			
Atlantic Hagfish	Seawater	24.2 ± 6.8 (6)	Alt et al. (1980),
Myxine glutinosa			Stolte and Schmidt-
Marine			Nielsen (1978)
Pacific Hagfish	Seawater	20.3 ± 2.13 (71)	Riegel (1978)
Eptatretus stoutii			
Marine			
Petromyzonta			
River lamprey	Freshwater	7.02 ± 0.027 (89)	Moriarty et al. (1978)
Lampetra fluviatilis	Seawater	2.9 ± 0.3 (9)	Logan et al. (1980c)
Freshwater, marine,			
or euryhaline			
Elasmobranchii			
Lesser spotted dogfish	Seawater	9.5 ± 1.4 (26)	Brown and Green
Scyliorhinus canicula			(1987)
Marine			
Telostei			
Rainbow trout	Freshwater	1.31 ± 0.20 (5)[a]	Brown et al. (1978)
Oncorhynchus mykiss		45 % filtering	Brown et al. (1980)
Euryhaline	Seawater	3.74 ± 1.12 (3)[a]	Brown et al. (1978)
		5 % filtering	Brown et al. (1978)
Amphibia			
Urodela			
Congo eel	Freshwater	17.5 ± 0.75 (18)	Persson (1978a)
Amphiuma means			
Aquatic, freshwater			
Mudpuppy	Freshwater	12.88 ± 1.56 (12)	Giebisch (1956)
Necturus maculosus			
Aquatic, freshwater			
Anura			
Frog	Control	13.44 ± 1.62 (3)	Walker and Hudson
Rana pipiens			(1937)
Semiaquatic, freshwater			
Reptilia			
Squamata			
Ophidia			
Garter snake	Control	0.6–5	Bordley and
Thamnophis sirtalis			Richards (1933)
Terrestrial, moist			Dantzler and
			Bradshaw (2009)
Aves			
Galliformes			
Gambel's quail	Control		
Callipepla gambelii	Mammalian-type	14.6 ± 0.79 (27)[a]	Braun and
Terrestrial, arid	nephrons		Dantzler (1972)
	Reptilian-type	6.4 ± 0.25 (41)[a]	Braun and
	nephrons		Dantzler (1972)
	Smallest	0.37 ± 0.082 (14)	W. H. Dantzler,
	Reptilian-type		unpublished

(continued)

Table 3.2   (continued)

	Conditions	SNGFR nl min^{-1}	References
Passeriformes			
European starling	Control		
Sturnus vulgaris	Mammalian-type	15.6 ± 0.75 (208)[a]	Braun (1978)
Terrestrial, moist	nephrons		
	Reptilian-type	7.0 ± 0.35 (185)[a]	Braun (1978)
	nephrons		
	Smallest	0.36 ± 0.040 (17)	Laverty and
	Reptilian-type		Dantzler (1982)
	nephrons		
Mammalia			
Laboratory rat	Control		
Rattus norvegicus	Superficial nephrons	30.1 ± 2.55 (7)	Brenner et al. (1971)
(Munich-Wistar strain)	Superficial	36.4 ± 3.5 (4)[a]	Trinh-Trang-Tan
Terrestrial	nephrons		et al. (1981)
	Juxtamedullary	51.7 ± 6.7 (4)[a]	Trinh-Trang-Tan
	nephrons		et al. (1981)

Values are means ± SE. Numbers in parentheses indicate number of determinations except in the case of the trout and rats where they indicate number of animals
[a]SNGFRs determined by constant infusion sodium ferrocyanide technique. All other determinations of SNGFR were made by micropuncture technique

No measurements of pressures involved in glomerular ultrafiltration have yet been made for birds, which, like mammals, are homeotherms with a relatively high and constant mean arterial pressure. However, the simplicity of the avian glomerular tuft, especially in the smallest, most superficial reptilian-type nephrons (vide supra; Fig. 2.10), should make it possible to measure the hydrostatic and colloid osmotic pressure profiles along the capillaries if any glomeruli can be found accessible to micropuncture (Braun 1982; Dantzler and Braun 1980; Laverty and Dantzler 1982). Such measurements would help greatly in understanding the regulation of filtration in all vertebrates.

3.2.2   Values for Single-Nephron Glomerular Filtration Rates

Direct measurements of filtration rates in individual nephrons have been made for mammals and for a number of species of nonmammalian vertebrates either by micropuncture methods or by the constant-infusion sodium-ferrocyanide method (deRouffignac et al. 1970). Values available for nonmammalian vertebrate species and for one mammalian species are shown in Table 3.2. A number of factors may account for differences among species in the values reported. First, they may be related, in part, to the status of the animals at the time the measurements were made and to differences in measurement techniques. Second, they may be related to differences in the population of nephrons sampled. Although, as noted above, no anatomically distinct populations of nephrons with differing filtration rates, like those found in the kidneys of birds and mammals (Table 3.2), have been found for the kidneys of fishes, amphibians, and reptiles, there is enough variation in glomerular size, at least within reptilian kidneys (Yokota et al. 1985a), that some heterogeneity in filtration rates among nephrons in an individual kidney may exist. Thus, it is quite possible that some of the variation in the micropuncture measurements of SNGFRs shown in Table 3.2 reflect inadvertent selection of accessible nephrons whose function was not representative of all nephrons in the kidney. In addition, in those species in which nephrons filter intermittently (vide infra), the selection of technically acceptable SNGFR measurements may have resulted in mean values greater than the true means for all nephrons. These possibilities are suggested for some species by marked differences between the mean measured values and the mean values predicted from allometric analyses (Yokota et al. 1985a).

Although the above factors may account for some interspecific differences shown in Table 3.2, the values for hagfish and all amphibians are very high compared to other poikilothermic nonmammalian vertebrates and even to the homeothermic birds. In fact, they are not far below those observed in many mammals. Because the net ultrafiltration pressure in amphibians is low, especially compared to that in birds and mammals (see, for example, Table 3.1; also Brenner et al. 1971; Persson 1981; Yokota et al. 1985a), the high SNGFRs may result from a large area available for filtration in glomeruli of these species (Renkin and Gilmore 1973). Also, the hydraulic conductivity of glomerular capillaries may be greater in these species than in others.

In the case of the hagfish, as noted above, net ultrafiltration pressure has even been reported to be absent or negative despite the high filtration rates (Riegel 1978, 1986a, b). How this occurs (if the measurements are reproducible) is difficult to understand no matter how large the area available for filtration or high the hydraulic conductivity in these animals. In any case, the physiological significance of the high filtration rates in amphibians and hagfish is not completely clear. The metabolic requirements for excretion in hagfish and amphibians are certainly lower than those in birds and mammals and, on that basis, the SNGFRs also should be lower (Yokota et al. 1985a). In amphibians, the high filtration rates may reflect requirements for excretion of water (Yokota et al. 1985a), but this is not the case for marine hagfish. Moreover, freshwater fish, which do need to get rid of excess water, have much lower SNGFRs than the marine hagfish (Table 3.2).

3.2.3   Changes in Whole-Kidney Glomerular Filtration Rates

Clearance measurements of whole-kidney GFRs have been made for many nonmammalian and mammalian species. Among nonmammalian vertebrates, whole-kidney GFRs often change with changes in the state of hydration or with osmotic stress. These functional patterns are shown in Table 3.3 for a number of species of fishes, amphibians, reptiles, and birds during acute adaptive changes in hydration or during intravenous administration of a salt load (hyperosmotic sodium chloride; usually 1 mol l^{-1}) or a water load (usually a hypoosmotic glucose solution).

For most wholly aquatic species that can be adapted to freshwater or seawater, whole-kidney GFR is much higher in freshwater than in seawater (Table 3.3). For the wholly aquatic sea snakes (*Aipysurus laevis*), however, the whole-kidney GFR, which is low under control circumstances, actually increases slightly with an acute salt load and decreases somewhat with an acute water load (Table 3.3). It does increase substantially with a chronic water load, either freshwater or seawater. Of course, these marine reptiles do have an extrarenal route (oral salt gland) for the excretion of sodium chloride (Dunson et al. 1971). If the salt gland removes sodium chloride rapidly enough, then an acute hyperosmotic salt load might be equivalent to an isosmotic plasma expansion or even an actual hypoosmotic fluid load. In this case, some increase, rather than decrease, in whole-kidney GFR might be expected. Data are not yet available on the simultaneous partitioning of ion excretion between salt gland and kidney in these wholly aquatic marine reptiles.

For terrestrial and semiaquatic nonmammalian vertebrates, the whole-kidney GFR tends to increase with a water load and to decrease with dehydration or a salt load (Table 3.3). As can be seen, however, some variation in this general pattern is found among animals of different species and from different habitats (Table 3.3). For example, among birds, the single passerine species studied (starling, *S. vulgaris*) does not show the same decrease in whole-kidney GFR with a salt load as the gallinaceous species studied. However, starlings cannot tolerate the same salt load that produces the decrease in whole-kidney GFR in gallinaceous birds (Braun 1978). Moreover, clearance studies on conscious, unrestrained starlings do indicate that moderate dehydration produces about a 60 % decrease in whole-kidney GFR (Table 3.3) (Roberts and Dantzler 1989). Thus, the whole-kidney GFR in all avian species studied does show considerable lability with changes in hydration.

Although all terrestrial and semiaquatic amphibians studied respond to changes in states of hydration with changes in whole-kidney GFR (Table 3.3), the response in a species of uricotelic, xerophilic South American tree frog (*Phyllomedusa sauvagii*) is quantitatively quite different from that of others (Table 3.3). The whole-kidney GFR in these animals can increase dramatically when they are placed in freshwater (Table 3.3). However, under normal circumstances, they can endure long periods without free drinking water, maintaining a very high GFR on a diet of insects (Table 3.3). This failure of the GFR to decrease in the absence of free drinking water, as it does in other amphibians, does not appear to be related in any simple way to the excretion of urates as the major end product of nitrogen metabolism (see Table 6.2), because a number of lizards and snakes, which are also uricotelic (Table 6.2), show a marked and rapid decrease in GFR with dehydration or a salt load (Table 3.3). The maintenance of a relatively high GFR with only the water obtained from an insect diet may relate to continued reabsorption of filtered water from the bladder as well as to low evaporative water loss across the skin (Shoemaker and Bickler 1979). In any case, whole-kidney GFR even in these amphibians can change substantially with hydration (Table 3.3).

Table 3.3 Changes in whole-kidney glomerular filtration rate for some fishes, amphibians, reptiles, and birds

	Condition	GFR ml kg^{-1} h^{-1}	References
Fishes			
Petromyzonta			
River lamprey,	Adapted to freshwater	25.07 ± 2.32	Logan et al. (1980b)
Lampetra fluviatilis	Adapted to seawater	4.66 ± 1.53	Logan et al. (1980c)
Freshwater, marine, or euryhaline			
Teleostei			
European eel,	Adapted to freshwater	4.6 ± 0.54	Sharratt et al. (1964)
Anguilla anguilla	Adapted to seawater	1.0 ± 0.22	
Euryhaline			
Rainbow trout,	Adapted to freshwater	8.6 ± 0.95	Brown et al. (1978)
Oncorhynchus mykiss	Adapted to seawater	1.2 ± 0.05	
Euryhaline			
Plains killifish,	Adapted to freshwater	25	Fleming and
Fundulus kansae	Adapted to seawater	1.4	Stanley (1965)
Euryhaline			
Winter flounder,	Adapted to brackish water	1.58 ± 0.29	Elger et al. (1987)
Pseudopleuronectes americanus	Adapted to seawater	0.61 ± 0.08	
Stenohaline, seawater			
Amphibia			
Anura			
Bullfrog	Control (in water)	34.2	Schmidt-Nielsen and
Rana clamitans	Dehydration	5.1	Forster (1954)
Semiaquatic, freshwater			
South American tree	Control (out of water	32.9 ± 4.6	Shoemaker and
frog,	3 days; fed cockroaches)		Bickler (1979)
Phyllomedusa sauvegei	Control (out of water	27.1 ± 3.0	
Terrestrial, arid	27 days; fed cockroaches)		
	Water load (in water)	92.3 ± 6.0	
South African clawed	Freshwater	30	McBean and
toad,	Hyperosmotic saline	12	Goldstein (1970)
Xenopus laevis			
Aquatic, freshwater			
Toad,			
Bufo boreas	Control (in water)	63.1 ± 9.2	Shoemaker and
Terrestrial, moist	Dehydration	1.8 ± 0.5	Bickler (1979)
Reptilia			
Testudinea			
Desert tortoise,	Control	4.7 ± 0.60	Dantzler and
Gopherus agassizii	Salt load (no urine flow	2.9 ± 0.91	Schmidt-Nielsen
Terrestrial, arid	when plasma osmolality		(1966)
	increased 100 mosmol)		
	Water load	15.1 ± 6.64	

(continued)

Table 3.3 (continued)

	Condition	GFR ml kg^{-1} h^{-1}	References
Freshwater turtle,	Control	4.7 ± 0.69	Dantzler and
Pseudemys scripta	Salt load (no urine flow	2.8 ± 0.90	Schmidt-Nielsen
Semiaquatic,	when plasma osmolality		(1966)
freshwater	increased 20 mosmol)		
	Water load	10.3 ± 2.00	
Crocodilia			
Crocodylus johnsoni	Control	6.0 ± 1.5	Schmidt-Nielsen and
Semiaquatic,	Dehydration	1.9 ± 0.2	Davis (1968)
freshwater	Water load	3.3 ± 1.1	
	Control	9.6 ± 1.0	Schmidt-Nielsen and
Crocodylus acutus	Dehydration	6.1 ± 0.6	Skadhauge (1967)
Semiaquatic,	Salt load	7.3 ± 0.6	
freshwater and	Water load	15.2 ± 2.0	
saltwater			
Crocodylus porosus	Control	1.5 ± 0.2	Schmidt-Nielsen and
Semiaquatic,	Salt load	2.8 ± 0.9	Davis (1968)
seawater	Water load	18.8 ± 2.3	
Squamata			
Ophidia			
Bull snake,	Salt load	16.1 ± 1.06	Komadina and
Pituophis melanoleucus	Water load	10.9 ± 1.07	Solomon (1970)
Terrestrial, arid			
Freshwater snake,	Salt load (no urine flow	13.1 ± 1.26	Dantzler (1967)
Nerodia sipedon	when plasma osmolality		
Semiaquatic,	increased more than		
freshwater	50 mosmol)		
	Water load	22.8 ± 1.75	Dantzler (1968)
Olive sea snake,	Control	0.78 (0.49–2.78)	Yokota et al. (1985b)
Aipysurus laevis	Salt load	2.24 (1.41–6.42)	
Aquatic, seawater	Chronic intraperitoneal		
	salt water load	7.05 (6.26–7.83)	
	Water load	0.17 (0.03–0.35)	
	Chronic intraperitoneal		
	water load	5.67 (4.40–6.20)	
Sauria			
Blue-tongued lizard,	Control	15.9 ± 1.0	Schmidt-Nielsen and
Tiliqua scincoides	Dehydration	0.7	Davis (1968)
Terrestrial, moist	Salt load	14.5 ± 0.5	
	Water load	24.5 ± 2.0	
Horned lizard	Control	3.5 ± 0.32	Roberts and Schmidt-
Phrynosoma cornutum	Dehydration	2.1 ± 0.20	Nielsen (1966)
Terrestrial, arid	Salt load	1.7 ± 0.40	
	Water load	5.5 ± 0.54	
Sand goanna,	Dehydration	10.99 ± 0.88[a]	Bradshaw and Rice
Varanus gouldii	Salt load	5.51 ± 1.10[a]	(1981)
Terrestrial, arid	Water load	15.98 ± 1.35[a]	

(continued)

Table 3.3 (continued)

	Condition	GFR ml kg^{-1} h^{-1}	References
Puerto Rico gecko	Control	10.4 ± 0.77	Roberts and Schmidt-
Hemidactylus sp.	Dehydration	3.3 ± 0.37	Nielsen (1966)
Terrestrial, moist	Salt load	11.0 ± 2.18	
	Water load	24.3 ± 1.67	
Rhynchocephalia			
Tuatara,	Control	3.9	Schmidt-Nielsen and
Sphenodon punctatus	Dehydration	3.4	Schmidt (1973)
Terrestrial, moist	Water load	4.8	
Aves			
Galliformes			
Chicken,	Control	73.8 ± 2.40	Ames et al. (1971)
Gallus gallus domesticus	Salt load	21.0 ± 2.40	Dantzler (1966)
Terrestrial, moist	Water load	190.8 ± 2.40	Skadhauge and
			Schmidt-Nielsen
			(1967)
Gambel's quail,	Control	52.8 ± 2.40	Braun and Dantzler
Callipepla gambelii	Salt load	9.0 ± 1.20	(1972), Braun and
Terrestrial, arid	Water load	83.4 ± 13.20	Dantzler (1975)
Passeriformes			
European starling,	Control	169.8 ± 4.80	Braun (1978)
Sturnus vulgaris	Dehydration	69.0 ± 5.40	Roberts and
			Dantzler (1989)
Terrestrial, moist	Salt load	168.6 ± 12.60	Braun (1978)

Values are means or means ± SE except for sea snakes for which, because the data did not show a normal distribution, the values are given as medians and interquartile ranges. All means with SE and medians are for four or more values. The values for the birds and some of the fishes were taken from the literature and converted to ml kg^{-1} h^{-1}

[a]Measurements of plasma levels of arginine vasotocin (AVT) were obtained simultaneously with these measurements of GFR

Among terrestrial and semiaquatic reptiles, the response to a salt load or dehydration is particularly variable for the crocodilians and saurians (Table 3.3). After the studies shown in Table 3.3 were performed, however, each of these crocodilian species was found to have functional extrarenal route (lingual salt gland) for the excretion of inorganic ions (Taplin and Grigg 1981), and, as in the case of the sea sakes (vide supra), its presence may account for the lack of decrease, or even increase, in GFR with an acute salt load. None of the other terrestrial or semiaquatic species listed in Table 3.3 has an effective extrarenal route for excreting inorganic ions.

Another possible reason for the variable response to an acute salt load, especially among ophidian and saurian reptiles, may be the state of hydration of the animal at the time of the experiment. Although experiments varied in the hands of different investigators, the acute salt load was always given in an attempt to increase plasma osmolality, especially in animals in which it was difficult to study dehydration. Unfortunately, in very well-hydrated animals this acute infusion would initially tend to lead to plasma expansion, increased renal blood flow, and increased whole-kidney GFR. With continued infusion, there would certainly be water and thus volume depletion, increased plasma osmolality, and decreased GFR, but most experiments were not run with long-term infusion. Other possible reasons for observed differences in responses are unknown (Dantzler 1976). Whatever the reasons for some of these variations in responses to changes in hydration and salt loads, it is clear that the whole-kidney GFR in nonmammalian vertebrates can change markedly, especially with acute changes in hydration, and therefore may contribute significantly to changes in the volume and composition of the final urine (vide infra).

The whole-kidney GFR in mammals, in contrast to that in nonmammalian vertebrates, is relatively stable during acute, but moderate changes in hydration (Yokota et al. 1985a). There is no increase with a water load and, although decreases do occur with severe dehydration, for most species these decreases appear more pathological than physiological (Yokota et al. 1985a). Therefore, in most mammals, physiological changes in whole-kidney GFR do not contribute significantly to changes in the volume and composition of the final urine (vide infra). However, in a few mammalian species from arid habitats, the whole-kidney GFR decreases significantly during chronic dehydration that the animals tolerate easily (Dantzler 1982). For example, the whole-kidney GFR of the spiny mouse (*Acomys cahirinus*) decreases about 55 % after 14 days of acclimation to a minimal water supply (Haines and Schmidt-Nielsen 1977) and that of the camel (*Camelus dromedarius*) decreases as much as 70 % after 10 days of dehydration (Yagil and Berlyne 1976). Since these animals are accustomed to these degrees of dehydration, these decreases in whole-kidney GFR appear to be physiological responses to the need to conserve water.

3.2.4   Changes in Single-Nephron Filtration Rates and in Number of Filtering Nephrons

The physiological decreases in whole-kidney GFR observed in a few mammalian species (e.g., the camel and spiny mouse, noted above) apparently involve a decrease in the filtration rates of all nephrons or populations of nephrons rather than a decrease in the number of nephrons actually filtering (Yokota et al. 1985a). In contrast, however, the changes in whole-kidney GFR observed in most nonmammalian vertebrates apparently result primarily from changes in the number of glomeruli filtering (Braun and Dantzler 1972, 1974, 1975; Brown et al. 1980; Dantzler 1966, 1967; Dantzler and Schmidt-Nielsen 1966; Forster 1942; Hickman 1965; Lahlou 1966; Mackay and Beatty 1968; Richards and Schmidt 1924; Schmidt-Nielsen and Forster 1954; Yokota and Dantzler 1990). Although changes in the individual filtration rates of glomeruli that continue filtering also apparently occur (Braun and Dantzler 1972, 1975; Brown et al. 1978, 1980; Richards and Schmidt 1924; Yokota and Dantzler 1990), changes in the number of glomeruli actually filtering appear to be quantitatively more important for the regulation of whole-kidney GFR in nonmammalian vertebrates.

An exception to this generalization appears to be the glomerular function in the primitive river lamprey (*L. fluviatilis*) discussed above. Apparently, when the whole-kidney GFR in these animals decreases with adaptation to brackish water (Table 3.3), the filtration rates of all individual nephrons decrease (Table 3.2), but all continue to filter (McVicar and Rankin 1985; Rankin et al. 1980). The decrease in SNGFR apparently results both from a decrease in mean arteriole pressure and from an increase in resistance at the afferent arterioles, thereby producing a decrease in the outwardly directed hydrostatic pressure along the glomerular capillaries (vide supra; Table 3.1). However, apparently because each afferent arteriole supplies more than one glomerulus and each glomerulus contains capillaries arising from more than one afferent arteriole (Hentschel and Elger 1987) (vide supra; Chap. 2), all glomeruli can continue filtering even if the increased resistance is much more marked in some afferent arterioles than others.

The concept that changes in the whole-kidney GFR result primarily from changes in the number of glomeruli filtering in most nonmammalian vertebrates is supported by a number of lines of evidence. The earliest evidence was the observation that for representative species of most major groups of nonmammalian poikilotherms—teleost fishes, amphibians, and reptiles—the maximum rate of transport (TM) of glucose or para-aminohippurate (PAH) by the renal tubules varies directly with the whole-kidney GFR (Brown et al. 1980; Dantzler 1967; Dantzler and Schmidt-Nielsen 1966; Forster 1942; Lahlou 1966; Mackay and Beatty 1968). If changes in whole-kidney GFR result from changes in the filtration rate of each glomerulus (or a population of glomeruli) but all continue to filter, the Tm for glucose or PAH transport should not change because the mass of tissue transporting these substances and contributing to the final urine would not have changed (Forster 1942; Ranges et al. 1939). Although this evidence for glomerular intermittency is indirect, it agrees rather well with direct visual observations of the activity of the glomerular circulation in amphibians and reptiles (Garland et al. 1975; Grafflin and Bagley 1952; Richards and Schmidt 1924; Sawyer 1951; Yokota and Dantzler 1990). Moreover, histologic studies of the kidneys of a number of reptiles show that the ratio of the number of open to closed proximal tubule lumina correlates roughly with the whole-kidney GFR (Schmidt-Nielsen and Davis 1968). Because a proximal tubule in most species collapses when the glomerulus ceases filtering, these observations also support the concept that changes in whole-kidney GFR reflect changes in the number of glomeruli filtering.

In addition, direct quantitative measurements of blood flow rates in single glomeruli in the kidney of a reptile (garter snake, *T. sirtalis*) now confirm the presence of intermittent blood flow and, presumably, intermittent filtration and indicate that the fraction of glomeruli with intermittent blood flow increases directly with increasing plasma osmolality (Fig. 3.1) (Yokota and Dantzler 1990). These studies also reveal highly variable blood flow in glomeruli that are continuously perfused, suggesting that variations in the filtration rates of glomeruli that are filtering may be more significant in reptiles than previously supposed. However, as noted above, the degree to which changes in blood flow along the glomerular capillaries influence the SNGFR depends on whether filtration equilibrium is reached along the glomerular capillaries, which may not be the case in these reptiles (S. D. Yokota personal communication).

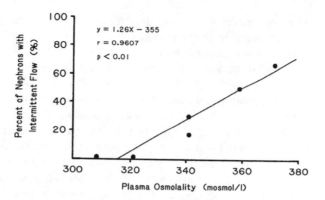

Fig. 3.1 Relation of percent of nephrons with intermittent flow to plasma osmolality in individual garter snakes (*Thamnophis sirtalis*). Each point represents a single animal. Plasma osmolality was measured in six animals. Line was fitted by linear regression (Yokota and Dantzler 1990, with permission)

Studies with the constant infusion sodium ferrocyanide technique (deRouffignac et al. 1970) and with Alcian blue injections have supported the concept of intermittent glomerular function and provided additional quantitative information on SNGFRs in teleost and elasmobranch fishes (Brown et al. 1978, 1980; Brown and Green 1987; Elger et al. 1984a). The sodium-ferrocyanide infusion technique, which relies on the formation of insoluble precipitates of radioactively labeled Prussian blue, permits the measurement of filtration rates of individual nephrons inaccessible to micropuncture at a single point in time. Such studies on a species of euryhaline teleost [rainbow trout, *Oncorhynchus mykiss* (formerly *Salmo gairdneri*)] indicate that about 45 % of the glomeruli are filtering in animals adapted to freshwater whereas about 5 % are filtering in animals adapted to seawater (Table 3.2). This difference between the numbers of glomeruli filtering in freshwater- and seawater-adapted animals corresponds to the difference between whole-kidney glomerular filtration rates under the same circumstances (Table 3.3). Of interest but of no clear physiological significance, the few glomeruli functioning during adaptation to seawater filter at an average rate greater than that of glomeruli functioning during adaptation to freshwater (Table 3.2).

These studies and others on a marine elasmobranch species (lesser spotted dogfish, *Scyliorhinus canicula*) also indicate that during changes in whole-kidney GFR, the filtration rates of individual functioning glomeruli can change (Brown et al. 1978; Brown and Green 1987). These observations provide quantitative support for the inferences drawn from visual observations of changes in blood flow through the glomerular capillaries of amphibians and reptiles (Bordley and Richards 1933; Grafflin and Bagley 1952; Richards and Schmidt 1924; White 1929) (W. H. Dantzler unpublished observations).

Finally, these sodium ferrocyanide infusion studies suggest that some of the nonfiltering glomeruli in both teleosts and elasmobranchs are still perfused with blood (Brown et al. 1978; Brown and Green 1987). Because this technique does not measure blood flow directly and distribution of the marker may sometimes be misleading (Brown et al. 1993), these observations require additional confirmation. If they are correct, however, they suggest that for some nephrons cessation of filtration results from a capillary hydrostatic pressure at the afferent end of the network equal to or even slightly below the opposing plasma colloid osmotic and capsular pressures or from an increase in the filtration barrier, not from a complete interruption of blood flow. In this regard, Brown et al. (1983) found structural changes in the podocytes and their processes in the glomeruli of seawater-adapted trout, suggesting that there may be changes in the filtration barrier and, thus, in the ultrafiltration coefficient under these circumstances.

Studies involving Alcian blue injections into stenohaline freshwater teleosts (Prussian carp, *Carassius auratus gibelio*) transferred to isosmotic seawater indicate that, during the first 2 h of adaptation, many glomeruli are no longer perfused with blood (Elger et al. 1984a). With prolonged adaptation to this medium, however, the surviving animals actually show atrophy and loss of many glomeruli (vide supra) (Elger and Hentschel 1981), but this is not a physiological form of adaptation.

Regulation of whole-kidney GFR by the regulation of the number of glomeruli filtering appears to be a practical adaptation for fishes, amphibians, and reptiles in which the nephrons empty at right angles into collecting ducts and are not arranged to function in concert to produce a urine hyperosmotic to the plasma. Moreover, most of those species in which glomerular intermittency has been well documented have renal portal systems that can continue to nourish the cells of nonfiltering nephrons in the absence of a postglomerular arterial supply. Indeed, this portal peritubular blood supply may permit the continued tubular secretion of inorganic ions, organic molecules, and possibly fluid by nonfiltering glomerular nephrons (vide infra).

In some fish species, however, in which there is some evidence for glomerular intermittency, a renal portal system is either not so clearly defined or even absent. For example, one freshwater teleost species (northern pike, *Esox Lucius*), in which there is some evidence for glomerular intermittency (Hickman 1965), apparently has a rudimentary renal venous portal system (Hickman and Trump 1969). However, there is no report of even a rudimentary renal venous portal system in another freshwater teleost (white sucker, *Catostomus commersoni*) in which a change in the number of filtering glomeruli with temperature appears to occur (Mackay and Beatty 1968). The possibility of a collateral postglomerular arterial blood supply from the efferent arterioles of filtering glomeruli to the capillary network surrounding the tubules of nonfiltering nephrons has yet to be explored in these animals.

However, the river lamprey, which spends part of its life cycle in freshwater and part in seawater and lacks any evidence of a renal portal venous system or of a postglomerular collateral blood supply (Logan et al. 1980a; McVicar and Rankin 1985), does not exhibit glomerular intermittency (vide supra) (McVicar and Rankin 1985; Rankin et al. 1980). Of course, as noted above, the lack of glomerular intermittency in these animals may also be related to the fact that each glomerulus is supplied by more than one afferent arteriole and that each afferent arteriole supplies more than one glomerulus.

Finally, as already pointed out above, the prolonged adaptation of the stenohaline freshwater Prussian carp to isosmotic seawater, which some animals do tolerate, leads to atrophy and complete disappearance of many nephrons in the surviving animals. This atrophy may possibly be a result of an inadequate nutrient blood supply in the absence of a renal venous portal system.

The function of the individual nephrons in birds is somewhat different from that in fishes, amphibians, and reptiles. Unlike other nonmammalian vertebrates and like mammals, birds are homeotherms with a relatively high and stable blood pressure. However, as discussed above, the avian kidney contains both reptilian-type (loopless) and mammalian-type (looped) nephrons (Figs. 2.1, 2.2 and 2.6). Also, like most other nonmammalian vertebrates but unlike mammals, birds have a renal portal system that contributes to sinuses surrounding the reptilian-type nephrons and the proximal and distal tubules of the mammalian-type nephrons (Fig. 2.7) (Wideman et al. 1981).

Because of the arrangement of the reptilian-type nephrons in a manner that should not contribute directly to the production of concentrated urine and because of the presence of a renal portal system, it appeared possible that avian nephrons could function intermittently. Indeed, initial studies on gallinaceous birds demonstrated that the Tm for glucose reabsorption or PAH secretion by the renal tubules varies directly with the whole-kidney GFR (Braun and Dantzler 1972; Dantzler 1966). These observations suggested that changes in whole-kidney GFR resulted from changes in the number of glomeruli filtering, as in most other nonmammalian vertebrates, but this approach was too indirect to indicate whether the apparent changes in the number of filtering glomeruli involved only the reptilian-type nephrons or both types of nephrons.

Studies employing the constant sodium ferrocyanide infusion technique, however, have provided an answer to this question and quantitative information on the filtration rates for both types of nephrons (Braun 1978; Braun and Dantzler 1972, 1974, 1975). In both the gallinaceous species (Gambel's quail, C. gambelii) and the passerine species (European starling, S. vulgaris) studied, the mean SNGFR for the mammalian-type nephrons is slightly more than twice that for the reptilian-type nephrons during a control diuresis (Tables 3.2 and 3.4). Moreover, the mean control values for the SNGFRs for each population are essentially the same for the two species (Tables 3.2 and 3.4). However, because there are more nephrons in the starling kidney (about 74,000) (Braun 1978) than in the quail kidney (about 47,000) (Braun and Dantzler 1972), the whole-kidney GFR for the starlings is greater than that for the quail under control conditions (Table 3.3). There is, of course, a range of SNGFRs within each nephron population in both species, apparently related to the range of nephron sizes noted above (Chap. 2), and both the sodium ferrocyanide infusion and micropuncture studies show that the smallest, most superficial reptilian-type nephrons have very low filtration rates (Table 3.2) (Laverty and Dantzler 1982, 1983) (J. R. Roberts and W. H. Dantzler unpublished observations).

When the whole-kidney GFR in Gambel's quail decreases during a salt load (Table 3.3), all reptilian-type nephrons apparently cease filtering, whereas the mammalian-type nephrons all continue filtering at a slightly reduced average rate (Table 3.4). Because about 90 % of the nephrons in this species are of the reptilian-type, the decrease in the number of these nephrons filtering accounts almost entirely

Table 3.4 Effects of salt load, water load, and arginine vasotocin administration on single nephron glomerular filtration rates (SNGFR) in Gambel's quail and European starlings

Treatment	Mammalian-type Nephrons		Reptilian-type Nephrons		References
	SNGFR (nl min^{-1})	Percent filtering	SNGFR (nl min^{-1})	Percent filtering	
Gambel's quail (*Callipepla gambelii*)					
Control (2.5 % mannitol)	14.6 ± 0.79 (27)	100	6.4 ± 0.20 (41)	71	Braun and Dantzler (1972)
Salt load (45 mEq kg^{-1})	12.7 ± 0.52 (70)	100	0	0	Braun and Dantzler (1972)
Water load	33.2 ± 1.57 (155)	100	11.4 ± 0.75 (146)	100	Braun and Dantzler (1975)
Arginine Vasotocin (10 ng kg^{-1})	11.3 ± 0.89 (102)	100	4.7 ± 1.05 (31)	52	Braun and Dantzler (1974)
(50 ng kg^{-1})	16.5 ± 0.75 (64)	100	6.9 ± 0.42 (26)	26	Braun and Dantzler (1974)
European Starling (*Sturnus vulgaris*)					
Control (2.5 % mannitol)	15.6 ± 0.75 (208)	?20	7.0 ± 0.35 (185)	?45	Braun (1978)
Salt load (34 mEq kg^{-1})	14.6 ± 0.51 (28)	?72	0	0	Braun (1978)

Values are means $\pm$ SE. Numbers in parentheses are sample sizes

for the decrease in whole-kidney GFR (Braun and Dantzler 1972). Studies in which the renal vasculature in Gambel's quail is filled with a silicone elastomer support the concept that during a salt load, reptilian-type nephrons cease filtering and suggest that this results from constriction of the afferent arterioles (Figs. 3.2 and 3.3) (also see Braun 1976). During a control diuresis, the vessels in the superficial areas of the kidney fill well and the afferent arterioles and glomerular capillaries of reptilian-type nephrons are clearly seen (Fig. 3.2). During the salt load that produces the decrease in whole-kidney GFR and apparent decrease in the number of filtering reptilian-type nephrons described above, few, if any, afferent arterioles or glomerular capillaries of reptilian-type nephrons in this same superficial area of the kidney are filled (Fig. 3.3).

It is also important to note that in Gambel's quail, only 71 % of the reptilian-type nephrons are filtering even during the control diuresis (Table 3.4). This finding is consistent with direct visual observations of the surface of the quail kidney in vivo during micropuncture experiments (W. H. Dantzler unpublished observations). It also resembles the situation in rainbow trout (*O. mykiss*) adapted to freshwater (Table 3.2) (Brown et al. 1980).

When whole-kidney GFR in Gambel's quail increases during a water load, all reptilian-type nephrons as well as all mammalian-type nephrons appear to be filtering and SNGFRs of both types approximately double (Table 3.4). In this species, therefore, the increase in whole-kidney GFR produced by a large intravenous

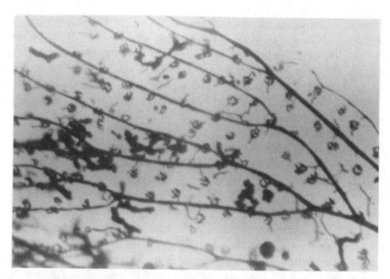

Fig. 3.2 Section of cleared tissue from superficial area of kidney of a bird (Gambel's quail, *Callipepla gambelii*) that had received only control mannitol infusion. Vasculature filled with silicone elastomer (Microfil) (Braun 1976, with permission)

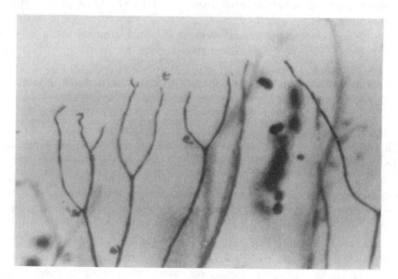

Fig. 3.3 Section of cleared tissue from superficial area of kidney of a bird (Gambel's quail, *Callipepla gambelii*) that had received a salt load. Vasculature filled with silicone elastomer (Microfil) (Braun 1976, with permission)

water load apparently results from a marked increase both in the number of filtering reptilian-type nephrons and in the SNGFRs of all nephrons (Braun and Dantzler 1972).

Unfortunately, no direct determination of the number of nephrons filtering in the starling kidney under any conditions is available. However, if all the mammalian-type and reptilian-type nephrons were filtering at the rates determined during the control diuresis (Table 3.4), the whole-kidney GFR would be far greater than that actually measured in these animals (Table 3.3) (Braun 1978). Braun, using his determination that about 70 % of the nephrons in the starling kidney are of the reptilian type and about 30 % are of the mammalian type, estimated that the control whole-kidney GFR shown in Table 3.3 could be attained if 45 % of the reptilian-type nephrons and 20 % of the mammalian-type nephrons were filtering at the control rates shown in Table 3.4 (Braun 1978). Of course, other combinations are also possible. It is not known whether any mammalian-type nephrons that are not filtering are small, transitional nephrons with short loops of Henle or larger nephrons with long loops.

During the maximum intravenous salt load tolerated by starlings, all the reptilian-type nephrons apparently cease filtering, but the SNGFRs of the filtering mammalian-type nephrons are unchanged from the control values (Table 3.4). Because the whole-kidney GFR does not change at this time (Table 3.3), about 72 % of the mammalian-type nephrons must be filtering (Table 3.4).

As discussed above for other nonmammalian vertebrates, regulation of the whole-kidney GFR in birds by altering the number of filtering reptilian-type nephrons appears practical because these nephrons empty at right angles into collecting ducts and apparently cannot function in concert to contribute directly to the urine-concentrating mechanism. However, such changes in the number of filtering reptilian-type nephrons may still influence the operation of the concentrating mechanism in less direct ways (vide infra; Chap. 7). The apparent changes in the number of filtering mammalian-type nephrons in the starling kidney may have direct effects on the operation of the concentrating process (vide infra; Chap. 7).

In the avian kidney, the viability of the proximal and distal tubules of the nonfiltering mammalian-type nephrons as well as all portions of the reptilian-type can be maintained by the renal portal system. However, the renal portal system of birds differs from that of other nonmammalian vertebrates in having a unique smooth muscle valve at the juncture of the external iliac vein and the efferent renal vein that determines whether blood from the posterior extremities bypasses the kidney and flows directly into the central circulation (open valve) or flows first through the renal portal system and supplies the peritubular sinuses (closed valve) (Sperber 1948). The fraction of flow in either direction, of course, depends on the extent to which the valve is opened or closed. This valve is subject to both adrenergic and cholinergic neural control (adrenergic stimulus producing relaxation and cholinergic stimulus producing contraction) (Akester and Mann 1969; Burrows et al. 1983) and, possibly, humoral control. However, it is not yet clear how, or even if, this control is coordinated with changes in the number of filtering nephrons (vide infra).

3.2.5 Regulation of Single-Nephron Glomerular Filtration Rates and Number of Filtering Nephrons

3.2.5.1 Hormonal Regulation

Neurohypophysial Peptides

These hormones apparently are important in regulating both the individual nephron filtration rates and the number of filtering nephrons. In mammals, the naturally occurring antidiuretic neurohypophysial peptide, arginine vasopressin (AVP), is a powerful vasoconstrictor, but there is no convincing evidence that it influences glomerular filtration rate in mammals under physiological conditions. It is still possible that AVP plays a role in the apparent physiological decrease in whole-kidney GFR observed in the spiny mouse or the camel (vide supra), but this possibility has yet to be adequately evaluated. If AVP does play such a role in these animals, it probably acts through mammalian V1a vascular receptors located in the afferent glomerular arterioles. These are G-protein coupled receptors with seven membrane-spanning domains that, in turn, act through a phosphatidyl inositol and calcium pathway to cause vasoconstriction (Morel et al. 1992).

Arginine vasotocin (AVT), which has been identified in the neurohypophysis of all those nonmammalian vertebrates examined (Follett and Heller 1964a, b; Heller and Pickering 1961; Munsick 1964, 1966; Sawyer et al. 1961; Sawyer and Pang 1975), clearly has a role in glomerular regulation. In fact, among the nonmammalian vertebrates, it may be the most important physiological regulator of the number of filtering nephrons and SNGFR with changes in hydration or the salinity of the aqueous environment. However, the details of this regulation can vary greatly among nonmammalian vertebrate classes and species.

For example, among the fishes, the administration of small, apparently physiological doses of AVT produces an increase in whole-kidney GFR and urine flow in the African lungfish (*Protopterus aethiopicus*) (Sawyer 1970), and initially this appeared to be true for teleosts as well. However, later work on euryhaline teleosts (European eel, *Anguilla Anguilla*) adapted to freshwater showed that small, nonpressor doses of AVT cause a decrease in whole-kidney and urine flow whereas higher, vasopressor doses cause an increase in whole-kidney GFR and urine flow (Babiker and Rankin 1978; Henderson and Wales 1974). In seawater-adapted eels, renal function responds only to the higher, vasopressor doses of AVT (Babiker and Rankin 1978). These observations suggest that the renal vessels, probably the afferent arterioles, are more responsive to the vasoconstrictor action of AVT than are the peripheral systemic arterioles. With sufficient AVT to produce an increase in systemic arterial pressure, however, the constriction of the afferent glomerular arterioles is overridden (Nishimura 1985; Nishimura and Imai 1982). If the hormone does function in this manner, then the seawater-adapted animals with their low whole-kidney GFR (Table 3.3) may already be responding maximally to low levels of AVT.

The cellular and molecular bases for these responses to AVT in fishes are far from completely understood, although some progress has been made in recent years. Homologues of mammalian V1a-type vascular receptors for AVT, with expression in the kidneys, have been cloned from African lungfish (*Protopterus annectens*) (Konno et al. 2009) and numerous teleosts (cave-dwelling fish, *Astyanax fasciatus*; euryhaline flounder, *Platichthys flesus*; Pacific salmon, *Oncorhynchus kisutch*; and white sucker, *Catostomus commersoni*) (Mahlmann et al. 1994; Warne 2001). Moreover, in the kidneys of the euryhaline flounder, these receptors have been localized to the endothelium of the glomerular afferent and efferent arterioles and the capillaries surrounding the collecting ducts (Weybourne et al. 2005). It is assumed that, as in mammals, these receptors act through the phosphatidyl inositol and calcium pathway to cause vasoconstriction, although this has yet to be demonstrated. Moreover, differences in degree of expression or sensitivity of these receptors in various vascular regions, which might account for the differences in responses to AVT noted above, have not been determined. In any case, the studies indicate that AVT may play a physiological role in the glomerular adjustment of euryhaline teleosts to seawater or freshwater (Table 3.3), probably largely by regulating the SNGFR (Table 3.2) and especially the number of filtering nephrons.

Arginine vasotocin also appears to have an important physiological role in the control of the glomerular filtration rate in amphibians and reptiles. Injections of small, apparently physiological doses of AVT that do not alter mean arterial pressure produce decreases in whole-kidney GFR that mimic those observed with a salt load or dehydration (Table 3.3) (Bradshaw and Rice 1981; Butler 1972; Dantzler 1967; Jard and Morel 1963). Although homologues of mammalian V1a vascular receptors with expression in the kidneys have been cloned from a newt (*Cynops pyrrhogaster*) (Hasunuma et al. 2007), frogs (*Rana catesbeiana* and *R. esculenta*) (Acharjee et al. 2004) and a toad (*Xenopus laevis*) (Mahlmann et al. 1994), nothing more is known about the sites, degrees of expression, or signaling pathways. Moreover, no AVT receptors have yet been cloned from any reptile species although it is likely that vascular V1-type receptors for AVT do exist.

There is, however, additional physiological data supporting the regulatory effects of AVT on glomerular filtration rate in reptiles. The plasma AVT level and plasma osmolality in the one lizard species in which they have been measured (sand goanna, *Varanus gouldii*) increase with dehydration or a salt load as the whole-kidney GFR decreases and decrease with a water load as the whole-kidney GFR increases (Table 3.3) (also Bradshaw and Rice 1981; Rice 1982). In addition, the diameter of the glomerular afferent arterioles and glomerular blood flow (and presumably filtration rate) of individual nephrons in the one species in which they have been measured quantitatively (garter snake, *T. sirtalis*) decrease with the administration of small doses of AVT (Fig. 3.4) (Yokota and Dantzler 1990). With the administration of larger but probably still physiological doses of AVT, glomerular blood flow and presumably filtration cease altogether (Fig. 3.4) (Yokota and Dantzler 1990). Thus, AVT appears to be physiologically important in amphibians and reptiles for the regulation of whole-kidney GFR, primarily by regulating

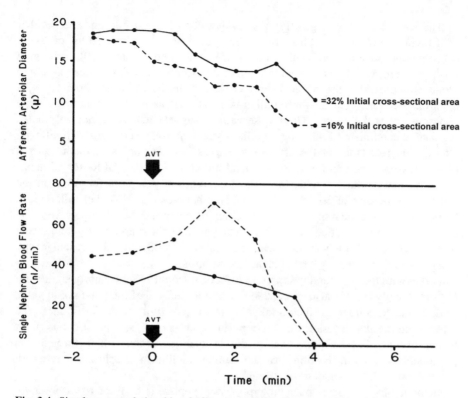

Fig. 3.4 Simultaneous relationship of afferent arteriolar diameter and blood flow for two representative nephrons in snake (*Thamnophis sirtalis*) kidney during continuous infusion of arginine vasotocin (AVT). Arrows mark start of AVT infusion at rate of 17 pg/100 g min^{-1} (Yokota and Dantzler 1990, with permission)

the number of glomeruli filtering but also by regulating the filtration rates of those glomeruli that are filtering.

Arginine vasotocin is also important physiologically in the regulation of glomerular filtration in birds. Small, apparently physiological doses that do not alter systemic blood pressure produce significant decreases in whole-kidney GFR in chickens and Gambel's quail (Ames et al. 1971; Braun and Dantzler 1974). In Gambel's quail, these decreases are explained quantitatively by decreases in the number of reptilian-type nephrons filtering without any significant changes in the average SNGFR of either nephron type (Table 3.4) (Braun and Dantzler 1974). The decreases in the number of filtering reptilian-type nephrons with the administration of AVT, as in the case of administration of a salt load (e.g., Fig. 3.3), appear to result from constriction of the afferent glomerular arterioles (Dantzler and Braun 1980). Furthermore, it should be noted, that a salt load comparable to that used in Gambel's quail produces a marked increase in the plasma level of AVT in chickens (Koike et al. 1979).

The role of AVT in regulating the glomerular filtration rate and the number of filtering nephrons in birds is further supported by the results of acute neurohypophysectomy in Gambel's quail (Dantzler and Braun 1980). Following this procedure, which is assumed to remove endogenous AVT, the mean systemic blood pressure falls, the whole-kidney GFR and the SNGFRs for both types of nephrons decrease, but all the reptilian-type as well as all the mammalian-type nephrons are still filtering. These observations suggest that AVT not only helps regulate the number of filtering reptilian-type nephrons and their SNGFRs by altering the resistance at the afferent arterioles but also helps maintain normal blood pressure and, therefore, normal renal blood flow and SNGFRs for all nephrons. In view of these results, there must be vascular receptors for AVT on the afferent glomerular arterioles and probably other vessels. However, although a homologue of the mammalian V1a vascular receptor has been cloned from the shell gland and brain of chickens, this particular receptor is apparently not present in the kidney or specific vascular locations (Tan et al. 2000). No other vascular-type receptors for AVT have yet been identified in birds.

Along with these general vasopressor effects, it would be particularly significant physiologically if AVT also stimulated the smooth muscle of the renal portal valve (vide supra) to contract. If the portal valve had a sensitivity to AVT similar to the afferent glomerular arterioles of the reptilian-type nephrons, this vasoconstrictor response could be very important in directing more venous blood through the peritubular sinuses at the time that the number of filtering nephrons is reduced. But this possibility has yet to be examined.

In mammals, oxytocin, the neurohypophysial peptide in addition to vasopressin, does not influence glomerular filtration under physiological conditions. However, the situation is less clear for mesotocin (8-isoleucine oxytocin), the oxytocin-like principle found in addition to vasotocin in the neurohypophysis of all nonmammalian tetrapods (amphibians, reptiles, and birds) as well as lungfish and marsupials. Very large, probably pharmacological doses of mesotocin produce a weak diuresis in lungfish (*Protopterus aethiopicus*). However, among amphibians, a number of studies suggest that it may have a physiological role as a glomerular diuretic agent in some anurans and urodeles. Small doses of mesotocin that have no systemic vasopressor effects produce an increase in whole-kidney GFR and urine flow without affecting the relative free-water clearance in amphibious bullfrogs (*R. catesbeiana*) and terrestrial Japanese toads (*Bufo bufo japonicas*) (Pang and Sawyer 1978; Uchiyama et al. 1985). Somewhat similarly, low doses of mesotocin restore whole-kidney GFR and urine flow in hypophysectomized larval and adult tiger salamanders (*Ambystoma tigrinum*) (Hartenstein and Stiffler 1990; Stiffler 1981; Stiffler et al. 1984). More recently, a mesotocin receptor with high expression in the kidney has been cloned from a North American terrestrial toad (*Bufo marinus*). However, many more detailed studies are required to be certain that mesotocin actually has a physiological role in maintaining or increasing whole-kidney GFR in amphibians and, if it does, whether it regulates primarily the SNGFR or the number of filtering nephrons.

Among reptiles, a relatively recent study on a chelonian species (Western painted turtle, *Chrysemis picta*) indicates that mesotocin has no effect on renal function except at very high pharmacological doses (Butler and Snitman 2005). Even at these doses, it has only a small effect that is antidiuretic rather than diuretic. And this antidiuretic effect involves a decrease in both whole-kidney GFR and tubular reabsorption of water. Thus, it seems unlikely that mesotocin has any physiological role in regulating SNGFR or the number of filtering nephrons in reptiles.

Among the birds, the plasma level of mesotocin in chickens (*Gallus gallus domesticus*) was found to increase with dehydration (Nouwen et al. 1984) and the infusion of an intravenous salt load (Koike et al. 1986). However, intravenous injections of mesotocin produced a biphasic response, depending on the dose (Takahashi et al. 1995). Low doses produced a decrease in urine volume and high, clearly pharmacological doses produced an increase in urine volume, but there is no information on whether these effects were glomerular or tubular or both. Mesotocin does bind to specific plasma membrane receptors in the chicken kidney, which appear to be different in the cortical and medullary regions (Takahashi et al. 1996, 1997), but these receptors have yet to be cloned. At this time, it is not clear whether mesotocin has any physiological role in regulating glomerular filtration in birds.

A receptor for isotocin, the oxytocin-like principle found in the neurohypophysis of all teleosts, has been cloned from the white sucker (*C. commersoni*). However, it is only present at rather low levels in the kidney. Moreover, although an early report suggested that mesotocin might be a glomerular diuretic in teleosts (Chester Jones et al. 1969), the physiological nature of this response has been challenged (Babiker and Rankin 1978), and it is simply not clear at this time that mesotocin is a physiological regulator of glomerular function.

Renin–Angiotensin System

Renin activity and granular cells in the kidneys and the production of angiotensin from homologous plasma and kidney extracts have now been demonstrated in representatives of all vertebrate classes (Nishimura 2001). Moreover, angiotensin II receptors have so far been demonstrated in elasmobranchs, teleosts, and all tetrapods (Nishimura 2001). Thus, the renin–angiotensin system appears to be widely distributed and probably to have evolved early in phylogeny (Nishimura 1980, 2001;Sokabe 1974). Any physiological renal effects of the renin–angiotensin system could result from effects on systemic blood pressure, the delivery of angiotensin II formed outside the kidney to renal structures, or the local action of angiotensin II formed within the kidney. Although angiotensin II could play a physiological role in regulating glomerular filtration rate, presumably by regulating the number of filtering glomeruli, the few studies on this problem have generally been negative or equivocal. Most of the experimental effects of angiotensin II on the glomerular filtration rate appear to be pharmacological in nature (Nishimura 1985). However, preliminary data on bullfrogs (*R. catesbeiana*) do indicate that during perfusion of renal arteries under constant pressure, even a nonpressor dose of angiotensin II reduces the whole-kidney GFR (Nishimura 1985). Also, a more complete study on anesthetized trout (*O. mykiss*), in which the systemic arterial pressure was maintained by an infusion of norepinephrine, indicates that infusions of angiotensin II can reduce whole-kidney GFR in freshwater-adapted animals by reducing the number of filtering glomeruli and in seawater-adapted animals by reducing the SNGFR of the few filtering nephrons (Brown et al. 1980). However, an additional, although less extensive, study on this same species in the absence of norepinephrine (see below for effects of catecholamines on GFR) suggests that regulation of GFR by angiotensin II in freshwater-adapted animals is accompanied by a systemic pressor effect (Brown et al. 1993; Gray and Brown 1985). Thus, although angiotensin II is certainly a naturally occurring peptide in both frogs and teleosts and its glomerular effects in trout suggest that it could play a physiological role in renal adaptation to seawater, it is still not clear whether the observed effects in either species are primarily physiological or pharmacological. Pharmacological effects of angiotensin II on the glomerular filtration rate of mammals are clearly documented (Brenner et al. 1981), but physiological regulation, in so far as it exists, may involve adjustments to the tubuloglomerular feedback system (vide infra).

Prolactin

The role of prolactin in regulating glomerular filtration rate, where it occurs, varies among the vertebrates. It appears to have no effect on whole-kidney GFR in mammals (Ibarra et al. 2005) or on whole-kidney GFR or SNGFR in birds (Roberts and Dantzler 1992). However, a few studies suggest that it could have a physiological role in determining the number of filtering nephrons in teleost fish and chelonian reptiles. A histological study of one euryhaline teleost species (the stickleback, *Gasterosteus aculeatus*) indicates that the administration of prolactin to prolactin-deficient, seawater-adapted animals increases the number of filtering glomeruli in a manner similar to adaptation to freshwater (Lam and Leatherland 1969). Also, administration of prolactin to some freshwater turtle species (at least, *Chrysemys picta* and, possibly, *Pseudemys scripta*) produces a significant increase in the low whole-kidney GFR produced by hypophysectomy (Brewer and Ensor 1980). Such changes in whole-kidney GFR in turtles are presumed to reflect changes in the number of filtering glomeruli (vide supra). These studies indicate that prolactin could help to determine the increase in GFR observed with adaptation to freshwater in teleosts or with the intake of a water load in some reptiles. However, much more information on the possible physiological versus pharmacological significance of this apparent glomerular action and on the mechanism involved is certainly required.

Epinephrine

The physiological role of circulating epinephrine, released from the adrenal glands under stress, on direct regulation of SNGFR and the number of filtering nephrons is unclear. However, in one study on a marine elasmobranch species (lesser spotted dogfish, *S. scyliorhinus*), infusions of epinephrine markedly increased whole-kidney GFR (Brown and Green 1987). Measurements of the number of filtering nephrons by the sodium-ferrocyanide infusion method and of SNGFRs by micropuncture during this epinephrine infusion indicate that the increase in whole-kidney GFR results primarily from an increase in SNGFR with a decrease in the number of filtering nephrons (Brown and Green 1987). This pattern is different from what might have been expected, but the exact effects on the local vasculature, probably the afferent arterioles, as well as the balance of local and systemic effects were not evaluated. Moreover, it is not clear that such an infusion of epinephrine adequately mimics a physiological situation. At present, there is no significant information about the possible glomerular regulatory effects of epinephrine in other non-mammalian vertebrates.

3.2.5.2   Neural Regulation

The true physiological significance of neural control of glomerular filtration is not completely clear. However, the afferent and efferent arterioles of mammalian glomeruli are richly innervated with sympathetic adrenergic nerves, and renal nerve stimulation increases both afferent and efferent resistances, thereby reducing glomerular blood flow and SNGFR (Munger et al. 2012). Thus, they could play a role in glomerular regulation.

Similarly, a number of studies suggest that sympathetic adrenergic innervation may play a role in regulating glomerular filtration in fishes, amphibians, and reptiles. Preglomerular sphincters at the start of the afferent arterioles and extensive sympathetic adrenergic innervation of both these sphincters and the afferent arterioles are found in the euryhaline rainbow trout (*O. mykiss*) (Elger et al. 1984b). Moreover, the administration of α-adrenergic blockers, bretylium and phentolamine, to these animals partially prevents the reduction in whole-kidney GFR observed with adaptation to salt water (Elger and Hentschel 1983). These observations suggest that α-adrenergic nerves may play a role in controlling resistance of afferent arterioles, and, thus, the number of filtering glomeruli, during adaptation to waters of varying salinity.

The autonomic nervous system also may be important in controlling the number of filtering glomeruli in amphibians during changes in hydration. A glomerular antidiuresis occurs during water deprivation in bullfrogs (*R. catesbeiana*) in the absence of any increase in circulating AVT and following the destruction of the hypothalamus, the source of AVT production (Gallardo et al. 1980). The glomerular response is eliminated when the animals are pithed or when an α-adrenergic blocker, phenoxybenzamine, is administered; it is mimicked by the arterial administration of norepinephrine to the pithed animals (Gallardo et al. 1980). Neural elements appear to exist in close proximity to the glomerular vessels (Gallardo et al. 1980), and, as in the case of the teleosts, it appears that α-adrenergic nerves may be involved in the control of the resistance of the afferent glomerular arterioles in amphibians. How, under normal, physiological conditions, this neural control may be integrated with control by AVT in the glomerular adaptation of teleost fishes to waters of varying salinity or of amphibians to varying degrees of hydration is completely unknown.

Preliminary data indicate that nerve endings exist near the glomerular arterioles of garter snakes (*Thamnophis* spp.) (S. D. Yokota, R. A. Wideman, and W. H. Dantzler unpublished observations) and that α-adrenergic inhibitors, phenoxybenzamine and phentolamine, block the decrease in whole-kidney GFR observed with high plasma concentrations of potassium in sea snakes (*A. laevis*) and garter snakes (Yokota et al. 1985b) (S. Benyajati, S. D. Yokota, and W. H. Dantzler unpublished observations). These data suggest that α-adrenergic agonists may be released by high plasma potassium levels and may play a role in regulating the resistance of the afferent arterioles (Yokota et al. 1985b) (S. Benyajati, S. D. Yokota, and W. H. Dantzler unpublished observations).

3.2.5.3 Renal Portal Influence

Although renal venous portal flow contributes to the blood supplying the tubule cells of the nephrons in most nonmammalian vertebrates, there is no evidence that it contributes directly to glomerular filtration under physiological conditions. However, for the avian kidney in which the amount of venous portal blood flow is regulated by the valve discussed above, there is some evidence that the amount of portal flow may indirectly influence glomerular filtration rate. When the venous portal blood flow in chickens is manipulated experimentally so that one kidney is completely perfused and the other kidney is completely bypassed, the whole-kidney GFR of the perfused kidney always exceeds that of the bypassed kidney (Braun and Wideman 1979). The mechanism underlying this observation is unknown. The method by which the manipulation is performed does not lead to backperfusion of the glomeruli themselves in the perfused kidney. However, it is possible that some increase in the pressure in the peritubular sinuses of the perfused kidney could lead to an increase in resistance at the efferent arterioles in that kidney, thereby leading to an increase in net filtration pressure. There may be other explanations as well. The important point, however, is that the amount of portal blood flow may have an influence on the filtration rate without contributing directly to the blood flow in the glomerular capillaries.

3.2.5.4   Renal Autoregulation

The Process of Autoregulation

In mammals, renal blood flow, whole-kidney GFR, and SNGFRs are relatively independent of the mean arterial perfusion pressure over a wide range of such pressures. This relative independence of renal blood flow and filtration from mean arterial perfusion pressure is termed "autoregulation" because it does not depend on systemic neural or humoral influences but is intrinsic to the kidney (Brenner et al. 1981; Munger et al. 2012). The control appears to involve primarily changes in resistance at the afferent glomerular arterioles (Brenner et al. 1981; Munger et al. 2012; Robertson et al. 1972). There are several mechanisms involved in this regulatory process, at least in the mammalian kidney (vide infra).

Although little is known about possible renal autoregulation in most nonmammalian vertebrates, Wideman and colleagues devised techniques that allowed them to reduce renal arterial perfusion pressure in a graded fashion while simultaneously measuring whole-kidney GFR, renal blood flow, and renal plasma flow in domestic fowl (*Gallus gallus domesticus*) (Wideman et al. 1992; Wideman and Gregg 1988). They found that whole-kidney GFR, renal blood flow, and renal plasma flow remain essentially constant as renal arterial perfusion pressure is reduced from 110 to 60 mmHg, thereby providing evidence of renal autoregulation in birds. They also obtained evidence that normal renal portal flow can contribute significantly to maintaining renal blood flow when arterial perfusion pressure is below the autoregulatory range for GFR (Wideman 1991; Wideman et al. 1992).

Renal autoregulation appears appropriate in the homeothermic birds with their high, generally stable systemic arterial pressures, but significant renal autoregulation appears far less likely in the poikilothermic nonmammalian vertebrates with their highly variable mean arterial pressures, variable SNGFRs, and glomerular intermittency. Indeed, in the one poikilothermic nonmammalian vertebrate in which this question has been addressed directly, a urodele amphibian (the Congo eel, *A. means*), micropuncture measurements show that the glomerular capillary pressure and SNGFR vary directly with the mean arterial pressure and that there is no autoregulation (Persson 1981).

However, there are other possibilities with regard to some form of renal autoregulation in poikilothermic nonmammalian vertebrates with highly variable SNGFRs. It may be that certain glomeruli in an individual kidney in these animals always have either high or low blood flow and, thus, either high or low filtration rate. This blood flow may be regulated in some specific fashion relative to the mean systemic arterial pressure. Indeed, Yokota and Dantzler (Yokota and Dantzler 1990) found that differences in mean single nephron blood flow rates between individual garter snakes (*T. sirtalis*) were not related to differences in mean systemic arterial pressures. Direct studies are needed to determine if a process different from, but similar to, classical autoregulation operates in the kidneys of poikilothermic nonmammalian vertebrates.

Renal Autoregulation Mediated by the Myogenic Mechanism

Myogenic control involves a mechanism intrinsic to the smooth muscle cells of arterioles whereby increased arterial pressure leads to constriction and decreased pressure leads to relaxation. This mechanism is well documented in the glomerular afferent arterioles in mammalian kidneys where it may account for as much as 50 % of the total autoregulatory response and reaches completion very rapidly (3–10 s). However, the cellular and molecular details of this intrinsic response, although certainly involving depolarization of the cell membrane and the entry of calcium, are far from completely clear and the current status is reviewed well elsewhere (Just 2006; Munger et al. 2012). It seems likely that a myogenic mechanism is involved in the renal autoregulation in birds, but this has yet to be examined.

Renal Autoregulation Mediated by Tubuloglomerular Feedback

A feedback mechanism from the distal tubule helping to control the filtration rate in individual mammalian nephrons is well supported by experimental evidence (Schnermann 2003). This tubuloglomerular feedback mechanism involves the cells of the macula densa sensing an increased load of solute (apparently sodium chloride) delivered to the early distal tubule, leading, in turn, via a paracrine factor (probably adenosine or possibly ATP) to constriction of the adjacent afferent arteriole (see Just 2006; Munger et al. 2012; Schnermann 2003 for reviews on additional details of this process). Although this feedback can respond to a decreased load of sodium chloride to lead to dilation of the afferent glomerular arteriole and an increase in SNGFR, it is poised to act primarily in response to an increased sodium chloride load to produce a decrease in SNGFR (Schnermann 2003). This tubuloglomerular feedback process is relatively slow, taking about 30–60 s to go to completion (Just 2006). The tubuloglomerular feedback mechanism and the myogenic mechanism act together in the autoregulation of renal blood flow and the filtration rate of individual nephrons. The myogenic mechanism begins and goes to completion rapidly whereas tubuloglomerular feedback begins rapidly but takes longer to go to completion, perhaps compensating for any initial increase in filtration rate before even the myogenic mechanism starts. However, the exact quantitative contribution of each at any given time is not clear. Moreover, it appears that a number of other factors, such as angiotensin II, can modify the tubuloglomerular feedback response or contribute more directly to autoregulation (Munger et al. 2012).

The possibility of such a single nephron feedback system in nonmammalian vertebrates is particularly intriguing because of the rapid changes that occur in the number of filtering nephrons and the single nephron glomerular filtration rates and because a macula densa appears to be present only in avian nephrons (vide supra; Chap. 2). Indeed such a mechanism appears very likely in avian kidneys which exhibit both autoregulation of renal blood flow and glomerular filtration rate and changes in the number of filtering nephrons. However, this possibility has yet to be examined in birds. Experiments to determine if such a mechanism exists in avian nephrons will be technically challenging, but the results could be important from an evolutionary viewpoint, especially because of other evidence of convergent evolution of avian and mammalian renal function (vide infra; Chap. 7).

Although there is no macula densa in reptilian and amphibian nephrons, the early distal tubule is closely associated with the glomerular vascular pole, particularly the afferent arteriole, in nephrons in all species examined of both these vertebrate classes (vide supra; Chap. 2). The possibility of tubuloglomerular feedback in reptile nephrons has yet to be examined. However, it has been examined in a urodele amphibian (the Congo eel, *A. means*) (Persson and Persson 1981). Surprisingly, microperfusion studies in these animals demonstrated a feedback system similar to that in mammals (Persson and Persson 1981). A depression in the glomerular filtration rate in an individual nephron occurs when the distal tubule is perfused with amphibian Ringer's solution at rates of 25 or 50 nl min^{-1} but not at 10 nl min^{-1}. The depression in SNGFR appears to result from an increase in resistance at the afferent arteriole. The lowest distal perfusion rate at which a decrease in SNGFR occurs is not known and the rates above appear somewhat high even for the high SNGFRs (Table 3.2) and relatively low proximal fluid reabsorption (Table 5.1) in these animals. Moreover, the variable or variables sensed and effector substance or substances that produce the increase in afferent arteriole resistance and reduction in SNGFR are unknown. Therefore, the physiological significance of this feedback process is far from clear, especially in the apparent absence of autoregulation in these same urodele amphibians (vide supra) (Persson and Persson 1981). Nevertheless, the very presence of such a process in the absence of a macula densa raises questions about the precise mechanism involved, its structural requirements, and its relationship to the process in mammals.

3.3 Filtration of Protein by Glomeruli

From the earliest micropuncture experiments of Richards and his colleagues on amphibians (both anurans, usually *Rana pipiens*, and urodeles, *Necturus maculosus*), the glomerular filtrate has been considered to be nearly protein-free (Richards 1934; Richards and Walker 1937; Wearn and Richards 1924), but it has long been known that some protein gets through the filter in mammalian and nonmammalian vertebrates and is reabsorbed along the proximal tubule (Christensen et al. 2009; Maunsbach 1976). Early studies in mammals (Maunsbach 1976) and lampreys (river lamprey, *L. fluviatilis* and sea lamprey, *Petromyzon marinus*) (Morris 1954; Youson and McMillan 1970) indicated that the reabsorptive process involves endocytosis, and more recent studies on mammals indicate that this endocytosis is mediated by two important receptor proteins, magalin and cubilin (Amsellem et al. 2010; Christensen et al. 2009). The reabsorbed proteins are then either transferred to lysosomes in the proximal tubule cells for degradation

(Christensen et al. 2009) or possibly transferred intact across the cells into the blood (Russo et al. 2007; Sandoval et al. 2012). No such detailed studies on the proximal endocytotic process for protein reabsorption have yet been made in nonmammalian vertebrates.

The questions of the actual amount of protein filtered by the glomerulus and the factors involved in determining that amount are far from completely answered for mammalian (Navar 2009; Sandoval et al. 2012; Tanner 2009) or nonmammalian vertebrates (Casotti and Braun 1996; Hausmann et al. 2010; Tanner et al. 2009). Recent technical developments involving two-photon confocal microscopy for in vivo analysis of protein filtration have led to conflicting results regarding the amount of serum protein (primarily albumin) filtered and thus the sieving coefficient at the mammalian glomerulus (Russo et al. 2007; Sandoval et al. 2012; Tanner 2009). One such study suggested that the amount of albumin filtered by rat glomeruli is many times greater (Russo et al. 2007) than that reported previously using micropuncture techniques (Tojo and Endou 1992). This study also indicated that much larger amounts of albumin than previously thought are reabsorbed by the proximal tubules because no albumin appears in the urine. However, later studies using the two-photon microscopy technique, but with careful analysis of the many technical factors involved, suggest that a number of factors could produce the high protein filtration or lead to technical errors and that the normal filtration of protein in rats is very low and comparable to that observed in micropuncture studies (Sandoval et al. 2012; Tanner 2009). Nevertheless, this topic remains controversial with regard to the mammalian kidney.

With regard to nonmammalian vertebrates, the two-photon microscopy technique applied to glomerular filtration in urodele amphibians (*N. maculosus*) indicates that the sieving coefficient for human or bovine albumin is exceedingly low, similar to that for rat albumin in the rat kidney, so that very little protein is filtered (Tanner et al. 2009; Tanner 2009). These recent data on amphibians are in accord with the earlier micropuncture studies (Richards 1934; Richards and Walker 1937; Wearn and Richards 1924). However, in the river lamprey (*L. fluviatilis*), micropuncture studies suggest that the sieving coefficient for molecules the size of rat or bovine serum albumin is much larger than that for rats or amphibians (Logan and Morris 1981). Nevertheless, the amount of protein actually filtered remains relatively low in these animals because most of their plasma proteins are much larger than bovine or rat serum albumin (Logan and Morris 1981). Although direct measurements of the quantity of protein filtered have yet to be made for reptilian or avian glomeruli, the amount either must be large or its reabsorption by the tubules must be minimal or both must occur because large amounts of protein are present in the tubular and ureteral urine of these vertebrates, where it plays an important role in the excretion of urates (vide infra; Chap. 6). Experiments determining the amount of filtered protein, its chemical form, and its degree of tubular reabsorption in reptiles and birds, although technically difficult, would be of considerable interest in understanding comparative renal function.

The role of each of the three layers of the glomerular filtration barrier—the fenestrated capillary endothelium, the basement membrane, and the visceral

epithelial cells or podocytes with filtration slits covered by a slit diaphragm
between them—in determining the degree of filtration of proteins is not completely
clear for any vertebrate species. However, the recent two-photon microscopy study
on *Necturus* provides strong evidence that the podocyte layer, apparently the slit
diaphragm, forms the major barrier (Tanner et al. 2009). A similar conclusion was
reached earlier in an electron micrographic study of ferritin movement across the
glomerular filtration barrier in bullfrogs (*R. catesbeiana*) (Schaffner and Rodewald
1978). Also, both the recent study on *Necturus* and an earlier electron micrographic
study on chickens suggest that the basement membrane does not form a significant
barrier to filtration in amphibians and birds (Casotti and Braun 1996; Tanner
et al. 2009), but this is far from certain (Navar 2009). Clearly, a great deal remains
to be learned about the specific significance of each of the three layers of the
glomerular filtration barrier for filtration of macromolecules in both mammalian
and nonmammalian vertebrates. In addition, although it is assumed that size, shape,
and deformability of plasma proteins all play a role in their possible filtration, much
remains to be learned about the plasma proteins in nonmammalian vertebrates,
particularly in reptiles and birds.

Finally, the role of the fixed anionic charges of the filtration barriers in excluding
negatively charged macromolecules, such as albumin, from the filter remains
unclear. Although charge selectivity has been shown to be important by numerous
studies on mammals over a number of years (Chang et al. 1975; Navar 2009), its
role has been challenged in recent studies (Rippe et al. 2007; Schaeffer et al. 2002).
Its importance has also been challenged in the recent two-photon microscopy study
on *Necturus* (Tanner et al. 2009). But if the charge is not important, how then are
relatively small negatively charged macromolecules (e.g., albumin) so markedly
excluded from the filter? (Navar 2009). The mechanism is far from clear. However,
both the recent study on *Necturus* (Tanner et al. 2009) and the earlier one on
chickens (Casotti and Braun 1996) indicate a lack of negative charges (apparently
low glycosaminoglycan concentration) in the glomerular basement membrane,
suggesting a lesser role for charge selectivity in these animals than in mammals.
Still, another recent study on *Necturus* indicates that filtration itself generates a
potential across the filtration barrier (negative within Bowman's space) that can
influence filtration of anionic macromolecules (Hausmann et al. 2010), but this
requires further study. In fact, glomerular filtration of macromolecules in general
and their reabsorption by the renal tubules remain fertile fields for future research
on nonmammalian vertebrates.

3.4   Secretion of Fluid by Tubules

Secretion of fluid by the proximal renal tubules is the essential process in the initial
formation of urine by aglomerular nephrons found in certain teleost fishes
(Berglund and Forster 1958; Beyenbach 2004; Hickman and Trump 1969) and
apparently a few reptiles (O'Shea et al. 1993; Regaud and Policard 1903). However,

it is now known to contribute to the initial formation of urine by the glomerular nephrons of marine teleosts (e.g., winter flounder, *Pseudopleuronectes americanus*; southern flounder, *Paralichthys lethostigma*; longhorn sculpin, *Myxocephalus octodecimspinosus*) (Beyenbach 1982; Forster 1953; Hickman 1968), marine elasmobranchs (e.g., dogfish shark, *Squalus acanthius*) (Beyenbach and Fromter 1985), euryhaline teleosts adapted to freshwater (e.g., American eel, *Anguilla rostrata*) (Schmidt-Nielsen and Renfro 1975), euryhaline teleosts adapted to either freshwater or seawater (e.g., *Fundulus heteroclitis*) (Beyenbach 1986; Cliff and Beyenbach 1992), and even marine snakes (e.g., olive sea snake, *Aipysurus laevis*) (Yokota et al. 1985b). Fluid secretion by both proximal tubules and collecting ducts may even contribute to urine formation in mammals (Grantham and Wallace 2002). The possible mechanisms involved in these secretory processes are discussed in Chaps. 4 and 5.

Among glomerular species, fluid secretion by the renal tubules appears to be most important when glomerular filtration is low, particularly perhaps when the number of filtering nephrons is reduced (Beyenbach 1982, 1986; Hickman 1968; Schmidt-Nielsen and Renfro 1975; Yokota et al. 1985b). In some species (e.g., flounders, dogfish sharks, and killifish), the rate of secretion of fluid by the tubules is about equal to the rate of filtration of fluid by the glomeruli (vide infra, Table 5.1) (see also Beyenbach 1982, 1986, 2004), whereas in others (e.g., olive sea snakes, Sprague-Dawley rats) it is much lower (Grantham and Wallace 2002; Wallace et al. 2001; Yokota et al. 1985b). In a number of marine species (e.g., longhorn sculpins, American eels, olive sea snakes), net fluid secretion appears to be most important when there is a need to excrete water (Forster 1953; Schmidt-Nielsen and Renfro 1975; Yokota et al. 1985a). Beyenbach (1986, 2004) and Grantham and Wallace (2002) suggest that the potential for net fluid secretion by renal tubules may be present as a primitive characteristic that can serve an important regulatory function in many, if not all, species with glomerular nephrons, not just in those species with aglomerular nephrons.

References

Acharjee S, Do-Rego J-L, Oh DY, Moon JS, Ahn RS, Lee K, Bai DG, Vaudry H, Kwon HB, Seong JY (2004) Molecular cloning, pharmacological characterization, and histochemical distribution of frog vasotocin and mesotocin receptors. J Mol Endocrinol 33:293

Akester AR, Mann SP (1969) Adrenergic and cholinergic innervation of the renal portal valve in the domestic fowl. J Anat 104:241–252

Alt JM, Stolte H, Eisenbach GM, Walvig F (1980) Renal electrolyte and fluid excretion in the atlantic hagfish *Myxine glutinosa*. J Exp Biol 91:323–330

Ames E, Steven K, Skadhauge E (1971) Effects of arginine vasotocin on renal excretion of Na$^+$, K$^+$, Cl$^-$, and urea in the hydrated chicken. Am J Physiol 221:1223–1228

Amsellem S, Gbureck J, Hamard G, Nielsen R, Willnow TE, Devuyst O, Nexo E, Verroust PJ, Christensen EI, Kozyraki R (2010) Cubilin is essential for albumin reabsorption in the renal proximal tubule. J Am Soc Nephrol 21:1859–1867

Babiker MM, Rankin JC (1978) Neurohypophysial hormonal control of kidney function in the european eel (*Anguilla anguilla* L.) adapted to sea-water or fresh water. J Endocrinol 76:347–358

Berglund F, Forster RP (1958) Renal tubular transport of inorganic divalent ions by the aglomerular marine teleost, *Lophius americanus*. J Gen Physiol 41:249–440

Beyenbach KW (1982) Direct demonstration of fluid secretion by glomerular renal tubules in a marine teleost. Nature 299:54–56

Beyenbach KW (1986) Secretory NaCl and volume flow in renal tubules. Am J Physiol Regul Integr Comp Physiol 250:R753–R763

Beyenbach KW (2004) Kidneys sans glomeruli. Am J Physiol Renal Physiol 286:F811–F827

Beyenbach KW, Fromter E (1985) Electrophysiological evidence for Cl secretion in shark renal proximal tubules. Am J Physiol Renal Physiol 248:F282–F295

Bordley J, Richards AN (1933) Quantitative studies of the composition of glomerular urine. VIII. The concentration of uric acid in glomerular urine of snakes and frogs, determined by an ultramicroadaptation of folin's method. J Biol Chem 101:193–221

Bradshaw SD, Rice GE (1981) The effects of pituitary and adrenal hormones on renal and postrenal reabsorption of water and electrolytes in the lizard, *Varanus gouldii* (gray). Gen Comp Endocrinol 44:82–93

Braun EJ (1976) Intrarenal blood flow distribution in the desert quail following salt loading. Am J Physiol 231:1111–1118

Braun EJ (1978) Renal response of the starling (*Sturnus vulgaris*) to an intravenous salt load. Am J Physiol Renal Physiol 234:F270–F278

Braun EJ (1982) Glomerular filtration in birds—its control. Fed Proc 41:2377–2381

Braun EJ, Dantzler WH (1972) Function of mammalian-type and reptilian-type nephrons in kidney of desert quail. Am J Physiol 222(3):617–629

Braun EJ, Dantzler WH (1974) Effects of ADH on single-nephron glomerular filtration rates in the avian kidney. Am J Physiol 226:1–8

Braun EJ, Dantzler WH (1975) Effects of water load on renal glomerular and tubular function in desert quail. Am J Physiol 229:222–228

Braun EJ, Wideman RF Jr (1979) Contribution of the renal portal system (RPS) to the variability in urine composition and renal function in birds. Fed Proc 38:902 (abstract)

Brenner BM, Ichikawa I, Deen WM (1981) Glomerular filtration. In: Brenner BM, Rector FC Jr (eds) The kidney, 2nd edn. Saunders, Philadelphia, pp 289–327

Brenner BM, Troy JL, Daugharty TM (1971) The dynamics of glomerular ultrafiltration in the rat. J Clin Invest 50:1776–1780

Brenner BM, Troy JL, Daugharty TM, Deen WM, Robertson CR (1972) Dynamics of glomerular ultrafiltration in the rat. II. Plasma-flow dependence of GFR. Am J Physiol 223:1184–1190

Brewer KJ, Ensor DM (1980) Hormonal control of osmoregulation in the chelonia. I. The effects of prolactin and interrenal steroids in freshwater chelonians. Gen Comp Endocrinol 42:304–309

Brown JA, Green C (1987) Single nephron function of the lesser spotted dogfish, *Scyliorhinus canicula*, and the effects of adrenaline. J Exp Biol 129:265–278

Brown JA, Jackson BA, Oliver JA, Henderson IW (1978) Single nephron filtration rates (SNGFR) in the trout, *Salmo gairdneri*. Validation of the use of ferrocyanide and the effects on environmental salinity. Pflugers Arch 377:101–108

Brown JA, Oliver JA, Henderson IW, Jackson BA (1980) Angiotensin and the single nephron glomerular function in the trout *Salmo gairdneri*. Am J Physiol Regul Integr Comp Physiol 239:R509–R514

Brown JA, Rankin JC, Yokota SD (1993) Glomerular haemodynamics and filtration in single nephrons of nonmammalian vertebrates. In: Brown JA, Balment RJ, Rankin JC (eds) New insights in vertebrate kidney function. Cambridge University Press, Cambridge, pp 1–44

Brown JA, Taylor SM, Gray CJ (1983) Glomerular ultrastructure of the trout, *Salmo gairdneri*. Glomerular capillary epithelium and the effects of environmental salinity. Cell Tissue Res 230:205–218

Burrows ME, Braun EJ, Duckles SP (1983) Avain renal portal valve: a reexamination of its innervation. Am J Physiol Heart Circ Physiol 245:H628–H634

Butler DG (1972) Antidiuretic effect of arginine vasotocin in the Western painted turtle (*Chrysemys picta belli*). Gen Comp Endocrinol 18:121–125

Butler DG, Snitman F (2005) Renal responses to mesotocin in Western painted turtles compared with antidiuretic response to arginine vasotocin. Gen Comp Endocrinol 144:101–109

Casotti G, Braun EJ (1996) Functional morphology of the glomerular filtration barrier of *Gallus gallus*. J Morphol 228:327–334

Chang RL, Deen WM, Robertson CR, Brenner BM (1975) Permselectivity of the glomerular capillary wall: III. Restricted transport of polyanions. Kidney Int 8:212–218

Chester Jones I, Chan DKO, Rankin JC (1969) Renal function in the European eel (*Anguilla anguilla* L.): effects of the caudal neurosecretory system, corpusclesof Stannius, neurohypophysial peptides and vasoactive substances. J Endocrinol 43:21–31

Christensen EI, Verroust P, Nielsen R (2009) Receptor-mediated endocytosis in renal proximal tubule. Pflugers Arch 458:1039–1048

Cliff WH, Beyenbach KW (1992) Secretory renal proximal tubules in seawater- and freshwater-adapted killifish. Am J Physiol Renal Physiol 262:F108–F116

Dantzler WH (1966) Renal response of chickens to infusion of hyperosmotic sodium chloride solution. Am J Physiol 210:640–646

Dantzler WH (1967) Glomerular and tubular effects of arginine vasotocin in water snakes (*Natrix sipedon*). Am J Physiol 212:83–91

Dantzler WH (1968) Effect of metabolic alkalosis and acidosis on tubular urate secretion in water snakes. Am J Physiol 215:747–751

Dantzler WH (1976) Renal function (with special emphasis on nitrogen excretion). In: Gans CG, Dawson WR (eds) Biology of Reptilia, vol 5, Physiology A. Academic, London, pp 447–503

Dantzler WH (1982) Renal adaptations of desert vertebrates. Bioscience 32(2):108–113

Dantzler WH, Bradshaw S (2009) Osmotic and ionic regulation in reptiles. In: Evans DH (ed) Osmotic and ionic regulation: cells and animals. CRC Press, Boca Raton, FL, pp 443–503

Dantzler WH, Braun EJ (1980) Comparative nephron function in reptiles, birds, and mammals. Am J Physiol Regul Integr Comp Physiol 239:R197–R213

Dantzler WH, Schmidt-Nielsen B (1966) Excretion in fresh-water turtle (*Pseudemys scripta*) and desert tortoise (*Gopherus agassizii*). Am J Physiol 210:198–210

Deen WM, Troy JL, Robertson CR, Brenner BM (1973) Dynamics of glomerular ultrafiltration in the rat. IV. Determination of the ultrafiltration coefficient. J Clin Invest 52:1500–1508

deRouffignac C, Deiss S, Bonvalet JP (1970) Détermination du taux individuel de filtration glomérulaire des néphrons accessibles et inaccessibles á la microponction. Pflugers Arch 315:273–290

Dunson WA, Packer RK, Dunson MK (1971) Sea snakes: an unusual salt gland under the tongue. Science 173:437–441

Elger B, Hentschel H (1983) Effect of adrenergic blockade with bretylium and phentolamine on glomerular filtration rate in the rainbow trout, *Salmo gairdneri* rich., adapting to saline water. Comp Biochem Physiol 75C:253–258

Elger E, Elger B, Hentschel H, Stolte H (1987) Adaptation of renal function to hypotonic medium in the winter flounder (*Pseudopleuronectes americanus*). J Comp Physiol B 157:21–30

Elger M, Hentschel H (1981) The glomerulus of a stenohaline fresh-water teleost, *Carassius auratus gibelio*, adapted to saline water. A scanning and transmission electron-microscopic study. Cell Tissue Res 220:73–85

Elger M, Kaune R, Hentschel H (1984a) Glomerular intermittency in a freshwater teleost, *Carassius auratus gibelio*, after transfer to salt water. J Comp Physiol B 154:225–232

Elger M, Wahlqvist I, Hentschel H (1984b) Ultrastructure and adrenergic innervation of preglomerular arterioles in the euryhaline teleost, *Salmo gairdneri*. Cell Tissue Res 237:451–458

Fleming WR, Stanley JG (1965) Effects of rapid changes in salinity on the renal function of a euryhaline teleost. Am J Physiol 209:1025–1030

Follett BK, Heller H (1964a) The neurohypophysial hormones in the lung fishes and cyclostomes. J Physiol (Lond) 172:92–106

Follett BK, Heller H (1964b) The neurohypophysial hormonesof bony fishes and cyclostomes. J Physiol (Lond) 172:74–91

Forster RP (1942) The nature of the glucose reabsorptive process in the frog renal tubule. Evidence for intermittency of glomerular function in the intact animal. J Cell Comp Physiol 20:55–69

Forster RP (1953) A comparative study of renal function in marine teleosts. J Cell Comp Physiol 42:487–510

Gallardo R, Pang PKT, Sawyer WH (1980) Neural influences on bullfrog renal functions. Proc Soc Exp Biol Med 165:233–240

Garland HO, Henderson IW, Brown JA (1975) Micropuncture study of the renal responses of the urodele amphibian *Necturus maculosus* to injections of arginine vasotocin and an anti-aldosterone compound. J Exp Biol 63:249–264

Giebisch G (1956) Measurements of pH, chloride, and inulin concentrations in proximal tubule fluid of *Necturus*. Am J Physiol 185:171–174

Grafflin AL, Bagley EH (1952) Glomerular activity in the frog's kidney. Bull Johns Hopkins Hosp 91:306–317

Grantham JJ, Wallace DP (2002) Return of the secretory kidney. Am J Physiol Renal Physiol 282: F1–F9

Gray CJ, Brown JA (1985) Renal cardiovascular effects of angiotensin II in the rainbow trout, Salmo gairdneri. Gen Comp Endocrinol 59:375–381

Haines H, Schmidt-Nielsen B (1977) Kidney functrion in spiny mice (*Acomys cahirinus*) acclimated to water restriction. Bull Mt Desert Isl Biol Lab 17:94–95

Hartenstein HR, Stiffler DF (1990) Renal responses to mesotocin in adult *Ambystoma tigrinum* and *Notophthalmus viridescens*. Exp Biol 48:373–377

Hasunuma I, Sakai T, Nakada T, Toyoda F, Namiki H, Kikuyama S (2007) Molecular cloning of three types of arginine vasotocin receptor in the newt, *Cynops pyrrhogaster*. Gen Comp Endocrinol 151:252–258

Hausmann R, Kuppe C, Egger H, Schweda R, Knecht V, Elger M, Menzel S, Somers D, Braun G, Fuss A, Uhlig S, Kriz W, Tanner G, Floege J, Moeller MJ (2010) Electrical forces determine glomerular permeability. J Am Soc Nephrol 21:2053–2058

Hayman JM (1927) Estimations of afferent arteriole and glomerular capillary pressures in the frog kidney. Am J Physiol 79:389–409

Heller H, Pickering BJ (1961) Neurohypophysial hormones of nonmammalian vertebrates. J Physiol (Lond) 155:98–114

Henderson IW, Wales NAM (1974) Renal diuresis and antidiuresis after injections of arginine vasotocin in the freshwater eel (*Anguilla anguilla* L.). J Endocrinol 61:487–500

Hentschel H, Elger M (1987) The distal nephron in the kidney of fishes. Adv Anat Embryol Cell Biol 108:1–151

Hickman CP Jr (1965) Studies on renal function in freshwater teleost fish. Trans R Soc Can 3:213–236

Hickman CP Jr (1968) Glomerular filtration and urine flow in the euryhaline southern flounder, *Paralichthys lethostigma*, in seawater. Can J Zool 46:427–437

Hickman CP Jr, Trump BF (1969) The kidney. In: Hoar WS, Randall DJ (eds) Fish physiology, vol I, Excretion, ion regulation, and metabolism. Academic, New York, pp 91–239

Ibarra F, Crambert S, Eklöf A-C, Lundquist A, Hansell P, Holtbäck U (2005) Prolactin, a natriuretic hormone, interacting with the renal dopamine system. Kidney Int 68:1700–1707

Jard S, Morel F (1963) Actions of vasotocin and some of its analogues on salt and water excretion by the frog. Am J Physiol 204:222–226

Just A (2006) Mechanisms of renal blood flow autoregulation: dynamics and contributions. Am J Physiol Regul Integr Comp Physiol 292:R1–R17

Koike TI, Neldon HL, McKay DW, Rayford PL (1986) An antiserum that recognizes mesotocin and isotocin: development of a homologous radioimmunoassay for plasma mesotocin in chickens (*Gallus domesticus*). Gen Comp Endocrinol 63:93–103

Koike TI, Pryor LR, Neldon HL (1979) Effect of saline infusion on plasma immunoreactive vasotocin in conscious chickens (*Gallus domesticus*). Gen Comp Endocrinol 37:451–458

Komadina S, Solomon S (1970) Comparison of renal function of bull and water snakes (*Pituophis melanoleucus* and *Natrix sipedon*). Comp Biochem Physiol 32:333–343

Konno N, Hyodo S, Yamaguchi Y, Kaiya H, Miyazato M, Matsuda K, Uchiyama M (2009) African lungfish, *Protopterus annectens*, possess an arginine vasotocin receptor homologous to the tetrapod V2-type receptor. J Exp Biol 212:2183–2193

Lahlou B (1966) Mise en evidence d'un "recrutement glomerulaire" dans le rein in des teleoseens d'apres la mesure du Tm glucose. C R Acad Sci 262:1356–1358

Lam TJ, Leatherland JF (1969) Effects of prolactin on the glomerulus of the marine threespine stickleback, *Gastersteus aculeatus* L., form *trachurus*, after transfer from seawater to fresh water, during the late autumn and early winter. Can J Zool 47:245–250

Landis EM, Pappenheimer JR (1963) Exchange of substances through capillary walls. In: Hamilton WF, Dow P (eds) Handbook of Physiology. Circulation. American Physiological Society, Washington, DC, pp 961–1034

Laverty G, Dantzler WH (1982) Micropuncture of superficial nephrons in avian (*Sturnus vulgaris*) kidney. Am J Physiol Renal Physiol 243:F561–F569

Laverty G, Dantzler WH (1983) Micropuncture study of urate transport by superficial nephrons in avian (*Sturnus vulgaris*) kidney. Pflugers Arch 397:232–236

Logan AG, Moriarty RJ, Morris R, Rankin JC (1980a) The anatomy and blood system of the kidney in the river lamprey, *Lampetra fluviatilis*. Anat Embryol 158:245–252

Logan AG, Moriarty RJ, Rankin JC (1980b) A micropuncture study of kidney function in the river lamprey, *Lampetra fluviatilis*, adapted to fresh water. J Exp Biol 85:137–147

Logan AG, Morris R (1981) The handling of macromolecules by the kidney of the river lamprey, *Lampetra fluviatilis*. J Exp Biol 93:303–316

Logan AG, Morris R, Rankin JC (1980c) A micropuncture study of kidney function in the river lamprey, *Lampetra fluviatilis*, adapted to sea water. J Exp Biol 88:239–247

Mackay WC, Beatty DD (1968) The effect of temperature on renal function in the white sucker fish, *Catostomus commersonii*. Comp Biochem Physiol 26:235–245

Mahlmann S, Meyerhof W, Hausmann H, Heierhorst J, Schönrock C, Zwiers H, Lederis K, Richter D (1994) Structure, function, and phylogeny of [Arg8]vasotocin receptors from teleost fish and toad. Proc Natl Acad Sci U S A 91:1342–1345

Maunsbach AB (1976) Cellular mechanisms of tubular protein transport. Int Rev Physiol 11:145–167

McBean RL, Goldstein L (1970) Renal function during osmotic stress in the aquatic toad *Xenopus laevis*. Am J Physiol 219:1115–1123

McVicar AJ, Rankin JC (1985) Dynamics of glomerular filtration in the river lamprey, *Lampetra fluviatilis* L. Am J Physiol Renal Physiol 249:F132–F139

Morel A, O'Carroll AM, Brownstein MJ, Lolait SJ (1992) Molecular cloning and expression of a rat V1a arginine vasopressin receptor. Nature 356:523–526

Moriarty RJ, Logan AG, Rankin JC (1978) Measurement of single nephron filtration rate in the kidney of the river lamprey, *Lampetra fluviatilis* L. J Exp Biol 77:57–69

Morris R (1954) Osmoregulation in cyclostomes with special reference to the river lamprey (*Lampetra fluviatilis* L.). Unpublished Dissertation

Munger KA, Kost CK Jr, Brenner BM, Maddox DA (2012) The renal circulations and glomerular ultrafiltration. In: Taal MW, Chertow GM, Marsden PA, Skorecki K, Yu ALS, Brenner BM (eds) Brenner and Rector's the kidney, 9th edn. Elsevier (Saunders), Philadelphia, pp 94–137

Munsick RA (1964) Neurohypophysial hormones of chickens and turkeys. Endocrinology 75:104–113

Munsick RA (1966) Chromatographic and pharmacologic characterization of the neurohypophysial hormones of an amphibian and a reptile. Endocrinology 78:591–599

Navar LG (1978) The regulation of glomerular filtration inmammalian kideys. In: Andreoli TE, Hoffman JF, Fanestil DD (eds) Physiology of membrane disorders. Plenum, New York, pp 593–625

Navar LG (2009) Glomerular permeability: a never-ending saga. Am J Physiol Renal Physiol 296: F1266–F1268

Navar LG, Bell PD, White RW, Watts RL, Williams RH (1977) Evaluation of the single nephron glomerular filtration coefficient in the dog. Kidney Int 12:137–149

Nishimura H (1980) Comparative endocrinology of renin and angiotensin. In: Johnson RA, Anderson RR (eds) The renin-angiotensin system. Plenum, New York, pp 29–77

Nishimura H (1985) Endocrine control of renal handling of solutes and water in vertebrates. Renal Physiol 8:279–300

Nishimura H (2001) Angiotensin receptors—evolutionary overview and perspectives. Comp Biochem Physiol [A] 128:11–30

Nishimura H, Imai M (1982) Control of renal function in freshwater and marine teleosts. Fed Proc 41:2355–2360

Nouwen EJ, Decuypere E, Kuhn ER, Michels H, Hall TR, Chadwick A (1984) Effect of dehydration, haemorrhage and oviposition on serum concentrations of vasotocin, mesotocin and prolactin in the chicken. J Endocrinol 102:345–351

O'Shea JE, Bradshaw SD, Stewart T (1993) Renal vasculature and excretory system of the agamid lizard, *Ctenophorus ornatus*. J Morphol 217:287–299

Pang PKT, Sawyer WH (1978) Renal and vascular responses of the bullfrog (*Rana catesbeiana*) to mesotocin. Am J Physiol Renal Physiol 235:F151–F155

Persson B-E (1978a) Driving forces for glomerular ultrafiltration in *Amphiuma means*. Acta Univ Ups 308(II):1–4

Persson B-E (1978b) Hydrostatic pressures in vascular and tubular structures of the *Amphiuma* kidney. Acta Univ Ups 308(I):1–13

Persson B-E (1981) Dynamics of glomerular ultrafiltration in *Amphiuma means*. Pflugers Arch 391:135–140

Persson B-E, Persson AEG (1981) The existence of a tubulo-glomerular feedback mechanism in the amphiuma nephron. Pflugers Arch 391:129–134

Ranges HA, Chasis H, Goldring W, Smith HW (1939) The functional measurement of the number of active glomeruli and tubules in the kidneys of normal and hypertensive subjects. Am J Physiol 126:P603

Rankin JC, Logan AG, Moriarty RJ (1980) Changes in kidney function in river lamprey, *Lampetra fluviatilis*, in response to changes in external salinity. In: Lahlou B (ed) Epithelial transport in the lower vertebrates. Cambridge University Press, Cambridge, England, pp 171–184

Regaud C, Policard A (1903) Sur l'existence de diverticules du tube urinpare sans relations avec les corpuscles de Malpighi, chez les serpents, et sur l'indépendence relative des fonctions glomérulaires et glandulaires dur rein en general. C R Soc Biol (Paris) 55:1028–1029

Renkin EM, Gilmore JP (1973) Glomerular filtration. In: Orloff J, Berliner RW (eds) Handbook of physiology 8: renal physiology. American Physiological Society, Washington, DC, pp 185–248

Rice GE (1982) Plasma arginine vascotocin concentrations in the lizard *varanus gouldii* (gray) following water loading, salt loading, and dehydration. Gen Comp Endocrinol 47:1–6

Richards AN (1934) Urine formation in the amphibian kidney. Harvey Lect 30:93–118

Richards AN, Schmidt CF (1924) A description of the glomerular circulation in the frog's kidney and observations concerning the action of adrenalin and various other substances upon it. Am J Physiol 71:179–208

Richards AN, Walker AM (1937) Methods of collecting fluid from known regions of the renal tubules of amphibia and of perfusing the lumen of a single tubule. Am J Physiol 118:111–120

Riegel JA (1978) Factors effecting glomerular function in the pacific hagfish *Eptatretus stouti* (Lockinton). J Exp Biol 73:261–277

Riegel JA (1986a) Hydrostatic pressures in glomeruli and renal vasculature of the hagfish, *Eptatretus stouti*. J Exp Biol 123:359–371

Riegel JA (1986b) The absence of an arterial pressure effect on filtration by perfused glomeruli of the hagfish, *Eptatretus stouti* (Lockington). J Exp Biol 126:361–374

Rippe C, Rippe A, Torffvit O, Rippe B (2007) Size and charge selectivity of the glomerular filter in early experimental diabetes in rats. Am J Physiol Renal Physiol 293:F1533–F1538

Roberts JR, Dantzler WH (1989) Glomerular filtration rate in concious unrestrained starlings under dehydration. Am J Physiol Regul Integr Comp Physiol 256:R836–R839

Roberts JR, Dantzler WH (1992) Micropuncture study of avian kidney: effect of prolactin. Am J Physiol Regul Integr Comp Physiol 262:R933–R937

Roberts JS, Schmidt-Nielsen B (1966) Renal ultrastructure and excretion of salt and water by three terrestrial lizards. Am J Physiol 211:476–486

Robertson CR, Deen WM, Troy JL, Brenner BM (1972) Dynamics of glomerular ultrafiltration in the rat. III. Hemodynamics and autoregulation. Am J Physiol 223:1191–1200

Russo LM, Sandoval RM, McKee M, Osicka TM, Collins AB, Brown D, Molitoris BA, Comper WD (2007) The normal kidney filters nephrotic levels of albumin retrieved by proximal tubule cells: Retrieval is disrupted by nephrotic states. Kidney Int 71:504–513

Sandoval RM, Wagner MC, Patel M, Campos-Bilderback SB, Rhodes GJ, Wang E, Wean S, Clendenon SS, Molitoris BA (2012) Multiple factors influence glomerular albumin permeability in rats. J Am Soc Nephrol 23:447–457

Sawyer WH (1951) Effect of posterior pituitary extracts on urine formation and glomerular circulation in the frog. Am J Physiol 164:457–466

Sawyer WH (1970) Vasopressor, diuretic, and natriuretic responses by lungfish to arginine vasotocin. Am J Physiol 218:1789–1794

Sawyer WH, Munsick RA, Van Dyke HB (1961) Evidence for the presence of arginine vasotocin (8-arginine oxytocin) and oxytocin in nerohypophysial extracts from amphibians and reaptiles. Gen Comp Endocrinol 1:30–36

Sawyer WH, Pang PKT (1975) Endocrine adaptation to osmotic requirements of the environment: endocrine factors in osmoregulation by lungfishes and amphibians. Gen Comp Endocrinol 25:224–229

Schaeffer RC Jr, Gratrix MI, Mucha DR, Carbajal JM (2002) The rat glomerular filtration barrier does not show negative charge selectivity. Microcirculation 9:329–342

Schaffner A, Rodewald R (1978) Glomerular permeability in the bullfrog *Rana catesbeiana*. J Cell Biol 79:314–328

Schmidt-Nielsen B, Davis LE (1968) Fluid transport and tubular intercellular spaces in reptilian kidneys. Science 159:1105–1108

Schmidt-Nielsen B, Forster RP (1954) The effect of dehydration and low temperature on renal function in the bullfrog. J Cell Comp Physiol 44:233–246

Schmidt-Nielsen B, Renfro JL (1975) Kidney function of the american eel *Anguilla rostrata*. Am J Physiol 228:420–431

Schmidt-Nielsen B, Schmidt D (1973) Renal function of *sphenodon punctatum*. Comp Biochem Physiol 44A:121–129

Schmidt-Nielsen B, Skadhauge E (1967) Function of the excretory system of the crocodile (*Crocodylus acutus*). Am J Physiol 212:973–980

Schnermann J (2003) The juxtaglomerular apparatus: from anatomical peculiarity to physiological relevance. J Am Soc Nephrol 14:1681–1694

Sharratt BM, Jones IC, Bellamy D (1964) Water and electrolyte composition of the body and renal function of the eel (*Anguilla anguilla* L.). Comp Biochem Physiol 11:9–18

Shoemaker VH, Bickler PE (1979) Kidney and bladder function in a uricotelic treefrog (*Phyllomedusa sauvagei*). J Comp Physiol 133:211–218

Skadhauge E, Schmidt-Nielsen B (1967) Renal function in domestic fowl. Am J Physiol 212:793–798

Sokabe H (1974) Phylogeny of the renal effects of angiotensin. Kidney Int 6:263–271

Sperber I (1948) Investigation on the circulatory system of the avian kidney. Zool Bidr Upps 22:429–448

Stiffler DF (1981) The effects of mesotocin on renal function in hypophysectomized *Ambystoma tigrinum* larvae. Gen Comp Endocrinol 45:49–51

Stiffler DF, Roach SC, Pruett SJ (1984) A comparison of the responses of the amphibian kidney to mesotocin, isotocin, and oxytocin. Physiol Zool 57:63–69

Stolte H, Schmidt-Nielsen B (1978) Comparative aspects of fluid and electrolyte regulation by cyclostome, elasmobranch, and lizard kidney. In: Barker Jorgenson C, Skadhauge E (eds) Osmotic and volume regulation. Alfred Benzon Symposium XI. Munksgaard, Copenhagen, pp 209–220

Takahashi T, Kawashima M, Yasuoka T, Kamiyoshi M, Tanaka K (1995) Diuretic and antidiuretic effects of mesotocin as compared with the antidiuretic effect of arginine vasotocin in the hen. Poult Sci 74:890–892

Takahashi T, Kawashima M, Yasuoka T, Kamiyoshi M, Tanaka K (1996) Mesotocin binding to receptors in hen kidney plasma membrane. Poult Sci 75:910–914

Takahashi T, Kawashima M, Yasuoka T, Kamiyoshi M, Tanaka K (1997) Mesotocin receptor binding of cortical and medullary kidney tissues of the hen. Poult Sci 76:1302–1306

Tan F-I, Lolait SJ, Brownstein MJ, Saito N, MacLeod V, Baeyens DA, Mayeux PR, Jones SM, Cornett LE (2000) Molecular cloning and functional characterization of a vasotocin receptor subtype that is expressed in the shell gland and brain of the domestic chicken. Biol Reprod 62:8–15

Tanner GA (2009) Glomerular sieving coefficient of serum albumin in the rat: a two-photon microscopy study. Am J Physiol Renal Physiol 296:F1258–F1265

Tanner GA, Rippe C, Shao Y, Evan AP, Williams JC (2009) Glomerular permeability to macro-molecules in the *Necturus* kidney. Am J Physiol Renal Physiol 296:F1269–F1278

Taplin LE, Grigg GC (1981) Salt glands in the tongue of the estuarine crocodile *Crocodylus porosus*. Science 212:1045–1047

Tojo A, Endou H (1992) Intrarenal handling of proteins in rats using fractional micropuncture technique. Am J Physiol Renal Physiol 263:F601–F606

Trinh-Trang-Tan M-M, Diaz M, Grunfeld J-P, Bankir L (1981) ADH-dependent nephron hetero-geneity in rats with hereditary hypothalmic diabetes insipidus. Am J Physiol Renal Physiol 240:F372–F380

Uchiyama M, Murakami T, Pang PKT (1985) Renal and vascular responses of the bullfrog, *Rana catesbeiana*, and the toad, *Bufo bufo japonicus*, to mesotocin. 10th Int Symp Comp Endocrinol, Abstracts, Copper Mountain, Colorado USA

Walker AM, Hudson CL (1937) The role of the tubule in the excretion of urea by the amphibian kidney. With an improved technique for the ultramicro determination of urea nitrogen. Am J Physiol 118:153–166

Wallace DP, Rome LA, Sullivan LP, Grantham JJ (2001) cAMP-dependent fluid secretion in rat inner medullary collecting ducts. Am J Physiol 280:F1019–F1029

Warne JM (2001) Cloning and characterization of an arginine vasotocin receptor from the euryhaline flounder *Platichthys flesus*. Gen Comp Endocrinol 122:312–319

Wearn JT, Richards AN (1924) Observations of the composition of glomerular urine, with particular reference to the problem of reabsorption in the renal tubules. Am J Physiol 71:209–227

Weybourne E, Warne JM, Hentschel H, Elger M, Balment RJ (2005) Renal morphology of the euryhaline flounder (*Platichthys flesus*). Distribution of arginine vasotocin receptor. Ann NY Acad Sci 1040:521–523

White HL (1929) Observations on the nature of glomerular activity. Am J Physiol 90:689–704

Wideman RF Jr (1991) Autoregulation of avian renal plasma flow: contribution of the renal portal
    system. J Comp Physiol B 160:663–669
Wideman RF Jr, Braun EJ, Anderson GL (1981) Microanatomy of the renal cortex in the domestic
    fowl. J Morphol 168:249–267
Wideman RF Jr, Glahn RP, Bottje WG, Holmes KR (1992) Use of thermal pulse decay system to
    assess avian renal blood flow during reduced renal arterial perfusion pressure. Am J Physiol
    Regul Integr Comp Physiol 22:R90–R98
Wideman RF Jr, Gregg CM (1988) Model for evaluating avian renal hemodynamics and glomer-
    ular filtration rate autoregulation. Am J Physiol Regul Integr Comp Physiol 254:R925–R932
Yagil R, Berlyne GM (1976) Sodium and potassium metabolism in the dehydrated and rehydrated
    bedouin camel. J App Physiol 41:457–461
Yokota SD, Benyajati S, Dantzler WH (1985a) Comparative aspects of glomerular filtration in
    vertebrates. Renal Physiol 8:193–221
Yokota SD, Benyajati S, Dantzler WH (1985b) Renal function in sea snakes I. Glomerular
    filtration rate and water handling. Am J Physiol Regul Integr Comp Physiol 249:R228–R236
Yokota SD, Dantzler WH (1990) Measurements of blood flow to individual glomeruli in the
    ophidian kidney. Am J Physiol Regul Integr Comp Physiol 258:R1313–R1319
Youson JH, McMillan DB (1970) The opisthonephric kidney of the sea lamprey of the great lakes,
    Petromyzon marinus L. II. Neck and proximal segments of the tubular nephron. Am J Anat
    127:233–258

Chapter 4
Transport of Inorganic Ions by Renal Tubules

Abstract This chapter considers tubular transport of sodium and chloride, potassium, hydrogen ion, calcium, phosphate, magnesium, and sulfate. For each inorganic ion, it examines the direction or directions of transepithelial transport, the tubule sites of transport, and the cellular and molecular mechanisms of transport, so far as they are known in nonmammalian vertebrates. For sodium and chloride, substantial quantitative information on net fluxes is provided. In addition, the electrical properties of the tubule segments measured to date in nonmammalian and mammalian vertebrates are provided and are related particularly to transport of sodium, chloride, and potassium. In so far as possible for each of these inorganic ions, the chapter describes and discusses the cellular and molecular characteristics of the transport steps at both the luminal and peritubular membranes of the renal tubule cells in the various tubule segments. These steps are very poorly defined for the renal tubule cells of many nonmammalian vertebrates, but it is often possible to suggest cellular and molecular processes by analogy with those found in mammals and a few nonmammalian vertebrates and to propose avenues of future research. Finally, the chapter considers hormonal, neural, and other regulatory factors of the transport steps of these inorganic ions, insofar as they are understood in nonmammalian vertebrates.

Keywords Inorganic ions • Tubular transport • Tubular reabsorption • Tubular secretion • Tubule transport sites • Molecular transporters • Transport regulation

4.1 Introduction

A large fraction of the inorganic ions filtered by glomerular nephrons does not appear in the ureteral urine and, therefore, must be reabsorbed by the renal tubules. In addition, because it has become clear that the variation in excretion of inorganic ions cannot be explained by filtration and reabsorption alone, tubular secretion must also play a role. The degree to which these processes have been studied in nonmammalian vertebrates varies greatly, but some major observations can be considered and compared and contrasted among the mammalian and nonmammalian vertebrates.

© The American Physiological Society 2016
W.H. Dantzler, *Comparative Physiology of the Vertebrate Kidney*,
DOI 10.1007/978-1-4939-3734-9_4

4.2   Sodium and Chloride

4.2.1   Direction, Magnitude, and Sites of Net Transport

Reabsorption of a major portion of the filtered sodium and chloride occurs along the renal tubules of mammals, all nonmammalian tetrapods, and most fishes under physiological conditions (Dantzler 1976; Dantzler and Bentley 1978; Hickman and Trump 1969; Jard and Morel 1963; Nishimura and Imai 1982; Sawyer 1970; Skadhauge 1973; Stolte et al. 1977a, b). However, only the renal tubules of mammals and birds are capable of reabsorbing almost all the filtered sodium, more than 99 % (Clark et al. 1976; Dantzler 1987; Skadhauge 1973). The renal tubules of most other vertebrates—reptiles, amphibians, glomerular teleosts, lung-fish, and freshwater lampreys—reabsorb between 35 % and 97 % of the filtered sodium and chloride, the largest amounts being reabsorbed by only a few freshwater forms (Dantzler 1976; Hickman and Trump 1969; Jard and Morel 1963; Long 1973; Nishimura and Imai 1982; Sawyer 1970; Stolte et al. 1977b). The renal tubules of marine and, possibly, freshwater elasmobranchs apparently reabsorb even less of the filtered sodium and chloride (Hickman and Trump 1969; Stolte et al. 1977a). And the archinephric ducts of the primitive marine hagfishes (e.g., *Myxine glutinosa*), which conform to their environment, reabsorb no filtered sodium (Stolte and Schmidt-Nielsen 1978).

It is important, however, to be aware of two factors regarding these values for fractional reabsorption by the renal tubules. First, they are obtained primarily from clearance studies that supply information only for the kidney as a whole. Such studies show only the net difference between the amount filtered and the amount excreted in the ureteral urine. Although net reabsorption is apparent, this approach does not distinguish between simple net reabsorption throughout the length of the tubules and net secretion in one portion and net reabsorption in another. In fact, as in the case of fluid, net secretion of sodium and chloride definitely can occur along the isolated, perfused proximal tubules from glomerular teleosts and elasmobranchs (Table 4.1) (Beyenbach 1982, 1986; Beyenbach and Fromter 1985). The proximal tubules of some other nonmammalian species also may secrete sodium and chloride to a lesser extent (Beyenbach 1986). In addition, net secretion of sodium and chloride apparently can occur along the isolated, perfused inner medullary collecting ducts of Sprague-Dawley rats (Wallace et al. 2001). Moreover, the ultrastructure of some of the collecting duct cells of marine catfish of the family Plotisidae is the same as that of chloride-secreting cells in the gills, suggesting that they too may secrete sodium and chloride (Hentschel and Elger 1987). Secretion of sodium chloride when it occurs may explain the apparent low fractional reabsorption of sodium chloride by the whole kidney in many nonmammalian vertebrates.

Second, the clearance determinations of fractional reabsorption of sodium and chloride are based solely on measurements of ion concentrations in the aqueous phase of the ureteral urine. In birds and uricotelic reptiles, significant quantities of

Table 4.1 Sodium and chloride transport by tubule segment

Tubule segment and species	J^{Na}_{Net} % filtered load	J^{Na}_{Net} pmol min^{-1} mm^{-1}	J^{Cl}_{Net} % filtered load	J^{Cl}_{Net} pmol min^{-1} mm^{-1}	References
Proximal					
Fishes					
Petromyzonta					
River lamprey, *Lampetra fluviatilis* (freshwater)	10	29.3	10	28.7	Logan et al. (1980), Moriarty et al. (1978)
Elasmobranchii					
Dogfish shark, *Squalus acanthias* (stimulated)		-10.3 ± 3.3 (4)		-8.5 ± 2.5 (4)	Beyenbach (1986), Sawyer and Beyenbach (1985)
Teleostei					
Winter flounder, *Pseudopleuronectes americanus*		-4.6 ± 0.5 (17)		-4.5 ± 0.4 (17)	Beyenbach et al. (1986a)
					Beyenbach et al. (1986a)
Killifish, *Fundulus heteroclitus* (seawater)		-6.1 ± 1.0 (16)		-7.7 ± 1.3 (16)	Beyenbach (1986)
(freshwater)		~ -5 (6)		~ 5 (6)	Cliff and Beyenbach (1992)
Amphibia					
Anura					
Bullfrog, *Rana catesbeiana*	17	35.7	17	30	Irish and Dantzler (1976), Long (1973)
Urodela					
Tiger salamander, *Ambystoma tigrinum*		26.9		24.6	Sackin and Boulpaep (1981b)
Mudpuppy, *Necturus maculosus*	29	39.3 ± 6.2 (47)			Boulpaep (1972), Garland et al. (1975)
Reptilia					
Squamata					
Ophidia					
Garter snake, *Thamnophis* spp.	45	130.5	45	115.4	Dantzler and Bentley (1978)
Sauria					
Blue spiny lizard, *Sceloporus Cyanogenys*	36.5	27.8			Stolte et al. (1977b)
Aves					
Passeriformes					
European starling, *Sturnus vulgaris* Reptilian-type nephrons	56	31.4	58	27.5	Laverty and Dantzler (1982)
Mammalia					
Rabbit, *Oryctolagus cuniculus*					
Convoluted segment	65	168.1			Kokko et al. (1971)
Straight segment	65	63.9			Schafer et al. (1977)

(continued)

Table 4.1   (continued)

Tubule segment and species	J_{Net}^{Na} % filtered load	J_{Net}^{Na} pmol min^{-1} mm^{-1}	J_{Net}^{Cl} % filtered load	J_{Net}^{Cl} pmol min^{-1} mm^{-1}	References
Early Distal "Diluting"					
Fishes					
Petromyzonta					
River lamprey,	1–2	17.6	1–2	17.6	Logan et al. (1980)
Lampetra fluviatilis					Moriarty et al. (1978)
(freshwater)					
Elasmobranchii					
Dogfish shark,					
Squalus acanthias					
Intermediate				47.7	Friedman and Hebert
segment IV					(1990)
Teleostei					
Rainbow trout,				65.5 ± 20.6 (12)	Nishimura
Oncorhyncus mykiss					et al. (1983a)
(freshwater)					
Amphibia					
Anura					
Leopard frog,		81.1 ± 10.1 (8)		56.6 ± 6.7 (3)	Stoner (1977)
Rana pipiens					
Bullfrog,	40				Long (1973)
Rana catesbeiana					
Urodela					
Tiger salamander,		85.6 ± 16.4 (5)			Stoner (1977)
Ambystoma tigrinum					
Mudpuppy,	32				Garland et al. (1975)
Necturus maculosus					
Congo eel,				33.4 ± 1.4 (12)	
Amphiuma means					Oberleithner
					et al. (1982)
Aves					
Galliformes					
Japanese quail,				271.8 ± 32.9 (13)	Miwa and Nishimura
Coturnix coturnix					(1986)
Mammalian-type					
nephrons, TAL					
Mammalia					
Rabbit,					
Oryuctolagus cuniculus					
Medullary TAL	25	28–276		19–94	Rocha and Kokko
					(1973),
					Fine and Trizna (1977),
					Stokes (1979)
Cortical TAL	25	60–70		39–60	Horster (1978),
					Burg and Green (1973),
					Shareghi and Agus
					(1982)

(continued)

Table 4.1 (continued)

Tubule segment and species	J_{Net}^{Na} % filtered load	J_{Net}^{Na} pmol min^{-1} mm^{-1}	J_{Net}^{Cl} % filtered load	J_{Net}^{Cl} pmol min^{-1} mm^{-1}	References
Late distal and collecting or connecting tubule					
Amphibia					
Urodela					
Tiger salamander, *Ambystoma tigrinum*		21.2 ± 4.2 (18)			Stoner (1977)
Mudpuppy, *Necturus maculosus*	22				Garland et al. (1975)
Congo eel, *Amphiuma means*	25				Wiederholt et al. (1971)
Reptilia					
Squamata					
Sauria					
Blue spiny lizard, *Sceloporus cyanogenys*	21.1	61.7			Stolte et al. (1977b)
Mammalia					
Rabbit, *Oryctolagus cuniculus*	5–7	62–82			Shareghi and Stoner (1978)
Collecting duct					
Fishes					
Petromyzonta					
River lamprey, *Lampetra fluviatilis* (freshwater)	80	153.3	80	150	Logan et al. (1980), Moriarty et al. (1978)
Amphibians					
Anura					
Bullfrog, *Rana catesbeiana*	47.1				Long (1973)
Urodela					
Tiger salamander, *Ambystoma tigrinum*		25.9 ± 4.9 (8)			Delaney and Stoner (1982)
Reptilia					
Squamata					
Sauria					
Blue spiny lizard, *Sceloporus cyanogenys*	37.2				Stolte et al. (1977b)
Mammalia					
Rabbit, *Oryctolagus cuniculus*					
Cortical	2–3	23		0	Stokes (1981, 1982)
Outer medullary	2–3	~0	~0	~0	Stokes (1981, 1982)
Papillary	3–3	~40	~80	~80	Rocha and Kudo (1982)

Values are means, ranges, or means ± SE. They are taken directly or calculated from the cited references. Numbers in parentheses indicate number of determinations. J_{Net}^{Na}, net transepithelial transport of sodium. J_{Net}^{Cl}, net transepithelial transport of chloride. Negative sign in front of value indicates net secretion rather than net reabsorption. TAL, thick ascending limb of loop of Henle in mammalian nephrons or mammalian-type avian nephrons

filtered sodium may be contained in urate precipitates (vide infra) so that the actual fraction of filtered sodium reabsorbed may be even lower than that usually measured (Dantzler 1978).

Even considering these cautionary notes about clearance measurements of net sodium and chloride transport, it is apparent that the fractional reabsorption of filtered sodium and chloride by the renal tubules of many nonmammalian species is low. Additional reabsorption, depending on the requirements for conserving sodium chloride may occur in regions distal to the kidney—cloaca, colon, or bladder—and must be integrated with the renal reabsorption in the maintenance of overall ionic balance.

The major sites of sodium and chloride transport in the renal tubules, determined by micropuncture and microperfusion experiments, vary among the vertebrates. Although the proximal tubule has generally been considered the primary site for reabsorption of filtered sodium and chloride, this is really only true for the mammals, 60–80 % (Giebisch and Windhager 1964), and birds, 50–60 % (Laverty and Dantzler 1982), among those vertebrates in which this problem has been studied (Table 4.1). Only some 20–45 % of the filtered sodium is reabsorbed along the proximal tubules of those amphibians and reptiles studied (Table 4.1) (also see Dantzler and Bentley 1978; Garland et al. 1975; Long 1973; Stolte et al. 1977b), and only 10 % is reabsorbed along the proximal tubules of the primitive river lamprey even in freshwater (Table 4.1) (also see Logan et al. 1980). In these nonmammalian vertebrates, much of the filtered sodium and chloride clearly is reabsorbed distal to the proximal tubule (Table 4.1). Some 50–70 % of the filtered sodium (and, presumably, chloride) is reabsorbed along the distal tubules or collecting ducts of those amphibians and reptiles studied, and about 80 % of the filtered sodium and chloride is reabsorbed along the collecting ducts of the river lamprey (Table 4.1).

The rate of sodium and chloride reabsorption per unit length of proximal tubule is about the same for most amphibians, reptiles, and birds studied (Table 4.1). It is somewhat higher for garter snakes (*Thamnophis* spp.) and considerably higher for mammals (Table 4.1). The fraction of filtered sodium and chloride reabsorbed in this tubule segment reflects the balance among the filtration rate, the reabsorption rate, and the length of the segment. The rate of sodium and chloride reabsorption per unit length of distal tubule and collecting duct segments is about the same for amphibians, reptiles, and mammals (Table 4.1). However, the rate of chloride reabsorption per unit length of early distal tubule (thick ascending limb) in the one avian species studied (Japanese quail, *Coturnix coturnix*) is considerably higher than that in amphibians and reptiles and as high as or even higher than that in mammals (Table 4.1). This high rate of reabsorption may reflect the primacy of sodium and chloride reabsorption in this tubule segment in the urine concentrating process in birds (vide infra; Chap. 7). Again, the fraction of filtered sodium and chloride reabsorbed in any of these distal segments reflects the balance among the filtered load reaching these regions, the rate of reabsorption per unit length, and the length of the segment. However, the rate of sodium and chloride reabsorption per unit length of collecting duct is much higher in the lampreys than in other vertebrates, apparently reflecting the extremely high fractional reabsorption in this region (Table 4.1).

The tubule sites, the direction at a given site, and the magnitude of sodium and chloride transport are not completely clear for the glomerular teleosts and elasmobranchs. Although net reabsorption is said to occur along the first segment of the proximal tubule (Beyenbach et al. 1986; Bulger and Trump 1968; Nishimura and Imai 1982), only one micropuncture study on a marine elasmobranch (little skate, *Raja erinacea*) supports this concept directly (Stolte et al. 1977a). Even this study provides no quantitative data on the magnitude of the reabsorptive transport in the first segment of the proximal tubule. Moreover, the data suggest that, in this marine species, substantially more sodium and chloride are reabsorbed in the second segment of the proximal tubule than in the first segment. To complicate matters further, studies with isolated, perfused tubules clearly demonstrate that net secretion of sodium and chloride can occur along the second segment of the proximal tubule of another marine elasmobranch (dogfish shark, *S. acanthias*), a marine teleost (winter flounder, *P. americanus*), and a euryhaline teleost (killifish, *F. heteroclitus*) adapted to either seawater (Beyenbach 1986; Sawyer and Beyenbach 1985) or, quite surprisingly, freshwater (Cliff and Beyenbach 1988a, 1992). Indeed, the rate of sodium chloride secretion in proximal tubules from fish adapted to freshwater is almost as great as it is in tubules from those adapted to seawater (Table 4.1) (Cliff and Beyenbach 1992). The rate of net secretion along the proximal tubules from those fish species studied is about one-fifth to one-third the rate of net reabsorption along the proximal tubules of other poikilothermic nonmammalian vertebrates and an even smaller fraction of the rate of net reabsorption along the proximal tubules of mammals (Table 4.1).

It is not yet known whether net secretion ever changes to net reabsorption in proximal tubules from most of the fish species studied. However, in killifish adapted to seawater, net secretion occurs in about 30–70 % of the proximal tubules whereas in killifish adapted to freshwater it occurs in only about 10 % of the proximal tubules (Cliff and Beyenbach 1992). The factors that regulate whether or not a given proximal tubule displays net secretion are unknown, but may relate to the mechanism involved in net secretion (vide infra). This difference in the number of tubules showing net secretion in seawater- versus freshwater-adapted animals certainly fits with the view that there should be a greater reliance on net secretion of sodium and chloride in seawater than in freshwater (Beyenbach 2004; Cliff and Beyenbach 1992). In this regard, it seems likely that net secretion always occurs along proximal tubules of stenohaline glomerular marine fish and euryhaline glomerular fish adapted to seawater and that it plays some role, in addition to extrarenal routes of ion secretion (gills and rectal glands), in eliminating excess sodium and chloride. For euryhaline teleosts, this secretory process may become particularly important for maintaining excretory function when the number of filtering glomeruli is reduced during adaptation to seawater (vide supra; Chap. 3) (Beyenbach 1986, 2004). It is apparently essential for aglomerular marine teleosts (vide infra; Chap. 5).

Studies of isolated, perfused early distal tubules from one euryhaline teleost species adapted to freshwater (rainbow trout, *O. mykiss*) (Nishimura et al. 1983a) show that net chloride reabsorption can occur at rates similar to those that occur under like circumstances in this segment of amphibian and mammalian nephrons (Table 4.1). Nishimura and Imai (Nishimura and Imai 1982) suggest that such net reabsorption, producing dilution of the tubular fluid, occurs in the early portion of the distal tubule of all freshwater and freshwater-adapted teleosts (vide infra; Chap. 7).

4.2.2   Mechanism of Transport

4.2.2.1   Introduction

Among the nonmammalian vertebrates, the mechanisms involved in net reabsorption of sodium and chloride have been studied in most detail in amphibians (primarily urodeles), although additional information has been contributed by measurements in scattered species from other nonmammalian classes. The studies on amphibians have also contributed to the understanding of these mechanisms in mammals, but they have been complemented by many additional physiological and molecular studies in mammalian species. The mechanisms involved in net secretion of sodium and chloride have been studied primarily in elasmobranchs and teleosts. The choice of species for the study of either reabsorption or secretion in nonmammalian vertebrates has depended on the significance of the transport process under study in that species, the accessibility of the tubules for in vivo micropuncture and microperfusion and in vitro microperfusion, and the suitability of tubule cells for microelectrode impalements. The current understanding of these transport processes in nonmammalian vertebrates, which is far from complete, is reviewed in the following discussions. The understanding in mammals, which has been reviewed frequently in recent years (e.g., Mount 2012), is touched upon only for critical comparisons. Some of the measurements of electrical potentials, electrical resistances, and ion fluxes from nonmammalian and mammalian vertebrates that are most important for a comparative discussion of transport mechanisms are summarized in Tables 4.1 and 4.2.

4.2.2.2   Proximal Tubules

Net Reabsorption

Although the data on sodium and chloride reabsorption in the proximal tubules of nonmammalian vertebrates are far from complete or comprehensive, certain general characteristics are apparent. Micropuncture and microperfusion studies reveal a lumen-negative transepithelial potential during net sodium reabsorption by

Table 4.2 Electrical properties by tubule segment

Tubule segment and species	V_T (mV)	V_{BL} (mV)	R_T (KΩcm)	R_T (Ωcm²)	References
Proximal					
Fishes					
Elasmobranchii					
Dogfish shark, *Squalus acanthias*					
(Unstimulated)	1.2 ± 0.6 (16)	−61.3 ± 1.6 (16)	2.87 ± 0.2 (16)	26.8 ± 3.1 (16)	Beyenbach (1986),
(Stimulated)	−1.6 ± 0.7 (7)	−48.0 ± 5.1 (17)		31.9 ± 3.3 (7)	Beyenbach and Fromter (1985)
Teleostei					
Rainbow trout, *Oncorhyncus mykiss* (Freshwater)	−5.0 ± 0.7 (15)				Nishimura et al. (1983a)
Winter flounder, *Pseudopleuronectes americanus*	−1.9 ± 0.2 (113)			25.6 ± 2.7 (28)	Beyenbach et al. (1986)
Amphibia					
Anura					
Leopard frog, *Rana pipiens*	−6.5 ± 1.6 (4)				Stoner (1977)
Toad, *Bufo marinus*	−6.6 ± 1.7 (4)				Stoner (1977)
Urodela					
Tiger salamander, *Ambystoma tigrinum*	−4.5 ± 0.2 (137)	−59.6 ± 2 (84)		52.1 ± 3.0 (81)	Sackin and Boulpaep (1981b)
Mudpuppy, *Necturus maculosus*	−15.4 ± 0.6 (93)	−61.0 ± 3.7 (10)		69.9 ± 8.97 (14)	Boulpaep (1972)
Reptilia					
Squamata					
Ophidia					
Garter snake, *Thamnophis* spp.	−0.49 ± 0.15 (14)	−60.1 ± 1.9 (13)			Dantzler and Bentley (1981), Kim and Dantzler (1995)
Aves					
Passeriformes					
European starling, *Sturnus vulgaris*					
Reptilian-type nephrons	−2.0				Laverty and Alberici (1987)
Mammalia					
Rabbit, *Oryctolagus Caniculus*					
Convoluted segment	−2.0 to 0	−51 ± 1.63 (24)		7.0	Biagi et al. (1981), Schafer and Andreoli (1979)
Straight segment	+2.0 to +3.1	−47.0 ± 0.97 (94)		8.2	Biagi et al. (1981), Schafer and Andreoli (1979), Schafer and Barfuss (1982)

(continued)

Table 4.2 (continued)

Tubule segment and species	V_T (mV)	V_{BL} (mV)	R_T (KΩcm)	R_T (Ωcm²)	References
Early distal "Diluting"					
Fishes					
Elasmobranchii					
Dogfish shark,					
Squalus acanthias					
Intermediate	8.7 ± 0.6 (21)	−57.5 ± 10.2 (21)		11.4 ± 0.8 (5)	Friedman and
segment IV					Hebert (1990),
					Hebert and Fried-
					man (1990)
Teleostei					
Rainbow trout,					
Oncorhyncus mykiss	+17.8 ± 1.4 (53)				Nishimura
					et al. (1983a)
(freshwater)					
Amphibia					
Anura					
Leopard frog,	+13.5 ± 1.1 (35)				Stoner (1977)
Rana pipiens					
Toad,					Stoner (1977)
Bufo marinus	+4.6 ± 1.4 (4)				
Urodela					
Tiger salamander	+13.6 ± 1.7 (14)	−54 ± 2			Sackin et al. (1981),
Ambystoma					Stoner (1977)
Tigrinum					
Congo eel,	+9.0 ± 0.5 (70)	−71.0 ± 0.5 (28)		10–30	Oberleithner
Amphiuma means					et al. (1982),
					Stanton
					et al. (1982)
Japanese newt,	+13.7 ± 1.0 (35)	−50.6 ± 1.6		29 ± 9 (4)	Hoshi et al. (1981)
Triturus sp.					
Aves					
Galliformes					
Japanese quail,					
Coturnix coturnix					
Reptilian-type	+4.5 ± 1.7 (7)				Nishimura
nephrons					et al. (1986)
Mammalian-type	+9.1 ± 0.72 (35)				Nishimura
nephrons, TAL					et al. (1986)
Mammalia					
Rabbit,					
Oryctolagus					
Caniculus					
Medullary TAL	+3 to +7				Burg (1982)
Cortical TAL	+3 to +10			25–34	Burg (1982)
Late distal and collecting or					
connecting tubule					
Amphibia					
Urodela					
Tiger salamander,	−41.2 ± 4.2 (25)				Stoner (1977)
Ambystoma					
Tigrinum					
Congo eel,					
Amphiuma means					
Late distal tubule	−0.2 ± 0.4 (11)	−71.3 ± 1.3 (11)			Stanton (1990)
Collecting tubule	−8.0 ± 2.0	−70 ± 4			Stanton
					et al. (1982)
Japanese newt	−6.5 ± 2,2 (11)			754	Hoshi et al. (1981)
Triturus sp.					

(continued)

Table 4.2 (continued)

Tubule segment and species	V_T (mV)	V_{BL} (mV)	R_T (KΩcm)	R_T (Ωcm^2)	References
Reptilia					
Squamata					
Ophidia					
Garter snake, *Thamnophis* sp.	-34.9 ± 2.1 (27)		23.4 ± 1.6 (27)	83.1	Beyenbach et al. (1980)
Aves					
Galliformes					
Japanese quail *Coturnix coturnix*	-3.2 ± 0.83 (7)				Nishimura et al. (1986)
Mammalia					
Rabbit *Oryctolagus cuniculus*	-5 to -40				Gross et al. (1975), Mount (2012), Shareghi and Stoner (1978)
Collecting duct					
Amphibia					
Urodela					
Tiger salamander *Ambystoma tigrinum*	-8.9 ± 1.9 (17)		31 ± 5	626	Delaney and Stoner (1982)
Japanese newt *Triturus* sp.	-23 ± 5				Hoshi et al. (1981)
Mammalia					
Rabbit *Oryctolagus Caniculus*					
Cortical	-23	96	15.2		Koeppen et al. (1983)
Outer medullary	-30 to $+10$				Stokes et al. (1978)
Papillary	~0	-15 to $+24$			Rocha and Kudo (1982), Terreros et al. (1981)

Values are means, ranges, or means $\pm$ SE. They are taken directly or calculated from the cited references. Numbers in parentheses indicate number of determinations. V_T indicates transepithelial potential difference; sign indicates tubule lumen relative to peritubular side. V_{BL} indicates potential difference across basolateral membrane; sign indicates inside of cell relative to outside of cell on basolateral side. R_T indicates transepithelial resistance. TAL indicates thick ascending limb of loop of Henle in mammalian nephrons or mammalian-type avian nephrons

proximal tubules of amphibians, reptiles, birds, and freshwater-adapted teleosts (Table 4.2), and indicate that transepithelial reabsorption occurs against an electrochemical gradient (Fig. 4.1) (Boulpaep 1972; Dantzler and Bentley 1981; Giebisch 1961; Nishimura et al. 1983b; Sackin and Boulpaep 1981a, b; Stoner 1977; Windhager et al. 1959). The fractional reabsorption of chloride is essentially the same as the fractional reabsorption of sodium along the proximal tubules of lampreys (Logan et al. 1980), amphibians (Giebisch 1956), reptiles (Dantzler and Bentley 1978), and birds (Laverty and Dantzler 1982) (Table 4.1). This also appears to be the case in the first portion of the proximal tubule of elasmobranchs (Stolte et al. 1977a) and teleosts (Nishimura and Imai 1982). Thus, it appears that transepithelial reabsorption of sodium and chloride by the proximal tubules of all

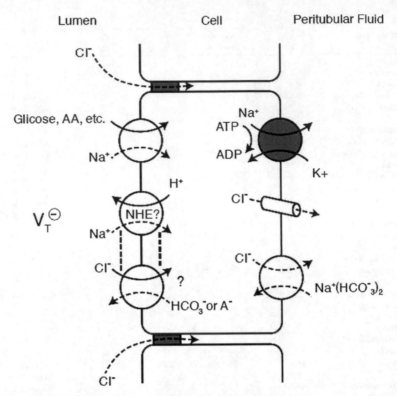

Fig. 4.1 Model for sodium and chloride reabsorption in proximal tubules of nephrons of nonmammalian vertebrates based mostly on studies with amphibians. Filled circle with solid arrows and breakdown of ATP to ADP indicates primary active transport, in this case, Na-K-ATPase. Open circles indicate carrier-mediated diffusion or secondary active transport. Broken arrows indicate movement down an electrochemical gradient. Solid arrows indicate movement against an electrochemical gradient. Tube crossing the cell membrane indicates channel for ions or other substances in the membrane. V_T indicates transepithelial potential difference with sign enclosed by a circle indicating whether lumen is negative or positive relative to peritubular fluid. Question marks indicate uncertainty about nature of transport steps. Modified from Braun and Dantzler 1997 with permission. See text for other information on which drawing is based

those nonmammalian vertebrates studied involves active reabsorption of sodium and passive reabsorption of chloride but with both ions being reabsorbed at essentially equivalent rates.

The transepithelial resistance of those proximal tubules of nonmammalian vertebrates in which it has been measured is low but not nearly so low as that of mammalian proximal tubules (Table 4.2). These data indicate that mammalian proximal tubules are leakier electrically than nonmammalian proximal tubules, a characteristic that also appears to be reflected in the generally lower transepithelial potentials in mammalian than in nonmammalian proximal tubules (Table 4.2).

This low transepithelial resistance in mammalian proximal tubules also reflects the fact that they are quite permeable to sodium, chloride, and other ions. Thus, in addition to transcellular active reabsorption of sodium and passive reabsorption of chloride, there is substantial paracellular reabsorption of these and other ions (Mount 2012; Weinstein and Windhager 1985). The combined transcellular and paracellular reabsorptive processes are reflected in transepithelial potentials in mammalian proximal tubules: slightly lumen-negative in the early portion and slightly lumen-positive in the late portion (Table 4.2). The initial lumen-negative potential reflects the presence of organic solutes and their electrogenic cotransport with sodium (vide infra; Chap. 6). The small lumen-positive potential in the late portion of the mammalian proximal appears to reflect the fact that chloride reabsorption lags behind sodium and bicarbonate reabsorption in the early portion so that its luminal concentration is greater than that of bicarbonate in the late portion. The lumen-negative potential then appears to be a chloride-diffusion potential generated by the transepithelial differences in chloride and bicarbonate concentrations and the greater paracellular permeability for chloride than for bicarbonate and other solutes (Mount 2012; Weinstein and Windhager 1985). This lumen-positive potential drives paracellular sodium reabsorption whereas the chemical gradient from lumen to bath drives paracellular chloride reabsorption in this segment, and overall about 40 % of the filtered sodium chloride is reabsorbed paracellularly in the mammalian late proximal tubule (Mount 2012). As noted above, however, in all portions of nonmammalian proximal tubules studied, chloride reabsorption does not lag behind sodium reabsorption, the transepithelial potential difference is significantly lumen-negative (Table 4.2), the transepithelial resistance is substantially higher than in mammalian proximal tubules (Table 4.2), and the paracellular pathway for sodium chloride reabsorption may be less important in some, but certainly not all, nonmammalian species than in mammals (vide infra).

A detailed analysis of ionic transport in proximal tubules of urodele amphibians (*N. maculosus, N. means, Ambystoma tigrinum*) involving intracellular recordings with ion-sensitive and conventional microelectrodes as well as transepithelial recordings has supplied information on the basolateral membrane potential (Table 4.2), the relative conductances across the basolateral and luminal membranes, the transepithelial ionic shunt pathways, and the intracellular ion activities (Abdulnour-Nakhoul and Boulpaep 1998; Boulpaep 1976; Guggino et al. 1982a, b, 1983; Kimura and Spring 1978; Sackin and Boulpaep 1981a, b; Spring and Kimura 1978). Studies with vesicles from brush-border membranes of *N. maculosus* have also supplied information on some ion transport steps across that membrane (Seifter and Aronson 1984). These data have permitted generation of a model for sodium and chloride reabsorption in amphibian proximal tubules (Fig. 4.1). In this model, sodium enters the cells across the luminal membrane down an electrochemical gradient by a saturable process that is rate-limiting for transepithelial transport (Fig. 4.1). This sodium entry step apparently involves primarily sodium-hydrogen exchange (Fig. 4.1), but other processes, e.g., coupled entry with organic solutes (Figs. 6.1 and 6.2), are certainly involved as well. Sodium is then transported out of the cells across the basolateral membrane by a rheogenic active process, apparently involving Na-K-ATPase, that is not saturable over a range of sodium concentrations greater than those normally found in the cells (Fig. 4.1). Saturation of the sodium entry step across the luminal membrane apparently results from decreases in the sodium permeability in response to increases in the intracellular sodium concentration.

Chloride appears to enter the cells across the luminal membrane in the transcellular reabsorptive process by an electroneutral, carrier-mediated process driven by the electrochemical gradient for sodium (Fig. 4.1). However, this process does not appear to involve a direct sodium-chloride cotransport step. Instead, it appears to involve an anion-exchange process, possibly chloride-bicarbonate exchange coupled to sodium-proton exchange (Fig. 4.1), but this is far from certain (Abdulnour-Nakhoul and Boulpaep 1998). In amphibians, chloride exits the cells across the basolateral membrane during transcellular reabsorption through chloride channels and also by an electroneutral process (Fig. 4.1) (Abdulnour-Nakhoul and Boulpaep 1998). The latter may involve some type of sodium-dependent chloride-bicarbonate exchange (Fig. 4.1) (Abdulnour-Nakhoul and Boulpaep 1998). In birds, inhibitor studies on regulation of intracellular pH also support the presence of such a sodium-dependent chloride-bicarbonate exchange process at the basolateral membrane of proximal tubules of intermediate and long-looped mammalian-type nephrons but not of loopless reptilian-type nephrons (Brokl et al. 1998; Kim et al. 1997; Martinez et al. 1997). In addition to the transcellular pathway in amphibian proximal tubules, there is a paracellular shunt pathway for chloride (at least in *Necturus*) that accounts for as much as two-thirds of the transepithelial reabsorption (Fig. 4.1) (Kimura and Spring 1978).

To what extent these cellular transport steps or paracellular shunts may be involved in proximal reabsorption of sodium and chloride in fishes, reptiles, and birds is not clear. However, given the importance of such reabsorption in the regulation of sodium and chloride balance in most species, the mechanisms involved certainly merit systematic study. Moreover, it should be noted that many of the cellular transport mechanisms described above for proximal tubules of amphibians are also found in proximal tubules of mammals (Mount 2012) and probably involve the same molecular families of transporters. However, no molecular studies of these tubule transporters have yet been made in nonmammalian vertebrates. Such studies need to be an important part of future studies to understand transepithelial sodium chloride reabsorption in renal proximal tubules of nonmammalian vertebrates.

Net Secretion

Beyenbach and his colleagues first defined the mechanism for net sodium chloride secretion in those fish proximal tubules in which it clearly occurs, helping to establish the basic pattern for secretion in other tubule segments (Beyenbach 2004; Beyenbach and Fromter 1985; Beyenbach et al.). They applied an electrophysiological analysis similar to that described above for amphibians to the isolated, perfused second segment of proximal tubules from elasmobranch (dogfish shark, *S. acanthias*) (Beyenbach and Fromter 1985) kidneys and a less rigorous analysis, because the tubule cells are too small to hold intracellular electrodes for long, to the second segment of proximal tubules from marine teleost (winter flounder, *P. americanus*) (Beyenbach et al. 1986) kidneys. This tubule segment from both species has a relatively low transepithelial resistance and, when secreting, a small, lumen-negative transepithelial potential (Table 4.2). The low transepithelial resistance appears to reflect a paracellular shunt pathway for sodium chloride that is more selective for sodium than chloride (Beyenbach 1986; Beyenbach and Frömter 1985; Beyenbach et al. 1986). Secretion of sodium and chloride by the elasmobranch tubule, and probably also the teleost tubule, is stimulated by cAMP and inhibited by furosemide and ouabain (Beyenbach and Fromter 1985).

The electrophysiological studies on secretion by elasmobranch tubules, involving cAMP stimulation and furosemide and ouabain inhibition, are consistent with a secondary active transport process for sodium chloride secretion, in which, as in the case of sodium chloride reabsorption, the primary energy is provided by the ouabain-sensitive Na-K-ATPase on the basolateral membrane (Fig. 4.2)

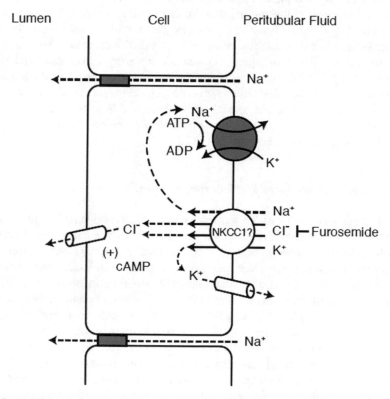

Fig. 4.2 Model for sodium and chloride secretion by proximal tubules based primarily on studies with elasmobranchs by Beyenbach and his colleagues (Beyenbach 1986, 2004; Beyenbach and Fromter 1985). Line with bar at the end indicates inhibition. Other symbols have the same meaning as in the legend for Fig. 4.1. See text for detailed description of the processes diagramed in the figure

(Beyenbach 1986, 2004; Beyenbach and Fromter 1985). The generation of a low intracellular sodium concentration by the primary active transport of sodium out of the cells across the basolateral membrane provides a substantial electrochemical gradient for the movement of sodium into the cells from the peritubular side (Fig. 4.2). Movement of chloride into the cells from the peritubular side is coupled to the energy provided by the sodium gradient apparently by means of a furosemide-sensitive Na-K-2Cl cotransporter (Fig. 4.2) (Beyenbach 2004). This should be the NKCC1 isoform of this transporter as in other secretory epithelia (Hass and Forbush 1998), but no molecular studies on the transporters in these secretory tubule segments have yet been performed. The sodium and potassium that enter the cells via this transporter are recycled out of the cells again via basolateral Na-K-ATPase and potassium channels, respectively (Fig. 4.2). The chloride, however, moves passively from cells to lumen through chloride channels in the luminal membrane, the conductance of which can be regulated by cAMP (Fig. 4.2). The transcellular transport of chloride from the peritubular side to the lumen is electrically balanced by the paracellular movement of sodium in the same direction (Fig. 4.2). In support of this model, stimulation of sodium chloride secretion (either spontaneously or with cAMP) in proximal tubules of *S. acanthias* decreases transepithelial resistance and resistance of the luminal membrane, depolarizes the potential across the basolateral membrane towards zero, and makes the transepithelial potential lumen-negative (Table 4.2) (Beyenbach 2004; Beyenbach and Fromter 1985). This mechanism for sodium chloride secretion in elasmobranch and probably teleost proximal tubules (Beyenbach and Fromter 1985; Beyenbach et al. 1986) resembles the process found in other secretory epithelia, e.g., reptilian and avian salt glands (Larsen et al. 2014).

4.2.2.3 Early Distal Tubules ("Diluting Segment")

As indicated above (Table 4.1), significant sodium and chloride reabsorption occurs along the early portion of the distal tubules of all those nonmammalian vertebrate species in which it has been studied, except freshwater lampreys. Such reabsorption also occurs to a significant extent along the comparable thick ascending limb of nephrons in mammals and of mammalian-type nephrons in birds. In addition, it occurs along the intermediate segment IV from the dorsal bundle zone of elasmobranch nephrons (Table 4.1). In all these cases, this segment appears to be impermeable to water and sodium and the chloride reabsorption appears to be involved in dilution of the tubular fluid (vide infra; Chap. 7). Moreover, as described above (Chap. 2), these early distal and thick ascending loop segments in nonmammalian and mammalian vertebrate nephrons as well as the intermediate segment IV of elasmobranch nephrons have certain common features of cell structure and, as discussed in the following, probably have similar reabsorptive mechanisms as well (Fig. 4.3).

The most detailed studies of sodium and chloride transport by the early distal tubule and thick ascending limb segments have been made in amphibians

(Bott 1962; Giebisch 1961; Hoshi et al. 1981, 1983; Oberleithner et al. 1982; Sackin et al. 1981; Stoner 1977; Sullivan 1968; Walker et al. 1937) and mammals (Greger 1985; Mount 2012), although less detailed studies have been made on freshwater teleosts (Nishimura, Imai, and Ogawa 1983a) and birds (Miwa and Nishimura 1986; Nishimura et al. 1984, 1986). Similar studies have been made on intermediate segment IV from elasmobranch kidneys (Friedman and Hebert 1990; Hebert and Friedman 1990). In vivo or in vitro microperfusions of these segments from each species studied reveal a lumen-positive transepithelial potential and a significant chloride reabsorptive flux (Tables 4.1 and 4.2), both of which are dependent on the presence of both sodium and chloride in the lumen and are eliminated by the addition of furosemide or bumetanide to the perfusate or ouabain to the peritubular bathing medium (Friedman and Hebert 1990; Greger 1985; Hebert and Friedman 1990; Hoshi et al. 1981, 1983; Miwa and Nishimura 1986; Nishimura et al. 1983a, 1984, 1986; Oberleithner et al. 1982; Stoner 1977).

No such measurements have been made on the early distal tubules of reptiles, but, as noted earlier (Chap. 2), the cells of this segment in reptiles do not necessarily have the structural characteristics described for other species. A few preliminary measurements on the thin intermediate nephron segment from one ophidian reptile species (garter snake, *Thamnophis* spp.) indicate that it has a significant lumen-positive transepithelial potential and, therefore, may reabsorb sodium chloride and dilute the urine by the same process as the early distal tubule of other nonmammalian vertebrates (S. D. Yokota and W. H. Dantzler unpublished observations). However, this is far from certain, and more definitive studies to determine the diluting site and mechanism of sodium chloride reabsorption in reptilian nephrons are certainly needed.

For the early distal tubules of amphibians (Congo eel, *A. means*), and, it is assumed for comparable tubule segments from other nonmammalian vertebrates, intracellular recordings with ion-sensitive and conventional microelectrodes as well as transepithelial recordings indicate that the presence of furosemide in the lumen or the absence of chloride or sodium from the lumen leads to hyperpolarization of the basolateral membrane to an extent equivalent to the decrease in the transepithelial potential (Oberleithner et al. 1982). At the same time, there is a significant decrease in the intracellular chloride activity from a control value that is about three times that expected at electrochemical equilibrium (Oberleithner et al. 1982). Removal of potassium from the lumen also reduces the intracellular activities of sodium and chloride (Oberleithner et al. 1983a, c). Greger and his colleagues obtained similar data on the thick ascending limb of mammals (Greger 1985). Taken together, these data support the general concept of chloride reabsorption by a secondary, sodium- and potassium-coupled mechanism, the energy for which is derived from the ouabain-sensitive Na-K-ATPase at the basolateral membrane (as in the case of the proximal tubule sodium chloride secretion discussed above) (Fig. 4.3) (Greger 1985; Oberleithner et al. 1982). The primary active transport of sodium out of the cells across the basolateral membrane establishes a chemical gradient for coupled electroneutral movement of sodium, chloride, and potassium into the cells across the luminal membrane by a

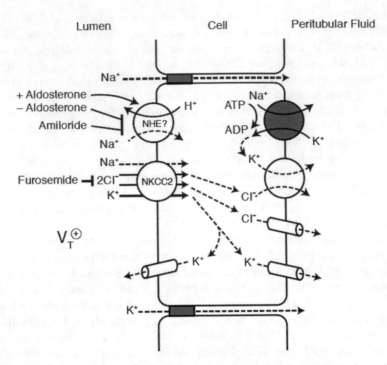

Fig. 4.3 Model for sodium, chloride, potassium, and hydrogen ion transport by early distal tubules of nephrons of nonmammalian vertebrates and thick ascending limbs of loops of Henle of mammalian-type avian nephrons and mammalian nephrons. Symbols have the same meaning as in the legends of Figs. 4.1 and 4.2. Although some details are based primarily on studies on amphibians and mammals, the steps for sodium and chloride reabsorption apparently hold for the early distal tubules of teleost nephrons, amphibian nephrons, and reptilian-type avian nephrons; the thick ascending limb of Henle's loop of mammalian-type avian nephrons and mammalian nephrons; intermediate segment IV of elasmobranch nephrons; and possibly the thin intermediate segment or early distal tubule of reptilian nephrons. See text for detailed description of the processes diagramed in the figure

furosemide- and bumetanide-sensitive Na-K-2Cl cotransporter (Fig. 4.3) (Greger and Schlatter 1983; Oberleithner et al. 1982). In mammals, this is the NKCC2 isoform of this transporter and, since this isoform has been found only in the kidney and, so far, only along the luminal membrane of the thick ascending limb and macula densa (Hass and Forbush 1998; Nielsen et al. 1998), it seems very likely that it is the transporter in the diluting segments of nonmammalian vertebrates. Indeed, NKCC2 has been identified in the distal tubules of euryhaline pufferfish (mefugu, *Takifugu obscurus*) (Kato et al. 2011). This transporter is also highly expressed in the early distal tubule segment of the pronephros in embryonic amphibians (African clawed toad, *Xenopus laevis*) (Hillyard et al. 2009; Tran et al. 2007; Zhou and Vize 2004), and it certainly seems likely that it is present in the early distal tubule of the mesonephros in the mature animals. In addition, immunocytochemical studies suggest that some isoform of NKCC is present in the diluting intermediate segment

IV of elasmobranch nephrons (Biemesderfer et al. 1996). However, NKCC2 was undetectable in the kidneys of five species of ophidian reptiles (Babonis et al. 2011), raising further questions about the site of the diluting region in reptiles.

Chloride that has entered the cells via the Na-K-2Cl cotransporter apparently leaves the cells across the basolateral membrane by both a chloride-conductive channel and an electroneutral cotransporter with potassium (Fig. 4.3) (Greger and Schlatter 1983; Oberleithner et al. 1982). In mammals, about half the transepithelial sodium reabsorption in the thick ascending limb occurs via a paracellular pathway (Fig. 4.3), thereby balancing the transcellular chloride reabsorption (Mount 2014). This is probably the case in the comparable tubule segments in nonmammalian vertebrates as well, but the process has yet to be studied. Potassium that has entered the cells via the Na-K-2Cl cotransporter is probably largely recycled back into the lumen via potassium-conductive channels in the luminal membrane, although some may undergo transepithelial reabsorption, exiting across the basolateral membrane by potassium-conductive channels or by the electroneutral cotransporter with chloride (Fig. 4.3) (vide infra; potassium transport).

In addition to this primary process, there is a sodium-proton exchanger in the luminal membrane of the early distal tubule of amphibian nephrons (vide infra; hydrogen ion transport) and the thick ascending limb of mammalian nephrons (Fig. 4.3) (Mount 2014; Oberleithner et al. 1984). In mammalian thick limbs, this exchanger is NHE3 (Mount 2014; Wang et al. 2001), but it has not been identified in amphibians. In amphibians, at least, this exchanger apparently functions primarily for the regulation of hydrogen ion transport and secondarily for the regulation of potassium secretion (vide infra; potassium and hydrogen ion transport). However, during chronic exposure of amphibians to a high potassium environment, this exchanger is stimulated (vide infra; potassium and hydrogen ion transport), and the Na-K-2Cl cotransporter is depressed so that sodium-proton exchange may account for some 30 % of the sodium entering the cells (Oberleithner et al. 1984). As in the case of sodium entry via the Na-K-2Cl cotransporter, the sodium that has entered the cells via sodium-proton exchange is transported out across the basolateral membrane by Na-K-ATPase. No information is currently available on the presence of such a sodium-proton exchanger in the comparable segment of other nonmammalian vertebrates.

4.2.2.4   Late Distal and Collecting or Connecting Tubules

As indicated earlier, about 20–25 % of the filtered sodium chloride is reabsorbed along the tubule segments between the diluting segment (early distal tubule or thick ascending limb) and the collecting duct in those nonmammalian vertebrates studied, and about 5–7 % is also reabsorbed along this region in mammals (Table 4.1). In mammals, the disappearance of NKCC2 shortly beyond the macular densa and the appearance of the neutral Na-Cl cotransporter (NCC) signals the end of the major diluting region and the start of the distal convoluted tubule (Mount 2012). This also may be the case in birds. Moreover, the disappearance of NKCC2 may well signal the end of the early distal tubule diluting segment and the start of the late distal tubule in other nonmammalian vertebrates. However, there is not enough detailed work on the presence, much less the exact localization, of NKCC2 in the renal tubules of nonmammalian vertebrates to be certain. Direct studies of the reabsorptive process are available only for the late distal tubules of amphibians, reptiles, and birds, the entire distal convoluted tubule of mammals, and the collecting tubules of amphibians and the connecting tubules of mammals. Because of the confusion about the relationship of these structures in one species to those in another (vide supra; Chap. 2), the data from the studies on nonmammalian verte-brates are to a large extent considered together as if they truly represented the analogous segment in all species (Table 4.2). This is undoubtedly an oversimplifi-cation, but much more information must be obtained on the relationship between structure and function in vertebrates other than amphibians and mammals before a clearer picture will emerge. However, it should be noted that the late distal segments and collecting tubules of many nonmammalian vertebrates and the connecting tubules of mammals as well as the collecting ducts of mammalian and many nonmammalian vertebrates have both principal cells (the sodium-reabsorbing cells) and intercalated cells (the proton or bicarbonate-secreting cells) (vide supra; Chap. 2) (also vide infra; current chap.).

In the segments studied by in vivo or in vitro microperfusion, the transepithelial potential varies from essentially zero to modestly lumen-negative (~3–10 mV) to highly lumen-negative (~40 mV), varying both by vertebrate class and by species within class (Table 4.2). Some of these differences may reflect the overlap between late distal tubule and collecting tubule in many nonmammalian vertebrate species. In the one amphibian species in which the two segments can be clearly differenti-ated (Congo eel, *Amphiuma means*) and in mammals, the transepithelial potential is not different from zero in the late distal tubule and significantly lumen-negative in the collecting tubule or connecting tubule (Table 4.2) (Mount 2012; Stanton 1988, 1990; Stanton et al. 1982, 1987b). There is a large intracellular-negative potential across the basolateral membrane in those tubule segments in which it has been measured (Table 4.2). In all tubules studied, regardless of species, the overall transepithelial reabsorption of sodium is against an electrochemical gradient and is inhibited by the presence of ouabain on the basolateral side, indicating that basolateral Na-K-ATPase is the primary energy-requiring step in this overall process (Fig. 4.4).

In those mammals studied, the absence of a transepithelial potential in the distal tubule apparently reflects the entry of sodium and chloride into the transporting cells across the luminal membrane by the electroneutral thiazide-sensitive passive Na-Cl cotransporter (NCC), with transport occurring down the electrochemical gradient for sodium established by Na-K-ATPase (Mount 2012) (Fig. 4.4). There may also be another parallel electroneutral DIDs-sensitive process involving the coupling of a luminal Na-H exchanger (NHE2) and a Cl-HCO$_3$ exchanger for sodium chloride entry into these cells (Mount 2012; Wang et al. 1993) (Fig. 4.4). The chloride apparently exits the cells across the basolateral membrane primarily via a chloride channel (possibly ClC-K2) or perhaps to a lesser extent by a K-Cl

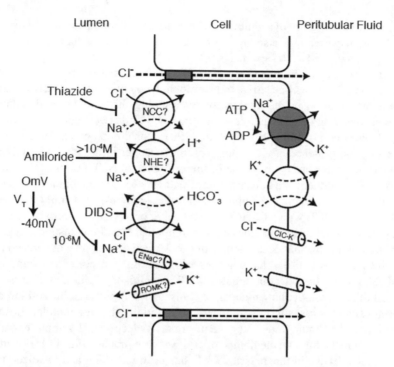

Fig. 4.4 Model for sodium and chloride reabsorption in principal cells or their equivalent in late distal tubules and collecting or connecting tubules of nonmammalian and mammalian vertebrates, based mainly on studies on amphibians, reptiles, and mammals. Symbols have the same meaning as in the legends of Figs. 4.1 and 4.2. See text for detailed description of the processes diagramed in the figure

cotransporter (apparently KCC4 in rabbit and mouse) (Mount 2012). The large, lumen-negative transepithelial potential in the connecting tubule apparently reflects the entry of sodium without chloride into the transporting cells (principal cells) down its electrochemical gradient via channels (ENaC) that are sensitive to micromolar concentrations of amiloride. Chloride largely moves across the epithelium in this region via a paracellular pathway (Mount 2012) (Fig. 4.4).

Similar, if not exactly the same, processes appear to be present in the late distal tubules and collecting tubules of those nonmammalian vertebrates studied, but in all species except for *Amphiuma* they may not be as clearly separated between tubule segments. In the late distal tubule of *Amphiuma*, in which there is no measurable transepithelial potential, electroneutral sodium chloride reabsorption is inhibited by thiazide (0.1 mM), by amiloride (1 mM), and by DIDS (0.5 mM) in the tubule lumen (Stanton 1990). The high concentration of amiloride is known to block sodium-proton exchange. All these maneuvers depolarize the basolateral membrane to a degree that can be explained entirely by a corresponding decrease in the intracellular chloride activity (Stanton 1990). These data all suggest that there is an electroneutral entry of sodium chloride across the luminal membrane by both a Na-Cl cotransporter and also by a process involving the coupling of a Na-H exchanger and a Cl-HCO$_3$$^-$ exchanger (Fig. 4.4).

In general, an inhibitory effect of millimolar luminal doses of amiloride is observed on the lumen-negative transepithelial potential and on transepithelial sodium reabsorption in the apparently overlapping distal tubule and collecting tubule segments in other amphibians, reptiles, and birds (Beyenbach and Dantzler 1978; Cohen et al. 1984; Hoshi et al. 1981; Nishimura et al. 1983b, 1986), suggesting the presence of a sodium-proton exchanger in the luminal membrane. In all species, this appears to be one of the orthologs of NHE (Brett et al. 2005). It seems likely that there is an electroneutral pathway for sodium chloride entry into the cells across the luminal membrane in all these species. This could involve the coupling of the NHE with a Cl-HCO$_3$$^-$ exchanger or a Na-Cl cotransporter or both (Fig. 4.4). An amphibian homolog of the mammalian Na-Cl cotransporter NCC has been identified in the late distal tubules of the African clawed toad (*X. laevis*) (Tran et al. 2007), and it is likely that other orthologs of this transporter are found in the late distal tubules or collecting tubules of all amphibians, reptiles, and birds.

In addition, the large lumen-negative transepithelial potential in reptilian distal tubules (*Thamnophis* spp.) is reduced substantially by micromolar doses of amiloride in the tubule lumen, suggesting that an electrogenic sodium channel is also present in the luminal membrane of cells in these tubules (Fig. 4.4) (Beyenbach and Dantzler 1978). In this regard, ENaC has been identified and localized to the luminal membrane of the collecting tubules of an anuran amphibian (*Bufo marinus*) (Uchiyama and Konno 2006), and it seems likely that this particular sodium channel is the one present in the late distal tubules and collecting tubules of reptiles and other nonmammalian vertebrates.

Chloride that has entered the cells via one or both of the electroneutral processes in the late distal tubules and collecting tubules of all these nonmammalian tetrapods apparently exits these cells across the basolateral membrane via a K-Cl cotransporter or a chloride channel or both (Fig. 4.4). Indeed, an amphibian homolog of the mammalian ClC-K chloride channel has been localized in the distal tubule of an amphibian (*X. laevis*) pronephros and is apparently present in the adult mesonephros (Vize 2003). It is likely that such a homolog is present in the distal tubule or collecting tubule of other nonmammalian tetrapods. However, given the electrogenic luminal sodium channel that also appears to be present in these tubule cells, additional chloride is probably reabsorbed across this tubule segment by a paracellular route (Fig. 4.4).

Although both the electroneutral and electrogenic pathways for sodium reabsorption could be present in all the cells (as illustrated in Fig. 4.4), it seems likely that only one path or the other is present in a given cell or that, when both pathways are present in the same cell, one or the other dominates. The size of the lumen-negative transepithelial potential (Table 4.2) then would reflect the degree of dominance of one pathway or the other, whether because of the dominance of one cell type or of one pathway within the same cell type. However, much work remains to be done both on the exact molecular structure of these channels and transporters and their locations in the late distal tubules and collecting tubules of nonmammalian vertebrates.

In vitro microperfusion studies on the late distal tubules of garter snakes (*Thamnophis* spp.) (Beyenbach 1984; Beyenbach and Dantzler 1978; Beyenbach et al. 1980) suggest that there may be some intrinsic cellular regulation of sodium reabsorption in this region. The substantial lumen-negative transepithelial potential (Table 4.2) and the calculated short-circuit current, apparently representing sodium reabsorption, although dependent on the presence of sodium in the lumen, decay rapidly when its concentration exceeds 30 mM. The decays in transepithelial potential and short-circuit current appear to result from an increase in resistance to sodium transport though the active pathway. Because sodium that enters the cells at this time cannot be extruded rapidly enough, cells swell. This response to an excessive sodium load may prevent the distal tubules from reabsorbing too much sodium when there is a need for additional sodium excretion. This adjustment of the transport system to operate effectively only at a low sodium concentration also permits further dilution of luminal fluid in which the sodium concentration is already low. Although the cell signaling pathway involved in this intrinsic regulation in snake distal tubules is unknown, it may involve inhibition of ENaC activity by high concentrations of sodium, possibly via a sodium-binding site in the acidic cleft of the molecule as suggested by studies with mammalian ENaC (Kashlan et al. 2015). Also unknown is whether or not such a process operates in the terminal regions of distal tubules in other reptiles or other nonmammalian vertebrates.

4.2.2.5 Collecting Ducts

Although substantial sodium reabsorption occurs along the collecting ducts of some fishes, amphibians, and reptiles (Table 4.1) and may even occur along the collecting ducts of birds, information bearing on the mechanism of transport exists only for urodele amphibians and mammals (Table 4.2). And, as noted above, the fraction of filtered sodium reabsorbed along mammalian collecting ducts is very low (Table 4.1).

Studies on *Amphiuma* indicate that the collecting tubule and collecting duct are structurally similar (vide supra; Fig. 2.16) (Stanton et al. 1984a) as discussed above. Indeed, micropuncture and microperfusion studies indicate that the sodium-dependent, ouabain-sensitive, lumen-negative transepithelial voltage in these segments (Table 4.2; shown for Collecting Tubule) is abolished by micromolar doses of amiloride in the luminal fluid, indicating the presence of an electrogenic sodium channel, probably ENaC, in the luminal membrane (Dietl and Stanton 1993; Horisberger et al. 1987; Hunter et al. 1987). In a somewhat similar fashion, micropuncture measurements on the collecting ducts of a newt (*Triturus* sp.), which appear to be separate from collecting tubules, in an in vitro kidney preparation (Hoshi et al. 1981) demonstrate a lumen-negative transepithelial potential difference (Table 4.2) that is abolished by the replacement of sodium with choline. The solute (apparently sodium) reabsorption is substantially reduced by the addition of amiloride to the incubation medium. The amiloride concentration in these experiments was low enough to suggest that its inhibitory effect reflects the presence of an electrogenic sodium channel, probably ENaC, in this amphibian species.

Studies on another urodele species, the tiger salamander (*A. tigrinum*), also generally agree with those on *Ambystoma* and *Triturus*. Segments of collecting ducts from this species, isolated and perfused in vitro, display a lumen-negative, ouabain-sensitive transepithelial potential difference (Table 4.2) (Delaney and Stoner 1982). In addition, these studies demonstrate that the collecting ducts have a substantial transepithelial specific resistance (Table 4.2) and a rate of sodium reabsorption (Table 4.1) equivalent to the measured short-circuit current (Delaney and Stoner 1982; Stoner 1977). Together, the data from all these studies indicate that the collecting ducts of urodele amphibians (and, in *Amphiuma*, collecting tubules as well) actively reabsorb sodium by an electrogenic process with basolateral Na-K-ATPase providing the driving force (Stoner 1985). They suggest that the sodium enters the cells across the luminal membrane through an amiloride-sensitive electrogenic channel, probably an amphibian homolog of ENaC (Uchiyama and Konno 2006). Chloride is apparently reabsorbed by a paracellular route. This transepithelial reabsorptive process for sodium and chloride in amphibian collecting duct cells corresponds to just the one electrogenic pathway for reabsorption observed in late distal tubule and collecting tubule cells in all nonmammalian vertebrates (Fig. 4.4). It also corresponds to the sodium chloride reabsorptive process in the principal cells of mammalian cortical collecting ducts (Mount 2012).

Although the process may be similar for the collecting ducts of other nonmammalian vertebrates, it has yet to be evaluated and, given the structural heterogeneity among species (vide supra; Chap. 2), there may be specific differences. Indeed, in the collecting ducts of euryhaline pufferfish (*Takifugu obscurus*), the electroneutral Na-Cl cotransporter NCC is highly expressed, whereas in the collecting ducts of stenohaline seawater pufferfish (*Takifugu rubripes*), mostly the Na-K-2Cl cotransporter NKCC2 is expressed (Kato et al. 2011). These molecular data suggest that the sodium and chloride reabsorptive processes in the collecting ducts of these teleosts are very different from the process discussed above for the collecting ducts of amphibians and mammals, but no physiological characterization of the transport process is available for collecting ducts of teleosts. Again, much work remains to be done to characterize sodium and chloride reabsorption in collecting ducts of nonmammalian vertebrates, especially in view of the very large fraction of filtered sodium and chloride that is reabsorbed in this nephron region in many species (Table 4.1).

Finally, Grantham and Wallace and their colleagues (Grantham and Wallace 2002; Wallace et al. 2001) have provided evidence that sodium and chloride can actually be secreted to a modest extent by mammalian inner medullary collecting ducts. Their data indicate that the secretory process is essentially the same as that described above for secretion in the proximal tubule of fishes (Fig. 4.2) and is a very primitive tubule function retained by mammals. These authors suggest that this solute and accompanying water secretion add to the fine regulation of solute and water excretion by the kidneys and thus to the fine regulation of extracellular fluid volume and composition.

4.2.3 Regulation of Transport

4.2.3.1 Hormonal Regulation

Antidiuretic Hormones

In view of the stimulatory effects of the antidiuretic neurohypophysial peptides on sodium transport by a number of nonrenal epithelia, e.g., toad urinary bladder, it has long appeared likely that these hormones might stimulate sodium reabsorption by the renal tubules. Such stimulation of sodium and chloride reabsorption along the thick ascending limb of Henle's loop in mammalian nephrons or in mammalian-type nephrons in birds could enhance the function of the countercurrent multiplier system and hence improve urine concentrating ability (vide infra; Chap. 7). Indeed, arginine vasopressin (AVP) stimulates adenylate cyclase activity and sodium chloride cotransport in isolated perfused thick ascending limbs of some mammalian species (mouse and rat) but not of others (rabbit) (Hall and Varney 1980; Hebert et al. 1981a, b; Sasaki and Imai 1980). The adaptive significance of this difference among mammalian species is not clear, although rabbits do not concentrate their urine very well. In mouse and rat thick ascending limbs, AVP, acting through the V2 receptors (Mutig et al. 2007) via the adenylate cyclase pathway, stimulates transepithelial sodium chloride reabsorption by increasing both the abundance and phosphorylation of the apical Na-K-2Cl cotransporter (NKCC2) (Ares et al. 2011; Mount 2014; Mutig et al. 2007). Given the presence of V2 receptors in the thick ascending limb of human nephrons (Mutig et al. 2007), it seems likely that the same process occurs in human nephrons as well.

In the case of birds, the situation is less clear. Early clearance studies on chickens indicate that arginine vasotocin (AVT) stimulates tubular reabsorption of sodium (Ames et al. 1971), but provide no information on the site of action. Although AVT stimulates adenylate cyclase activity in isolated thick ascending limbs of house sparrow (*Passer domesticus*) nephrons (Goldstein et al. 1999), it has no effect on either transepithelial voltage or chloride reabsorption in isolated, perfused thick ascending limbs of Japanese quail (*Coturnix coturnix*) nephrons (Miwa and Nishimura 1986). Certainly, more work needs to be done to determine to what extent, if any, AVT regulates sodium chloride reabsorption in the thick ascending limbs of avian mammalian-type nephrons and whether there are any other sites of its natriferic action.

In the mammalian kidney, AVP, acting through V2 receptors, stimulates sodium reabsorption in both the distal convoluted tubule and the connecting tubules and cortical collecting ducts (Mount 2012; Subramanya and Ellison 2014). In the distal convoluted tubules, this process involves an increase in the activity of the neutral sodium-chloride cotransporter (NCC) in the luminal membrane (Subramanya and Ellison 2014), whereas in the connecting tubules and cortical collecting ducts, it involves activation of the sodium channel (ENaC) in the luminal membrane by increasing its open time (Mount 2012).

AVT appears to have a variable effect on sodium reabsorption by renal tubules of those fishes that have been studied. It apparently has no direct effect on tubular sodium reabsorption in teleost fishes (Nishimura 1985). This is perhaps not surprising, because although renal V1 receptors have been cloned in teleosts (vide supra; Chap. 3) and AVT has been found to increase cyclic AMP content of suspensions of rainbow trout (*O. mykiss*) renal tubules (Perrott et al. 1993), no V2 receptors have been found (An et al. 2008). In lungfish (*P. aethiopicus*), in contrast to teleosts, AVT produces a natriuresis out of proportion to the increase in glomerular filtration that it also produces (vide supra; Chap. 3) (Sawyer 1970). However, it appears that this natriuretic effect reflects not a direct inhibition of tubular sodium reabsorption but a failure of sodium reabsorption to keep pace with the increasing filtered load (Sawyer 1972). The direct effects of AVT in lungfish demand further study because V2 receptors have been cloned from another species (*P. annectens*) and localized to the late distal tubule by immunohistochemistry (Konno et al. 2009). The direct effects of AVT on tubular sodium reabsorption in other groups of fishes have yet to be examined.

Clearance studies suggest that AVT stimulates tubular sodium reabsorption in frogs (*R. esculenta*) (Jard and Morel 1963), and studies of the short-circuit current across monolayers of the A6 cell line derived from the kidneys of the African clawed toad (*Xenopus laevis*) indicate that physiological concentrations of AVT activate ENaC sodium channels as well as a chloride channel (possibly CFTR) (Shane et al. 2006). Although immortal cell lines, such as A6, no longer exactly represent the tissues from which they were derived, ENaC is highly conserved throughout evolution and thus it seems likely that AVT activates it in the late distal tubule and collecting tubule of amphibians. How the chloride channel, if CFTR, may relate to the natural tubule segments is not as clear. In any case, it seems reasonable that an antidiuretic hormone would stimulate sodium reabsorption in the late distal tubule and collecting tubule of amphibian nephrons. Stimulation of NKCC2 in the diluting segment of amphibian nephrons, as in the thick ascending limb of mammalian and possibly avian nephrons, would not be appropriate because it would only lead to production of more dilute urine and a diuretic, not an antidiuretic effect.

The effect of physiological doses of AVT on renal tubular sodium transport in reptiles appears to be highly variable among species. For example, clearance studies indicate that AVT stimulates tubular sodium reabsorption in water snakes (*Nerodia (=Natrix) sipedon*) (Dantzler 1967), inhibits tubular sodium reabsorption in freshwater turtles (*Chrysemys picta bellii*) (Butler 1972), and has no effect on tubular sodium reabsorption in lizards (*V. gouldii*) (Bradshaw and Rice 1981). AVT also has no effect on the amiloride-sensitive transepithelial voltage in isolated, perfused segments of late distal tubules from garter snakes (*Thamnophis* spp.) (Beyenbach 1984), suggesting that it does not affect ENaC in this species. The tubule site or sites of any direct effect of AVT on sodium reabsorption in reptiles remain unknown. Moreover, whereas it is assumed that tubular actions of AVT involve interaction with some reptilian homolog of the mammalian V2 receptor and V2-like receptors have been identified in the thin intermediate segment and collecting duct of nephrons of the ornate dragon lizard (*Ctenophorus ornatus*) (Bradshaw and Bradshaw 1996), no specific AVT receptors have yet been cloned from reptiles.

Angiotensin II

Angiotensin II has a clear, but variable, effect on tubular sodium reabsorption in mammals, but its possible effect on tubular sodium reabsorption in nonmammalian vertebrates is far from clear. Early micropuncture and peritubular microperfusion studies showed that it has a dose-dependent, biphasic effect on sodium reabsorption by mammalian proximal tubules (Harris and Young 1977; Steven 1974). Low doses of angiotensin II (10^{-10} to 10^{-12} M) stimulate reabsorption whereas high doses ($>10^{-7}$) inhibit reabsorption. In mammalian proximal tubules, AT_1 angiotensin receptors are located on both the luminal and peritubular membranes, and the biphasic response can occur from applications at either membrane (Harrison-Bernard et al. 1997; Li et al. 1994). Moreover, angiotensin II is produced and secreted by mammalian proximal tubule cells and can have a significant autocrine effect on sodium reabsorption (Quan and Baum 1996).

Much less information is available about any possible effects of angiotensin II on renal tubular transport of sodium in nonmammalian vertebrates, although some forms of angiotensin receptors have been cloned in teleosts, amphibians, and birds (Nishimura 2001). Moreover, autoradiographic studies demonstrated significant angiotensin II binding to proximal tubules in teleosts (rainbow trout, *O. mykiss*) with much greater binding in freshwater-adapted than in seawater-adapted animals (Cobb and Brown 1992). Because retention of sodium is important in freshwater species, this study suggests that angiotensin II might stimulate sodium reabsorption in proximal tubules. However, studies of a hormone effect in living teleosts provide no clear evidence of a tubule effect. For example, pressor doses of angiotensin II produce a natriuresis in some teleost species (e.g., American eel, *A. rostrata*), but this response, which may be pharmacological, appears to result primarily from an increase in GFR (Nishimura 1985; Nishimura and Sawyer 1976). Moreover, even high doses of angiotensin II have little or no effect on sodium excretion by aglomerular teleosts (goosefish, *Lophius americanus*; toadfish, *Opsanus tau*), strongly suggesting that there actually is no direct effect on the renal tubules of teleosts (Churchill et al. 1979; Nishimura 1985; Zucker and Nishimura 1981).

In contrast to the teleosts, infusions of angiotensin II into the renal portal system of amphibians, reptiles, and birds suggest that the hormone may have a direct effect on sodium transport by the renal tubules. In an anuran amphibian (Chilean toad, *Calyptocdephallela cardiverbela*), infusions of low, nonpressor doses produce an antidiuresis and antinatriuresis without a change in GFR, whereas higher doses produce a diuresis and natriuresis despite an apparent reduction in GFR (Galli-Gallardo and Pang 1978; Nishimura 1985). Likewise, in a chelonian reptile (freshwater turtle, *Pseudemys scripta elegans*), a renal portal infusion of a nonpressor dose produces an antinatriuresis (Nishimura 1985), and, in chickens, such an infusion of a moderate dose produces a diuresis and natriuresis predominately on the ipsilateral side (Nishimura 1985; Stallone and Nishimura 1985). Together, these data from renal portal infusions in amphibians, reptiles, and birds suggest that, as in mammals, angiotensin II may act directly on the tubules at low levels to enhance reabsorption of sodium and at high levels to inhibit reabsorption of sodium. However, these studies on nonmammalian vertebrates are more indirect than those on mammals and there is no information on possible tubule sites or mechanisms of action involved in any possible hormonal effect.

Adrenocorticosteroids

Among mammals, the importance of aldosterone, the natural mineralocorticoid, in stimulating the reabsorption of sodium by the renal tubules is well documented and subject of continuing study and frequent reviews (e.g., Mount 2012). The primary site of aldosterone action in mammals is the distal nephron (the distal convoluted tubule, the connecting tubule, and the collecting duct) where it increases the abundance of NCC in the luminal membrane of the distal convoluted tubule (Subramanya and Ellison 2014) and the expression of both Na-K-ATPase in the basolateral membrane and ENaC in the luminal membrane of principal cells of the connecting tubules and collecting ducts (Mount 2012). It also activates the ENaC channels in these latter structures. A great deal is now known about the molecular steps in these processes and possible additional effects of aldosterone (e.g., modifying claudins in tight junctions to increase paracellular permeability) (Le Moellic et al. 2005) and are reviewed elsewhere (Mount 2012).

Among nonmammalian vertebrates, natural mineralocorticoids are not even clearly defined for many species and their effects, if any, on sodium reabsorption by the renal tubules are even less clear. In fact, as Nishimura (Nishimura 1985) pointed out three decades ago, individual adrenocorticosteroids often have both glucocorticoid and mineralocorticoid effects in nonmammalian vertebrates. This may be particularly true of fishes where, in contrast to tetrapod vertebrates, few, if any, species appear to synthesize aldosterone (Henderson and Kime 1986; Prunet et al. 2006; Sandor et al. 1976). However, other corticosteroids, primarily cortisol, are synthesized and released by the intrarenal tissue of all teleosts and cyclostomes studied (Atlantic hagfish, *M. glutinosa*; marine lamprey, *P. marinus*) (Henderson 1998; Henderson and Kime 1986; Sandor et al. 1976). Moreover, cortisol plays a role in both energy metabolism and hydromineral balance in teleosts (Prunet et al. 2006; Wendelaar Bonga 1997). In this regard, both glucocorticoid and mineralocorticoid receptors, for which cortisol is a potent agonist, have been cloned in teleosts. Thus, the action of the steroid depends on the receptor with which it interacts. Moreover, the observation that partial adrenalectomy of freshwater European eels (*A. Anguilla*) results in an increased fractional excretion of sodium (Chan and Chester Jones 1969) suggests that mineralocorticoid receptors could be present on renal tubules and that cortisol could stimulate sodium reabsorption. At present, however, there is no information on localization of mineralocorticoid receptors to renal tubules in teleosts.

Aldosterone is present in amphibians and clearly stimulates sodium transport in nonrenal epithelium [e.g., frog (*R. catesbeiana*) skin, toad (*Bufo marinus*) urinary bladder, etc.] (Hillyard et al. 2009). As far as the kidney is concerned, physiological concentrations of aldosterone activate ENaC channels via the phosphatidylinositol pathway in monolayers of the A6 cell line, thereby enhancing transepithelial sodium transport (Blazer-Yost et al. 1999; Shane et al. 2006). As mentioned above, this immortal cell line will not exactly reflect the function of the anuran kidney tissue from which it is derived. However, because ENaC is so highly conserved throughout evolution, it seems very likely that aldosterone activates ENaC in the late distal and collecting tubules of amphibians to stimulate sodium reabsorption. There is at present no information on whether aldosterone also increases the abundance and activity of Na-K-ATPase in these regions of the amphibian renal tubules. Nevertheless, it appears likely that in amphibians, as in mammals, aldosterone plays an important role in regulating sodium excretion by regulating its reabsorption by the distal nephron.

Aldosterone also apparently stimulates the sodium-proton exchanger in the luminal membrane of amphibian early distal tubules during chronic exposure to a high potassium environment (Oberleithner et al. 1987; Weigt et al. 1987) (vide infra; potassium and hydrogen ion transport) and thereby stimulates sodium reabsorption by this process. However, this may only compensate for decreased reabsorption by the Na-K-2Cl cotransporter and thus may not produce any increase in net sodium reabsorption.

The observed renal effects of adrenocorticoids in reptiles are so variable that a clear picture has yet to emerge. Indeed, the data suggest that the effects in some species are exactly opposite to those in other species. For example, clearance studies suggest that aldosterone, a naturally occurring adrenal steroid in reptiles, stimulates sodium reabsorption by renal tubules in sodium-chloride-loaded snakes (*Nerodia cyclopion*) and turtles (*C. picta*), but has no effect on water-loaded or control animals (Brewer and Ensor 1980; Elizondo and Lebrie 1969; Lebrie and Elizondo 1969). Although administration of aldosterone to desert iguanas (*Dipsosaurus dorsalis*) appears to have no effect on sodium excretion, hypophysectomy and dexamethasone administration in these animals and in a species of Australian agamid lizard (ornate dragon lizard, *C. ornatus*) suggests that the adrenocorticoids may actually inhibit reabsorption by renal tubules (Bradshaw et al. 1972; Bradshaw 1975). More recent studies on a varanid lizard (*V. gouldii*), however, indicate that both plasma levels of aldosterone and fractional reabsorption of sodium decrease during administration of a salt load (Bradshaw and Rice 1981). The fractional reabsorption of sodium in *V. gouldii* is also reduced following adrenalectomy and restored to control level by the administration of aldosterone (Rice et al. 1982). Together, these conflicting data suggest that in some reptiles aldosterone has the same stimulatory effect on sodium reabsorption as in mammals whereas in others it, or some other adrenocorticoid, has no effect or the opposite effect (see more detailed review in Dantzler and Bradshaw 2009). Although it is not necessary that adrenocorticoids have the same effects in all species of reptiles, direct studies of tubular effects and possible sites of action, as well as of mineralocorticoid receptors are needed to begin to understand these complex effects.

Finally, in birds, at least ducks, *Anas platyrhynchos*, adrenalectomy and administration of corticosteroids suggest that both aldosterone and corticosterone, two naturally occurring adrenal corticosteroids, have physiological roles in stimulating sodium reabsorption by renal tubules (Holmes and Phillips 1976). However, these studies were made four decades ago and since then there have still been no direct studies of tubular effects, sites of action, or mechanisms of action in birds.

Prolactin

There is no current evidence of a direct effect of prolactin on tubular reabsorption of sodium chloride in mammals (Mount 2012). However, in nonmammalian vertebrates there may be some effects.

Although prolactin clearly affects sodium transport in the gills and intestines of teleost fishes (Sakamoto and McCormick 2006), the fragmentary data on renal effects are far from clear. For example, administration of the hormone to goldfish (Lahlou and Giordan 1970; Lahlou and Sawyer 1969) appears to stimulate sodium reabsorption by the renal tubules, but the hormone preparation may well have been contaminated with neurohypophysial peptides and these may have played some role in the observed effects. Nothing further has been done on direct tubular effects in teleosts since that time.

In anuran amphibians, prolactin increases sodium transport across the skin by stimulating ENaC in the luminal membrane and Na-K-ATPase in the basolateral membrane (Hillyard et al. 2009; Takada and Hokari 2007), and specific binding sites for prolactin have been described along the renal proximal tubules (Gono 1982; Nishimura 1985; White and Nicoll 1979). Although the presence of these binding sites suggests that the hormone has an effect on sodium reabsorption in the proximal tubule, there are as yet no studies to determine whether this is the case or the mechanism by which it might occur.

In reptiles, an early study indicated that the administration of prolactin to freshwater turtles (*C. picta*) (Brewer and Ensor 1980) can stimulate renal sodium reabsorption, but, as in the early studies on teleosts, this may not have been a pure preparation. In any case, there have been no further studies on possible effects of the hormone on renal tubule sodium reabsorption in these vertebrates.

In the case of birds, a simultaneous renal clearance and micropuncture study on European starlings (*S. vulgaris*) during molt and in the absence of molt suggested that prolactin, which is elevated during molt, might influence tubular reabsorption of sodium and water (Roberts and Dantzler 1992). A follow-up study on animals in the absence of molt but in the presence and absence of an infusion of prolactin designed to mimic high physiological levels during periods such as molt indicated that prolactin increases the fractional excretion of sodium and chloride (Roberts and Dantzler 1992). However, micropuncture measurements performed at the same time on the superficial reptilian-type nephrons (the only ones accessible to micropuncture) showed no significant reduction in sodium or chloride reabsorption up to the point of micropuncture (Roberts and Dantzler 1992). Therefore, it appears that any reduction in tubular reabsorption of sodium chloride in the presence of prolactin, indicated by the whole-animal clearance measurements, occurs in the distal portions of the reptilian-type nephrons or in some portion of the mammalian-type nephrons.

Natiuretic Peptides

One or more of the known mammalian natriuretic peptides (atrial natriuretic peptide, ANP; brain natriuretic peptide, BNP; and C-type natriuretic peptide, CNP) have been identified in fish, amphibians, reptiles, and birds, and binding sites have often been found in the kidneys, especially in the glomeruli (Toop and Donald 2004). However, their effects on renal function have not been well studied and are far from clear (Toop and Donald 2004). In teleost fishes (rainbow trout, *O. mykiss*), heterologous mammalian peptides apparently produce a diuresis and natriuresis primarily by producing an increase in GFR (Duff et al. 1997). This also appears to be the case for arterial injections of salmon cardiac natriuretic peptide into salmon (*Salmo salar*) (Tervonen et al. 2002). However, infusions of ANP have no effect on renal excretion of sodium in seawater eels (Tsukada and Takai 2006). Although possible tubule effects generally have not been examined in most fishes, one interesting study on the aglomerular toadfish, *Opsanus tau*, indicated that heart extracts and heterologous ANP are both diuretic and natriuretic, demonstrating that natriuretic peptides are capable of an effect on tubular secretion. However, there is no information on the cellular and molecular mechanisms involved in this effect.

Among the amphibians, ANP binding sites have been found in the glomeruli of a number of anurans (*R. temporaria, X. laevis*, and *B. marinus*) (Kloas and Hanke 1992a, b; Meier et al. 1999) and urodeles (*Ambystoma mexicanum* and *Triturus alpestris*) (Kloas and Hanke 1993). However, they have so far only been found in the renal tubules of some urodeles (*A. mexicanum* and *T. alpestris*) (Kloas and Hanke 1993). There is, however, some evidence for expression of ANP in the proximal tubules of anurans (De Falco et al. 2002). In any case, studies of the effects of ANP on renal function have been performed only in anurans (*B. marinus*), and they have revealed no evidence of an effect on renal tubular sodium transport (Donald et al. 2002; Toop and Donald 2004).

Among the reptiles, ANP has been found in the atria of one turtle species (*Amyda japonica*) (Cho et al. 1988) and a similar peptide, probably BNP, has been found in the atria of another turtle species (*Pseudemys scripta*) (Reinhart and Zehr 1994). A cDNA for BNP, but not for ANP, has been obtained from the atria of the longneck tortoise (*Chelodina longicollis*) and the saltwater crocodile (*Crocodylus porosus*) (Toop and

Donald 2004). However, although renal binding sites for ANP have been found in *A. japonica*, no renal effects of injections of heterologous ANP or turtle atrial extracts in this chelonian species were found (Toop and Donald 2004). No other studies of the renal effects of natriuretic peptides have yet been reported for reptiles.

In birds, binding sites for chicken BNP, the avian cardiac natriuretic factor, are found in glomeruli and collecting ducts of reptilian-type nephrons and for chicken CNP in the glomeruli of mammalian-type nephrons in Pekin ducks (*Anas platyrhynchos*) (Schütz et al. 1992). Infusions of ANP in chickens and of BNP in ducks produced both a diuresis and natriuresis, suggesting that these peptides increase GFR (at least in reptilian-type nephrons) and block sodium reabsorption in collecting ducts (Gray 1993; Gregg and Wideman 1986; Schütz et al. 1992). However, there are as yet no direct studies in birds that demonstrate that natriuretic peptides actually inhibit tubular sodium reabsorption, much less the mechanism by which such inhibition might occur. In summary, although there are some studies that suggest that natriuretic peptides can inhibit tubular sodium reabsorption in nonmammalian vertebrates, there is no evidence that this actually is the case.

4.2.3.2 Neural Regulation

In mammalian kidneys, efferent adrenergic sympathetic nerves terminate on epithelial cells of proximal tubules, thick ascending limbs, distal tubules, and collecting ducts where they function to stimulate sodium reabsorption (Johns et al. 2011). This process is the subject of continuing study and has been frequently reviewed (see, e.g., DiBona and Kopp 1997; Johns and Kopp 2013; Johns et al. 2011). Briefly, norepinephrine released from the nerve terminals interacts with G-coupled α_1-adrenoreceptors on proximal tubule epithelial cells and with both α_1- and especially α_2-adrenoreceptors on more distal epithelial cells (Johns et al. 2011). Most study has concentrated on the proximal tubule cells where interaction with the α_1-adrenoreceptors activates both phospholipase C and MAP kinase signaling pathways to increase the activity of sodium-hydrogen exchangers (both NHE1 and NHE3), thereby stimulating sodium reabsorption (Johns et al. 2011). The details in the distal parts of the nephron are less clear and much study is continuing on this process as well as the involvement of cotransmitters (e.g., neuropeptide Y and ATP) released during sympathetic nerve depolarization in the regulation of tubular sodium reabsorption (Johns et al. 2011).

These studies on neural regulation of sodium transport in mammals, although certainly not complete, are far more extensive than anything that has been done on nonmammalian vertebrates. In fact, only one study in amphibians has any direct bearing on this problem. In bullfrogs (*R. catesbeiana*), the systemic arterial infusion of phenoxybenzamine, an α-adrenergic blocker, initiates a bilateral natriuresis, whereas renal portal venous infusion of the same drug produces a natriuresis only from the ipsilateral kidney (Gallardo et al. 1980). Neural elements in close proximity to the renal tubules are also observed in these animals (Gallardo et al. 1980).

Thus, it is possible that α-adrenergic nerves play some role in stimulating tubular sodium reabsorption in bullfrogs. However, no additional, more direct information is available on amphibians or any other nonmammalian vertebrates.

4.3 Potassium

4.3.1 Direction, Magnitude, and Sites of Transport

Clearance studies indicate that either net reabsorption or net secretion of potassium by the renal tubules can occur in mammals (Giebisch 1975), representatives of all classes of nonmammalian tetrapods (Dantzler 1976; Holmes et al. 1968; Shoemaker 1967; Skadhauge and Schmidt-Nielsen 1967; Stolte, Schmidt-Nielsen, and Davis 1977b; Wiederholt et al. 1971), and freshwater teleosts (Hickman and Trump 1969). This may also be the case for marine elasmobranchs (Hickman and Trump 1969; Stolte et al. 1977b). However, net tubular reabsorption generally occurs in euryhaline teleosts adapted to seawater and net tubular secretion in euryhaline teleosts adapted to freshwater (Hickman and Trump 1969; Schmidt-Nielsen and Renfro 1975). Only net tubular reabsorption has been observed in freshwater lampreys (Logan et al. 1980) and only net tubular secretion in hagfishes (Alt et al. 1980; Stolte and Schmidt-Nielsen 1978).

As in the case of sodium excretion (vide supra), these measurements of net secretion and net reabsorption have been made only with the aqueous phase of the ureteral urine. In birds and uricotelic reptiles, some filtered or secreted potassium may be combined in urate precipitates in the tubular fluid (vide infra) so that true values for the magnitude and, in some cases, even the direction of net transport may be different from those generally reported (Dantzler 1978).

The sites of net transport along the renal tubules and the magnitude of such transport are more variable among nonmammalian vertebrates than among mammals. In all mammals studied, about 50–75 % of the filtered potassium is reabsorbed along the proximal tubules whether or not overall net secretion by the tubules is occurring (Mount 2012). However, among those nonmammalian species in which micropuncture measurements have been made, potassium transport along the proximal tubules is much less consistent. About 25–35 % of filtered potassium is reabsorbed along the proximal tubules of freshwater lampreys (river lamprey, L. fluviatilis) (Logan et al. 1980), anuran amphibians (bullfrog, R. catesbeiana) (Long and Giebisch 1979), and lizards (blue spiny lizard, S. cyanogenys) (Stolte et al. 1977b), but little or no net transepithelial transport occurs along the proximal tubules of urodele amphibians (mudpuppy, N. maculosus, and Congo eel, A. means) (Garland et al. 1975; Oken and Solomon 1963; Wiederholt et al. 1971). Transepithelial transport can vary from net reabsorption to no net transport to possible net secretion along the proximal tubules of the superficial reptilian-type nephrons of birds (starlings, S. vulgaris) (Laverty and Dantzler 1982). However, a much greater fraction of filtered potassium may undergo net reabsorption along the proximal tubules of the short-looped transitional and long-looped mammalian-type nephrons of birds, as suggested by the much higher rate of net fluid reabsorption along the proximal tubules of transitional nephrons than of reptilian-type nephrons (vide infra; Table 5.1) (Brokl et al. 1994; Laverty and Dantzler 1982).

About 25 % of the filtered potassium is reabsorbed along the thick ascending limbs of the loops of Henle in mammalian nephrons (Mount 2012), and a significant fraction of the filtered potassium may also be reabsorbed along the thick ascending limbs of avian mammalian-type nephrons. However, this has yet to be studied.

In both mammalian and nonmammalian vertebrates, overall net secretion of potassium by the renal tubules, when it occurs, results from net secretion along the distal portions of the nephrons (Dantzler 1967; Dietl and Stanton 1993; Giebisch 1975; Mount 2012; Stolte et al. 1977a, b; Stoner 1977, 1985; Wiederholt et al 1971). In mammals, the late distal tubule, connecting tubule, and the cortical collecting duct are the principal sites of potassium secretion with the connecting tubule being the most important quantitatively (Mount 2012). In amphibians, however, net potassium secretion can occur along the early distal tubule (the diluting segment) as well as along the late distal tubule and collecting tubule (Dietl and Stanton 1993; Oberleithner et al. 1983c, d; Stoner 1977, 1985). The specific distal nephron sites of net potassium secretion are not known for other nonmammalian vertebrates.

4.3.2   Mechanism of Transport

4.3.2.1   Proximal Tubules

The mechanism of potassium reabsorption by mammalian proximal tubules is not completely clear. However, it appears that it is passive and that most of it occurs paracellularly, probably through claudin 2, in the late proximal tubule driven by the small, lumen-positive transepithelial potential found in this region of the nephron (Fig. 4.5) (Mount 2012; Yu 2015). This movement may involve some component of solvent drag, not simply diffusion, but to what extent this occurs is subject to considerable debate (Mount 2012; Yu 2015). In amphibian proximal tubules, the substantial lumen-negative transepithelial potential (vide supra; Table 4.2) and lack of evidence for any substantial paracellular water flux (vide infra; Chap. 5) suggest that there is little or no passive paracellular potassium reabsorption, but this has not been examined directly. If potassium is reabsorbed by a transcellular route, it must be transported into the cells against an electrochemical, probably by a primary active transport step (Fig. 4.5), and this appears to be the major step in transepithelial reabsorption (Fujimoto et al. 1980; Wright and Giebisch 1985). Nothing further is known about the reabsorptive mechanism for potassium in the proximal tubules of amphibians or any other nonmammalian vertebrates.

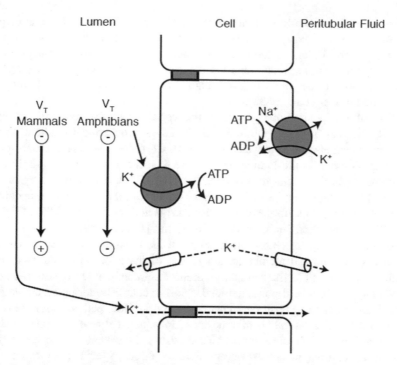

Fig. 4.5 Model for potassium reabsorption by renal proximal tubules based on studies on mammals and amphibians. Symbols have the same meaning as in the legends of Figs. 4.1 and 4.2. Note that the transepithelial potential difference changes from lumen-negative in the early proximal tubule to lumen-positive in the late proximal tubule in mammals but remains lumen-negative throughout the proximal tubule in amphibians. See text for detailed description of the processes diagramed in the figure

4.3.2.2 Distal Nephrons

Studies of the mechanisms of potassium transport along the distal portions of the nephrons have been made only for mammals and amphibians. The studies on amphibians include micropuncture and microperfusion of tubules in intact and doubly perfused kidneys and microperfusion of isolated renal tubules. They involve transepithelial electrical potential measurements and intracellular measurements with ion-sensitive and conventional microelectrodes and kinetic analyses of tracer washout. Similar microelectrode studies have been made for portions of mammalian distal nephrons. These studies indicate, as noted above, that either net reabsorption or net secretion of potassium can occur along the early distal (diluting) segment of both anuran and urodele amphibian nephrons (Oberleithner et al. 1983b; Stoner 1977, 1985) whereas only net reabsorption occurs along the thick ascending limb of Henle's loop of mammalian nephrons (Mount 2012, 2014). Even in amphibians, however, net reabsorption occurs along this tubule segment during control conditions (Dietl and Stanton 1993). The initial step in this reabsorptive process in both the early distal segment of amphibian nephrons and the thick ascending limb of Henle's loop in mammalian nephrons involves entry into the cells across the luminal membrane via the Na^+-K^+-$2Cl^-$ cotransporter (NKCC2) discussed above (vide supra; Fig. 4.3; also Dietl and Stanton 1993; Mount 2012, 2014; Oberleithner et al. 1983b).

Some of the potassium that enters the cells across the luminal membrane normally diffuses back into the lumen down its electrochemical gradient through potassium channels and is recycled (Fig. 4.3). Currently, two channels for potassium diffusion across the luminal membrane have been identified in early distal tubules of amphibian nephrons: a maxi-channel of 120–200 ps activated by calcium (Congo eel, *A. means*) and a smaller channel of about 30 ps with multiple conductances (*A. means* and a frog, *Rana temporaria*) (Dietl and Stanton 1993; Giebisch et al. 1990; Hunter and Giebisch 1988; Hurst and Hunter 1989).

Normally, during net reabsorption, some of the potassium that has entered the cells across the luminal membrane moves out of the cells across the basolateral membrane down its electrochemical gradient via a conductive potassium channel and via the electroneutral potassium and chloride cotransporter noted above (vide supra; Fig. 4.3), thereby resulting in net reabsorption. Both pathways for potassium exit across the basolateral membrane exist in single cells of mammalian thick ascending limbs of Henle's loop (Mount 2014). However, in the early distal segment of nephrons in at least one urodele amphibian species (*A. means*), there appear to be two cell types, one with one pathway for potassium exit and the other with the other pathway (Guggino 1986). Even during the process of net reabsorption in both amphibian early distal tubules and mammalian thick ascending limbs, active transport of potassium back into the cells, involving Na-K-ATPase, occurs across the basolateral membrane.

Net secretion of potassium in the early distal region of nephrons in amphibians and, possibly, in other nonmammalian vertebrates is produced by a high potassium environment or a high potassium intake and by alkalosis or administration of carbonic anhydrase inhibitors (Dantzler 1976; Oberleithner et al. 1983b; Wiederholt et al. 1971). This secretory process is influenced by the peritubular potassium concentration (Oberleithner et al. 1983c; Wiederholt et al. 1971). Electrophysiological and kinetic studies indicate that in both anurans (*R. catesbeiana*) and urodeles (*A. means*), net secretion results from three changes in the transport processes: (1) increased potassium conductance across the luminal membrane via an increase in the number of active potassium channels; (2) increased potassium uptake across the basolateral membrane via Na-K-ATPase; and (3) possibly decreased potassium transport into the cells across the luminal membrane via Na-K-2Cl (NKCC2) cotransport (Fig. 4.3) (Dietl and Stanton 1993; Giebisch et al. 1990; Hurst and Hunter 1989, 1990; Oberleithner et al. 1983b; Sullivan et al. 1977; Wiederholt et al. 1971; Wilkinson et al. 1983). Of these, increased potassium conductance across the luminal membrane appears to be of primary importance in producing increased potassium secretion in response to a high potassium environment or other secretory stimuli (vide infra; Hormonal Regulation) (Dietl and Stanton 1993; Oberleithner et al. 1985b).

In amphibians, there is no potassium reabsorption or secretion by the cells of the late distal portion of the nephrons, which can be clearly differentiated from collecting tubules and collecting ducts (vide supra; Fig. 2.16) (Dietl and Stanton 1993; Stanton et al. 1984b). However, as in the connecting tubules and cortical collecting ducts of mammalian nephrons, potassium is secreted by the collecting tubules and collecting ducts of amphibian nephrons (Dietl and Stanton 1993; Hillyard et al. 2009). Also, as in the connecting tubules and cortical collecting ducts of mammals, potassium secretion in the collecting tubules and collecting ducts of amphibians is driven by the lumen-negative transepithelial potential generated by sodium movement into principal cells through ENaC channels in the luminal membrane (Dietl and Stanton 1993; Horisberger et al. 1987; Hunter et al. 1987; Mount 2012). In at least one terrestrial anuran amphibian (toad, *Bufo bufo*) this secretion, as in mammals, occurs through potassium channels in the luminal membrane of the principal cells (Mobjerg et al. 2002). However, in contrast to mammals and apparently anurans, potassium secretion in urodele amphibians (at least in *A. means*) occurs between the principal cells via a potassium-selective paracellular pathway (Dietl and Stanton 1993). Nothing is known about the mechanism of potassium secretion by the distal nephrons of other nonmammalian vertebrates, although both principal and intercalated cells appear to exist in the late distal tubules and collecting ducts of reptiles and birds (vide supra; Chap. 2).

4.3.3 Hormonal Regulation

4.3.3.1 Antidiuretic Hormones

Arginine vasopressin has a well-known stimulatory effect on potassium secretion by the distal segments of mammalian renal tubules. This appears to involve two major effects on principal cells (Mount 2012). First, it activates ENaC, permitting increased entry of sodium across the luminal membrane, resulting in an increased lumen-negative transepithelial membrane potential and stimulation of potassium secretion. Second, it activates potassium channels in the luminal membrane.

Although arginine vasotocin appears to have some effect on potassium transport by renal tubules of some nonmammalian vertebrates, it is opposite to the effect in mammals and its physiological significance is unclear. The administration of apparently physiological doses of the hormone appears to reduce the fraction of filtered potassium excreted by some reptiles (water snakes, *N. sipedon*, but not sand goannas, *V. gouldii*) (Bradshaw and Rice 1981; Dantzler 1967), but not by amphibians (Jard and Morel 1963) or birds (Ames et al. 1971; Bradley et al. 1971). If all filtered potassium is reabsorbed by reptilian renal tubules and that excreted is derived solely from tubular secretion, then a reduction in fractional excretion suggests that the hormone inhibits secretion. If all filtered potassium is not reabsorbed, then a reduction in fractional excretion could mean that the hormone stimulates reabsorption, inhibits secretion, or both. However, it is not clear that the effective doses of the hormone are truly physiological and no direct studies of hormone action on the potassium transport process have yet been made.

4.3.3.2   Adrenocorticosteroids

In mammals, aldosterone and similar mineralocorticoids stimulate net potassium secretion by the distal nephron. As discussed above with regard to tubular sodium transport, aldosterone activates Na-K-ATPase in the basolateral membrane and ENaC in the luminal membrane of principal cells of connecting tubules and cortical collecting ducts, both of which actions enhance potassium secretion (Mount 2012). There is also an increase in the number of potassium channels in the luminal membrane of these cells but much of this increase may be the result of the high plasma potassium concentration itself, rather than from the aldosterone whose release from the adrenals it induces (Mount 2012). Over time, the hormone can actually lead to an increase in the area of the basolateral membrane and, with it, an increase in the total Na-K-ATPase activity (Wright and Giebisch 1985).

A considerable amount of information is also now available on the effects of adrenocorticosteroids on potassium transport by the renal tubules of amphibians. The plasma level of aldosterone increases with chronic exposure to a high potassium environment at the same time that the potassium conductance of the luminal membrane of the early distal tubule cells is increased and net potassium secretion by this tubule segment is stimulated (vide supra) (Oberleithner et al. 1983b, 1985b). Intracellular and transepithelial recordings with conventional as well as ion-sensitive microelectrodes and whole-cell patch clamp studies indicate that aldosterone (but not glucocorticoids) stimulates hydrogen ion secretion by enhancing sodium-hydrogen exchange at the luminal membrane (Fig. 4.6), thereby alkalinizing the interior of the cells. This, in turn, leads to an increase in the potassium conductance of the luminal membrane and the flux of potassium out of the cells across this membrane (Dietl and Stanton 1993; Hunter et al. 1988; Hurst and Hunter 1989; Oberleithner et al. 1985a, 1987, 1988; Wang et al. 1989; Weigt et al. 1987). The increase in the potassium conductance of the luminal membrane by intracellular alkalinization results from an increase in the number of potassium channels, not from an increase in the open time of each channel (Giebisch et al. 1990; Hurst and Hunter 1989, 1990). It also appears that aldosterone can produce a small increase in the potassium conductance by some unknown mechanism even in the absence of cellular alkalinization (Wang et al. 1989). The increase in potassium flux across the luminal membrane that occurs from the increase in the number of potassium channels converts net reabsorption by the cells to net secretion. However, for net secretion to continue, there must be increased cellular potassium uptake at the basolateral membrane via Na-K-ATPase, a process that is apparently stimulated by the alkaline intracellular pH (Dietl and Stanton 1993).

Chronic exposure to a high potassium environment, which will increase plasma aldosterone levels, has no effect on the structure of the early distal tubules of amphibians, but, in urodeles (A. means) at least, it leads to hypertrophy of the

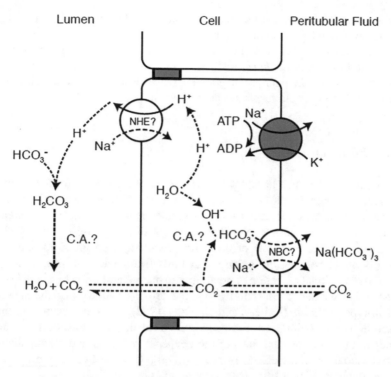

Fig. 4.6 Model for hydrogen ion secretion and bicarbonate reabsorption by the proximal and distal tubules of amphibians. C.A. indicates carbonic anhydrase. Other symbols have the same meaning as in the legends of Figs. 4.1 and 4.2. See text in both Chaps. 4 and 6 for detailed description of the processes diagramed in the figure

basolateral membrane of principal cells of the collecting ducts (Stanton et al. 1984b). This is assumed to reflect an increase in Na-K-ATPase. By itself, this is unlikely to enhance potassium secretion by the collecting ducts because, as noted above, there are no potassium channels in the luminal membrane of these cells. Instead, the increased activity of Na-K-ATPase hyperpolarizes the basolateral membrane leading to an increase in potassium conductance in this membrane and recycling of potassium back into the peritubular fluid (Dietl and Stanton 1993). There are no direct measurements of aldosterone effects on potassium secretion by the collecting ducts in *A. means*. However, there is evidence that aldosterone increases ENaC activity in the luminal membrane of the principal cells (vide supra; Hormonal Regulation of sodium transport). As noted above, this will increase the lumen-negative transepithelial potential and could enhance paracellular secretion of potassium by the amphibian collecting ducts.

The effects of adrenocorticoids on potassium transport by the tubules of other nonmammalian vertebrates are far from clear. There is no evidence of any effect in fishes. Aldosterone and corticosterone apparently increase the fractional excretion of potassium in birds (Holmes 1972), but the mechanism involved in this effect is unknown. The effects of these hormones in reptiles are highly variable. They appear to reduce fractional potassium excretion in sodium-chloride-loaded water snakes (Lebrie and Elizondo 1969), to increase fractional excretion in freshwater turtles (*C. picta* and *Pelomedusa subrufa*) (Brewer and Ensor 1980), and to stimulate net potassium secretion in lizards (*V. gouldii*) (Rice et al. 1982). The mechanisms involved in these variable responses are unknown.

4.3.3.3   Prolactin

The administration of this hormone to one species of freshwater turtle (*P. subrufa*) increases potassium excretion (Brewer and Ensor 1980), but whether this results from an increase in GFR, a decrease in tubular reabsorption, or an increase in tubular secretion is unknown. Nor is it clear that this response is even of physiological significance. Infusions of prolactin producing high physiological plasma levels (as might be seen during molt) in birds (European starlings, *S. vulgaris*) reduced the fractional excretion of potassium in the ureteral urine by about 40 % but had no effect on the reabsorption along the proximal tubules of the superficial reptilian-type nephrons (Roberts and Dantzler 1992). Thus any change in tubular reabsorption or secretion must have occurred in the distal tubules of these superficial nephrons and/or any portions of the deeper transitional and mammalian-type nephrons. No data on the effects of prolactin on potassium excretion are available for other nonmammalian vertebrates or mammals.

4.4   Hydrogen Ion

4.4.1   Magnitude and Sites of Net Transport

Very little is known about hydrogen ion transport or even about the role of the kidneys in the regulation of acid–base balance in nonmammalian vertebrates. In many, extrarenal structures, e.g., gills in teleost fishes and bladder in chelonian reptiles, are more important than the renal tubules for the elimination of hydrogen ions and have been subjected to much more intensive study than the kidneys in this regard. The tubular transport of hydrogen ions may be related to the excretion of the end products of nitrogen metabolism (vide infra; Chap. 6) and to the function of these extrarenal routes for ion excretion or for the modification of ion excretion.

Ureteral urine of elasmobranchs, at least in little skate (*R. erinacea*) and spiny dogfish (*S. acanthias*), is normally always acidic with a fixed pH of about 5.8 (Smith 1939) and that of alligators (*Alligator mississippiensis*) is normally alkaline regardless of dietary intake (Coulson and Hernandez 1964), whereas that of other reptiles, as well as that of amphibians, birds, and mammals, normally can vary from alkaline to acid depending on dietary intake. Moreover, ureteral urine of mammals, birds, and amphibians can be greatly acidified (pH about 4.5 with blood pH about 7.4) during an acid load, whereas that of some reptiles (water snakes, *N. sipedon*) can only can only be acidified to a modest extent (pH 5.8 even with a blood pH of 7.0) (Dantzler 1968, 1976; Koeppen et al. 1985). These data suggest that the renal tubules of some reptiles may not be able to secrete hydrogen ions against a steep gradient and that significant acidification may occur in regions distal to the kidneys, e.g., bladder or cloaca. As noted above, hydrogen ion secretion by the turtle bladder has been well documented and studied in some detail (Steinmetz 1967, 1974).

In elasmobranchs (e.g., *R. erinacea* and *S. acanthias*), micropuncture studies involving the injection of buffer dyes indicate that acidification occurs in the earliest portion of the proximal tubules and can go to completion at any point along the proximal tubule (Deetjen and Maren 1974; Kempton 1940). The rate of net acid secretion by the proximal tubules of *R. erinacea* amounts to about 40 μmol liter (of cells)$^{-1}$ min^{-1} (Deetjen and Maren 1974), a rate that compares favorably with that determined for intact, free-swimming *S. acanthias* (Hodler et al. 1955). Moreover, the acid secretory process is nearly inexhaustible (Swenson and Maren 1986).

In contrast, micropuncture studies on both anuran (*R. pipiens; R. esculenta*) and urodele (*N. maculosus*) amphibians indicate the pH of the tubular fluid only falls below that of the blood along the distal tubules, beginning with the early distal diluting segment (Giebisch 1956; Montgomery and Pierce 1937; O'Regan et al. 1982; Oberleithner et al. 1984). These observations also indicate that the bicarbonate concentration does not change along the proximal tubules of amphibians. In birds, the micropuncture measurements on the proximal tubules of superficial reptilian-type nephrons indicating that sodium and chloride are reabsorbed at equivalent rates (vide supra; Laverty and Dantzler 1982) suggested that the bicarbonate concentration does not fall and that the tubular fluid is not acidified along this region of the avian nephron. Lack of acidification along the proximal tubules was confirmed and the presence of acidification at the level of the superficial cortical collecting ducts was demonstrated by direct micropuncture measurements of pH along these nephron segments (Laverty and Alberici 1987). These observations on amphibians and birds differ from those on mammals that show significant acidification along the proximal tubules (Koeppen et al. 1985). Although acidification of tubular fluid does occur along the proximal tubules of elasmobranchs, the process involved appears to be quite different from that in mammals (vide infra).

4.4.2 Mechanism of Transport

4.4.2.1 Proximal Tubules

As in the case of sodium and chloride transport discussed above, detailed information bearing on the mechanism of hydrogen ion transport by proximal renal tubules of nonmammalian vertebrates is only available for urodele amphibians, and this information is far from complete. The amphibian studies involve intracellular as well as transepithelial pH measurements, generally with doubly perfused kidneys of *N. maculosus* and *A. means* (Boron and Boulpaep 1983a, b; O'Regan et al. 1982), and transport measurements with brush-border membrane vesicles from kidneys of *N. maculosus* (Seifter and Aronson 1984). The data indicate that the hydrogen ion concentration within the cells is maintained below electrochemical equilibrium apparently primarily by extrusion across the luminal membrane by the same sodium-hydrogen exchanger discussed above (Figs. 4.1 and 4.6). The energy for this hydrogen extrusion is derived from the movement of sodium into the cells down its electrochemical gradient (Figs. 4.1 and 4.6). Extrusion of hydrogen ions across the basolateral membrane also may occur, although the mechanism of such extrusion is unknown (Boron and Boulpaep 1983a, b).

Although the fluid in the lumen of the proximal tubules of urodele amphibians is not acidified under control conditions, a reduction in luminal pH does occur when the lumen is perfused with high concentrations of bicarbonate (O'Regan et al. 1982). However, at steady state, no significant bicarbonate or pH gradient can be established across the epithelium (O'Regan et al. 1982). The rate of bicarbonate reabsorption in these experiments cannot be explained by passive movement (O'Regan et al. 1982). These observations suggest that, in view of hydrogen ion secretion across the luminal membrane, most bicarbonate reabsorption is driven by hydrogen ion secretion. Apparently, as in the mammalian kidney, such reabsorption involves conversion of bicarbonate to carbon dioxide and water in the tubule lumen and then reconversion to bicarbonate within the renal tubule cells (Fig. 4.6). It also seems likely that, as in mammalian proximal tubules, carbonic anhydrase plays an essential role in this process (Fig. 4.6), especially since the failure to develop any significant transepithelial pH or bicarbonate gradient along amphibian proximal tubules during control, free-flow conditions indicates that net bicarbonate reabsorption is proportional to net fluid reabsorption. However, the presence of carbonic anhydrase in amphibian proximal tubules is not well documented. Although histochemical evidence for carbonic anhydrase is conflicting, it suggests that, if the enzyme is present, there is much less of it than in the distal tubules (vide infra) (Lonnerholm and Ridderstrale 1974; Ridderstrale 1976; Rosen 1972). Finally, the exit step for bicarbonate across the basolateral membrane during the transepithelial reabsorptive process is unknown, although it could well involve a sodium-bicarbonate cotransporter similar to the one found in mammalian proximal tubules.

Although the exact mechanism of acidification along the proximal tubules of elasmobranchs is unknown, preliminary data indicate that it is very different from that in the proximal tubules of mammals. First, these animals have no renal carbonic anhydrase (Maren 1967) (Maren 1967). Second, in isolated perfused kidneys of *R. erinacea*, transepithelial voltage and pH measurements with micro-electrodes in the presence and absence of inhibitors on the basolateral side of the tubules suggest that both the basolateral Na-K-2Cl cotransporter (probably NKCC1) discussed above (Fig. 4.2) and a basolateral chloride-bicarbonate exchange system play a role in the acidification process (Silbernagl et al. 1986).

No cellular details of transepithelial hydrogen ion secretion or bicarbonate reabsorption are known for proximal renal tubules of other nonmammalian vertebrates, but the observation that the administration of carbonic anhydrase inhibitors to water snakes (*N. sipedon*) results in an alkaline urine, an increased excretion of sodium and potassium, and an unchanged excretion of chloride suggest that the mechanisms in these reptiles may be similar to those in amphibians and mammals (Dantzler 1968). However, one group of reptiles, the crocodilians, is clearly different from mammals, amphibians, and other reptiles (Coulson and Hernandez 1964, 1983; Lemieux et al. 1985). The kidneys of alligators (*A. mississippiensis*) normally produce an alkaline urine containing bicarbonate and ammonia, despite a high protein, acid-ash diet (Coulson and Hernandez 1964, 1983). Net tubular secretion of bicarbonate occurs and the blood pH is low (around 7.1–7.2) compared to that of other nonmammalian vertebrates or mammals at comparable temperatures (Coulson and Hernandez 1964, 1983; Lemieux et al. 1985). In these animals, the administration of carbonic anhydrase inhibitors results in a mildly acid urine and the conversion of net secretion of bicarbonate to net reabsorption (Coulson and Hernandez 1964; Lemieux et al. 1985). It appears that carbonic anhydrase plays a role in bicarbonate production by renal tubule cells and that hydrogen ions may be extruded from the cells into the blood as bicarbonate is secreted into the lumen (Lemieux et al. 1985). However, nothing is known about the tubular sites or cellular and molecular mechanisms involved.

4.4.2.2   Distal Tubules

Among the nonmammalian vertebrates, specific information on the hydrogen ion secretory process in distal tubules is available only for amphibians. In the case of the early distal diluting segment, this information involves only anurans (*R. esculenta* and *R. pipens*). Studies on this process have involved transepithelial and intracellular recordings with conventional and ion-sensitive microelectrodes in isolate, doubly perfused frog kidneys (Oberleithner et al. 1984, 1987; Weigt et al. 1987) and intracellular recordings on fused "giant cells" from frog diluting segments (Oberleithner et al. 1987) (Oberleithner et al. 1987). These studies all indicate that hydrogen ion secretion involves an amiloride-sensitive secondary active sodium-hydrogen exchange process at the luminal membrane (Fig. 4.3). As described above for the proximal tubules, the energy for the hydrogen extrusion from the cells into the lumen is derived from the movement of sodium down its electrochemical gradient into the cells. This electrochemical gradient for sodium is maintained, in turn, by the primary energy-consuming Na-K-ATPase at the basolateral membrane (Figs. 4.1 and 4.6). As indicated above for proximal tubules, the sodium-hydrogen exchanger, as in mammals, probably belongs to the NHE transporters (Figs. 4.1 and 4.6), but no molecular studies are available for this process.

Activity of the sodium-hydrogen exchanger in these amphibians appears to depend on the metabolic state of the animal (Oberleithner 1985). Adaptation to a

high potassium environment stimulates sodium-hydrogen exchange, thereby raising the intracellular pH and lowering the luminal pH, whereas adaptation to a high
sodium environment suppresses sodium-hydrogen exchange, thereby leaving
hydrogen ions distributed at electrochemical equilibrium (Oberleithner 1985;
Oberleithner et al. 1984). As noted above, aldosterone secretion is stimulated by
adaptation to a high potassium environment and suppressed by a high sodium
environment, suggesting that this hormone may play a major role in the observed
changes in hydrogen ion secretion (Oberleithner et al. 1983b). Direct evaluation of
the effects of aldosterone indicates that it does, indeed, stimulate sodium-hydrogen
exchange and that such stimulation can be inhibited by spironolactone (Fig. 4.3)
(Oberleithner et al. 1987; Weigt et al. 1987). In fact, aldosterone appears to be
required for activation of the exchange process (Weigt et al. 1987). These data all
support the concept that this hormone is responsible for the enhanced hydrogen ion
secretion observed when these amphibians are adapted to a high potassium
environment.

In isolated, perfused late distal tubules of urodele amphibians (A. means), direct
pH measurements indicate that there is significant electrically silent hydrogen ion
secretion that is inhibited by acetazolamide and luminal amiloride (Stanton
et al. 1987a). These data indicate that the enzyme carbonic anhydrase is important
for the intracellular generation of hydrogen ions, an idea supported by histochemical evidence that significant amounts of carbonic anhydrase are found in amphibian late distal nephrons (Lonnerholm and Ridderstrale 1974). The data also suggest
that hydrogen ions are secreted into the lumen via a sodium-hydrogen exchanger
like that discussed above for the urodele proximal tubule and anuran diluting
segment (Fig. 4.4). However, Stanton and colleagues (Stanton 1988; Stanton
et al. 1987a) have described a second, minority cell type in the late distal tubules
of *Amphiuma* that has the sodium-hydrogen exchanger in the luminal membrane
but lacks the chloride-bicarbonate exchanger found in that same membrane in the
majority of cells (Fig. 4.4). They suggest that these cells alone account for the net
hydrogen ion secretion in urodele late distal tubules. Clearly, as in all areas of renal
function in nonmammalian vertebrates, more work needs to be done on the hydrogen ion secretion by amphibian renal tubules.

As discussed in Chap. 2, the collecting tubules and collecting ducts of both
urodele (e.g., Congo eel, A. means) and anuran (e.g., toad, B. bufo) amphibians
contain large numbers of mitochondrial-rich intercalated cells (Mobjerg et al. 1998;
Stanton et al. 1984b). Similar cells may be found in the collecting tubules or
collecting ducts of other nonmammalian vertebrates. In analogy with some types
of intercalated cells in mammalian collecting ducts (Roy et al. 2015), these cells
may actively secrete hydrogen via some form of H^+-ATPase or an H^+-K^+-ATPase,
but this has yet to be studied directly. Moreover, because active hydrogen ion
secretion has been well documented in the bladder of a freshwater turtle (*P. scripta*)
(Steinmetz 1967, 1974), it is likely that postrenal structures (bladder or cloaca) may
be more important than the renal tubules in regulating acid secretion in many
nonmammalian vertebrates.

4.5 Calcium

4.5.1 Direction, Magnitude, and Sites of Transport

Clearance studies indicate that overall net tubular secretion of calcium occurs in glomerular and aglomerular marine teleosts and euryhaline teleosts adapted to seawater (Hickman and Trump 1969). Because marine teleosts lack distal tubules (vide supra; Chap. 2), this early observation suggests that net secretion definitely can occur in the proximal tubules of teleosts (vide infra). In freshwater teleosts and euryhaline teleosts adapted to freshwater, overall net tubular reabsorption occurs but final excretion is relatively high (about 40 % of that filtered) (Hickman and Trump 1969; Schmidt-Nielsen and Renfro 1975). Because freshwater teleosts have distal tubules, these observations suggest that distal tubules are the site of most reabsorption in teleosts.

Overall net tubular secretion has also been observed in one marine elasmobranch species (little skate, *R. erinacea*) (Stolte et al. 1977a), whereas overall net tubular reabsorption of calcium has been observed in the only other marine elasmobranch species studied (dogfish shark, *S. acanthias*) (Hickman and Trump 1969). Whether or not net secretion always occurs in skates and net reabsorption always occurs in dogfish sharks is unknown. However, it would seem advantageous for marine fishes always to eliminate excess calcium via the kidneys. Micropuncture studies on the little skate indicate that the net tubular secretion in this species occurs primarily along the second portion of the proximal tubule (Stolte et al. 1977a).

In all other nonmammalian vertebrates and in mammals, clearance studies indicate that overall net tubular reabsorption of calcium always occurs (Bindels et al. 2012; Clark and Dantzler 1972, 1975; Dantzler 1976; Laverty and Clark 1981). In most of these species, the fraction of filtered calcium excreted in the ureteral urine is about equal to the fraction of filtered sodium excreted. Thus, the excretion of calcium is quite low (usually about 1–2 % of the filtered load).

Micropuncture and microperfusion studies on mammals indicate that about 65 % of the filtered calcium is reabsorbed along the proximal tubule, about 20 % along the thick ascending limb of Henle's loop, and most of the rest along the distal convoluted tubule and connecting tubule (Bindels et al. 2012; Blaine et al. 2015). Similar studies on urodele amphibians (*N. maculosus*) indicate that all significant net reabsorption occurs along the distal tubules, not the proximal tubules (Garland et al. 1975). A complete description of tubular transport sites is not available for other nonmammalian vertebrates. However, micropuncture studies in birds (European starlings, *S. vulgaris*) (Laverty and Dantzler 1982) do show that, as in mammals, net reabsorption can occur along the proximal tubules of the superficial, reptilian-type nephrons. But, in contrast to mammals, the rate of reabsorption of calcium along these avian proximal tubules exceeds the rate of reabsorption of sodium and water (Laverty and Dantzler 1982). In some of these birds, however, for reasons that are not yet clear, net secretion of calcium occurs along the proximal tubules of the superficial, reptilian-type nephrons (Roberts and Dantzler 1990,

1992). Because almost all filtered calcium, even in these particular birds, is still reabsorbed before the final urine is reached (Roberts and Dantzler 1990, 1992), substantial net reabsorption must take place in the distal portion of these nephrons and in the collecting ducts as well as in the mammalian-type nephrons. However, as noted above, no information is available about other possible reabsorptive sites for calcium in the renal tubules of birds or other nonmammalian vertebrates.

4.5.2 Mechanism of Transport

Much is now known about the mechanisms of calcium transport by renal tubules of mammals, but very little is known about the process in renal tubules of nonmammalian vertebrates. In mammals, the reabsorption of calcium along the proximal tubules (about 65 % of the filtered load) and thick ascending limb (about 20 % of the filtered load) is passive and moves from lumen to peritubular fluid by a paracellular route (Bindels et al. 2012). In the proximal tubule, this reabsorption involves passive diffusion driven by the lumen-to-peritubular fluid calcium concentration gradient and the small lumen-positive transepithelial potential in the late proximal tubule (vide supra; Table 4.2) and possibly also by solvent drag (Yu 2015). Claudin-2 appears to account primarily for the permeability of the paracellular pathway to calcium (Yu 2015). Similarly, paracellular calcium reabsorption in the thick ascending limb is driven by the lumen-positive transepithelial potential (vide supra; Table 4.2) with claudins 16 and 19 being primarily responsible for the paracellular calcium permeability (Yu 2015). This pathway in mammalian thick ascending limbs is regulated by the interaction of plasma calcium with a calcium-sensing receptor (CaSR) on the basolateral membrane of the cells (Bindels et al. 2012; Yu 2015). Activation of the CaSR by increased plasma calcium levels leads to the upregulation of claudin 14, which physically interferes with the paracellular cation channels formed by claudins 16 and 19 to reduce reabsorption (Yu 2015 for more details on the intracellular pathways).

In birds, calcium reabsorption, when it occurs, along the proximal tubules of the superficial reptilian-type nephrons cannot involve simple passive paracellular movement because the calcium concentration in the tubular fluid is below that in the ultrafiltrate of the plasma (Laverty and Dantzler 1982), and there is a small, lumen-negative, not lumen-positive, transepithelial potential (vide supra; Table 4.2). It appears that some form of active transcellular transport must occur, but this process has yet to be examined.

However, if calcium reabsorption occurs along the thick ascending limb of avian mammalian-type nephrons and the distal diluting segment of reptilian-type nephrons, as appears almost certain, it could well be driven by the lumen-positive transepithelial potential (vide supra; Table 4.2) via a passive paracellular pathway as in mammals. This also could occur in the distal diluting segment of nephrons in other nonmammalian vertebrates. In support of this concept, measurements with conventional and ion-sensitive microelectrodes on the diluting segment in nephrons of the isolated, perfused frog kidney (*R. pipiens*) suggest that all reabsorption is passive via the paracellular pathway and is driven by the lumen-positive transepithelial potential (Dietl and Oberleithner 1987).

In mammals, reabsorption of calcium by the distal convoluted tubules and connecting tubules (where, as noted above, most of the remaining calcium is reabsorbed) occurs solely by a transcellular route. This route for calcium reabsorption involves passive entry across the luminal membrane via the transient receptor potential vanilloid 5 (TRPV5) calcium channel, binding by calbindin-D28k for diffusion through the cytoplasm to the basolateral membrane, and active transport out of the cells across the basolateral membrane (Bindels et al. 2012; Blaine et al. 2015). The active transport out of the cells involves two basolateral membrane processes: secondary active transport via a sodium-calcium exchanger (NCX1) and primary active transport via calcium-ATPase isoform 1b (PMCA1b) (Bindels et al. 2012; Blaine et al. 2015).

It seems likely that calcium reabsorption by the distal nephron beyond the diluting segment in all nonmammalian vertebrates also occurs via a transcellular route. If so, this would involve, as in mammals, calcium entry into the cells down its electrochemical gradient across the luminal membrane and transport out of the cells against its electrochemical gradient across the basolateral membrane. These steps may involve channels and transporters similar to those found in mammals, but as yet there is no information available on these transport processes in nonmammalian vertebrates.

A bit more information is available about net secretion of calcium in the proximal tubules of marine or seawater-adapted teleosts. Transport of calcium across the basolateral membrane into the cells of proximal tubule cells isolated from the kidneys of seawater-adapted winter flounder (*P. americanus*) involves both fast and slow components (Renfro 1978). These studies and others with isolated tubules and plasma membranes indicate that the slow component, which may be more important than the fast component for net calcium secretion, is saturable, ATP-dependent, sodium-dependent (but apparently not directly linked to the sodium gradient), and partially inhibited by magnesium (Renfro 1978; Renfro et al. 1982). Although a calcium-ATPase is easily identified in teleost renal plasma membranes, it is not clear whether it is located on the luminal membrane, the basolateral membrane, or both (Flik et al. 1996). Clearly, for transcellular secretion to occur, calcium must be transported out to the cells against an electrochemical gradient at the luminal membrane, almost certainly by a calcium-ATPase, but the details of such transport are far from clear. In addition, net transepithelial secretion has long been known to respond directly to the concentration of calcium in the peritubular fluid, increasing with an increase in concentration and decreasing with a decrease in concentration (Hickman and Trump 1969). It seems likely that there is some sort of calcium-sensing receptor on the basolateral membrane of marine teleost proximal tubules, perhaps similar to that found in the thick ascending limb of mammalian nephrons. However, there is as yet no information available on such a possible sensor or how it might function to alter transcellular calcium secretion. Certainly, a great deal remains to be learned about calcium transport by the renal tubules of nonmammalian vertebrates.

4.5.3   Hormonal Regulation

4.5.3.1   Parathyroid Hormone

In mammals and in most birds studied (starling, *S. vulgaris*; Japanese quail, *C. coturnix*; domestic chickens, *G. gallus domesticus*), parathyroid hormone (PTH) appears to be required for normal reabsorption of calcium by the renal tubules because parathyroidectomy reduces reabsorption, and the administration of PTH to parathyroidectomized animals restores reabsorption to control levels (Agus and Goldfarb 1985; Clark and Sasayama 1981; Clark and Wideman 1977; Mok 1978). In mammalian kidneys, PTH receptors have been localized to proximal convoluted tubules, proximal straight tubules, cortical thick ascending limbs, and distal convoluted tubules, but the action of PTH on calcium transport and the mechanisms involved in the different tubule segments are not completely clear and are the subject of ongoing study (Bindels et al. 2012). In the proximal tubule, PTH enhances production of Vitamin D_3, but does not itself directly affect the calcium transport process (Bindels et al. 2012). In thick ascending limbs, it appears to enhance the both the lumen-positive transepithelial potential (the driving force for calcium reabsorption) and the calcium permeability of the paracellular pathway (Wittner et al. 1993). In distal convoluted tubules, it clearly stimulates active reabsorption, but this may involve insertion of more TRPV5 calcium channels in the luminal membrane, stimulation of calcium-ATPase (PMCA1b) in the basolateral membrane, increased expression of sodium-calcium exchange (NCX1) in the basolateral membrane, increased expression of calbindin-D28k, or any combination of these (Bindels et al. 2012).

Preliminary micropuncture studies on starlings indicate that PTH does not stimulate calcium reabsorption along the proximal tubules of the superficial reptilian-type nephrons (G. Laverty and W. H. Dantzler unpublished observations) (Laverty and Dantzler 1983). Parathyroid hormone may, of course, stimulate calcium reabsorption along the proximal tubules of the deeper mammalian-type nephrons. However, in view of the lack of a direct effect on calcium reabsorption in mammalian proximal tubules, any proximal effect in avian nephrons seems unlikely. The stimulatory effect is more likely to occur in the more distal regions of the avian nephrons. In fact, because the response of the whole kidney in starlings is similar to that in mammals and because PTH can stimulate calcium reabsorption in the thick ascending limb of Henle's loops in mammals, a substantial part of the whole-kidney response in birds may reflect the presence of thick ascending limbs of Henle's loops in the mammalian-type nephrons. In this regard, it is particularly noteworthy that PTH does not appear to stimulate renal calcium reabsorption in reptiles, where the nephrons lack loops of Henle. Neither parathyroidectomy nor the administration of PTH to intact or parathyroidectomized snakes (*Nerodia* spp.) or turtles (*Chrysemys* spp.) has any observable effect on reabsorption of calcium by the renal tubules (Clark and Dantzler 1972; Laverty and Clark 1981). However, it should also be noted that in two species of birds (herring gull, *Larus argentatus* and great black-backed gull, *L. marinus*), which, of course, do have mammalian-type nephrons with loops of Henle, neither parathyroidectomy nor the administration of PTH to intact or parathyroidectomized animals appears to affect net tubular reabsorption of calcium (Clark and Mok 1986). These differences among species of birds do not appear to be the result only of differences in age or sex and, thus, of differences in the reproductive state and in the requirement for calcium for egg shell formation (Clark and Mok 1986). They also do not appear to be the result of differences in the maturity of the calcium transport system (Clark and Mok 1986). At present, it is not clear from whole-animal studies whether parathyroidectomy or PTH administration actually has any direct effect on tubular reabsorption of calcium in amphibians (Stiffler 1993).

Fish were long thought to lack PTH because they apparently lack the equivalent of the parathyroid gland found in tetrapod vertebrates. Also, the hormone appeared to be unnecessary to mobilize or retain calcium because it is readily available to both freshwater and marine fish. However, in recent years two forms of the hormone PTH and two forms of the paracrine factor PTH-related protein (PTHrP) have been identified and cloned in teleosts (Guerreiro et al. 2007). PTHrPs have also been identified in elasmobranchs and receptors that work for both PTH and PTHrPs have been identified in teleosts and elasmobranchs (Guerreiro et al. 2007). Nevertheless, despite evidence for the presence of PTHrPs and receptors in teleost kidneys, good evidence for a direct effect on tubular calcium reabsorption is still lacking. PTHrP has a minimal effect on calcium fluxes across a monolayer of a primary culture of winter flounder (*P. americanus*) proximal tubule cells (Guerreiro et al. 2004), but since calcium reabsorption probably occurs mainly in teleost distal tubules and winter flounder lack distal tubules, this may mean little. With the current molecular knowledge of PTHrPs and PTH in fish, it should be possible to learn much more about the possible renal effect of these factors in the near future.

4.5.3.2 Calcitonin

The effects of calcitonin on calcium reabsorption by specific segments of mammalian nephrons are not yet clear and may vary among species. In rats, calcitonin appears to stimulate calcium reabsorption by the thick ascending limb of Henle's loops but not by the proximal or distal tubules of mammalian nephrons (Quamme 1980). In rabbits, it appears to stimulate reabsorption by the distal tubules, apparently by activating calcium channels in the luminal membrane (presumably TPV5) and the sodium-calcium exchanger (presumably NCX1) in the basolateral membrane (Zuo et al. 1997). However, it is not clear that these effects are important physiologically. Among the nonmammalian vertebrates, the administration of calcitonin to intact or parathyroidectomized snakes (*Nerodia* spp.) has no effect on calcium transport by the renal tubules (Clark and Dantzler 1975). Also, there is no evidence to date of an effect on calcium transport by the renal tubules of any other nonmammalian vertebrates.

4.5.3.3   Stanniocalcin

Stanniocalcin from the corpuscles of Stannius clearly plays a role in regulating calcium balance in teleost fishes, apparently by reducing calcium uptake by gills and intestine (Guerreiro et al. 2007; Tseng et al. 2009). However, there is currently no evidence that it reduces calcium reabsorption by the renal tubules. In fact, studies of the action of stanniocalcin on transepithelial calcium transport by monolayers of cultured flounder (*P. americanus*) proximal tubule cells mounted in Ussing chambers showed no direct effect at all (Lu et al. 1994b). Moreover, although stanniocalcin has now been identified in mammals, including humans (Olsen et al. 1996; Wagner et al. 1997), there is again no evidence that it directly affects the steps in calcium transport by the renal tubules.

4.6   Phosphate

4.6.1   *Direction, Magnitude, and Sites of Net Transport*

Clearance and micropuncture studies have revealed both net reabsorption and net secretion of phosphate by the renal tubules of birds, reptiles, and fishes (Clark et al. 1976; Clark and Dantzler 1972; Hickman and Trump 1969; Knox et al. 1973; Laverty and Dantzler 1982; Levinsky and Davidson 1957; Schneider et al. 1980). Among amphibians, net tubular reabsorption definitely occurs and net tubular secretion also may occur, but the data are not convincing (Schneider et al. 1980; Walker and Hudson 1937). Among mammals, however, only net tubular reabsorption has been documented (Bindels et al. 2012; Knox and Haramati 1985; Mizgala and Quamme 1985; Schneider et al. 1980), and tubular secretion appears to play no role at all in overall renal excretion (Greger et al. 1977).

Among the fishes, hagfishes, lampreys, marine elasmobranchs, and stenohaline marine teleosts exhibit net tubular secretion (Cliff et al. 1986; Hickman and Trump 1969; Stolte et al. 1977a). Net tubular reabsorption always appears to occur among true stenohaline freshwater teleosts (Hickman and Trump 1969). In general, euryhaline teleosts exhibit net reabsorption when adapted to freshwater and net secretion when adapted to seawater (Hickman and Trump 1969), but the mechanism involved in the change in the dominant direction of transport is, at best, only partially understood (vide infra). Moreover, net tubular secretion has been observed in euryhaline eels (*A. Anguilla*) adapted to freshwater (Chester Jones et al. 1969).

Among reptiles, clearance studies reveal net tubular secretion requiring the presence of parathyroid hormone in water snakes (*Nerodia* spp.) and freshwater turtles (*C. picta* and *C. picta elegans*) (vide infra) (Clark and Dantzler 1972; Laverty and Clark 1981). This may also be the case for alligators (*A. mississippiensis*) (Dantzler 1976). Under control conditions in water snakes, the rate of excretion of phosphate can actually be 2.5 times the rate of filtration, a relative rate of net tubular secretion higher than that generally found in other vertebrate species (Clark and Dantzler 1972).

Among birds, clearance studies normally reveal net tubular reabsorption, but this changes to net tubular secretion with phosphate loading or the administration of parathyroid hormone (Clark et al. 1976; Levinsky and Davidson 1957). Moreover, as among the reptiles, net tubular secretion always requires the presence of parathyroid hormone (vide infra).

Descriptions of the tubular sites of net transport are not truly complete for any vertebrates. Among the mammals, about 80 % of the filtered phosphate is reabsorbed by the renal tubules, and micropuncture and microperfusion studies suggest that 90 % (if not all) of the reabsorption occurs along the proximal tubules (Knox and Haramati 1985). The degree of net reabsorption, if any, along the distal convoluted tubules and collecting ducts is not yet resolved (Bindels et al. 2012; Knox and Haramati 1985; Mizgala and Quamme 1985). Micropuncture measurements on a marine elasmobranch species (little skate, *R. erinacea*) indicate that the major site of secretion is the second segment of the proximal tubule (Stolte et al. 1977a). In vitro microperfusions of tubules from a marine teleost species (winter flounder, *P. americanus*) indicate that net secretion also occurs primarily in the second segment of the proximal tubules of these animals (vide infra; Mechanism of Transport) (Cliff et al. 1986). Nothing further is known about the sites of transport in fishes. Although micropuncture and microperfusion studies indicate that net reabsorption of phosphate can occur along the proximal tubules of amphibians and reptiles, they do not clearly demonstrate the site of the marked phosphate secretion observed in reptiles (Dantzler 1992; Walker and Hudson 1937). Micropuncture studies on starlings indicate that in intact control animals either net reabsorption or net secretion or no net transepithelial transport can be observed along the proximal tubules of superficial reptilian-type nephrons (Laverty and Dantzler 1982; Roberts and Dantzler 1990). It is not clear whether this variation in direction of net transport is time-dependent or indicates simultaneous differences among the individual nephrons in transport of phosphate. Nothing more is known about the tubule sites of phosphate transport in birds.

4.6.2 Mechanism of Transport

The cellular and molecular mechanisms involved in phosphate reabsorption by mammalian renal tubules have been and still are the subject of intensive investigation, whereas, except for significant work on teleost fishes, much less research effort has been expended on the mechanisms of phosphate transport by the renal tubules of nonmammalian vertebrates. This latter situation is particularly unfortunate with regard to reptiles and birds because clear evidence of net secretion under the control of parathyroid hormone is found only among these nonmammalian vertebrates.

The many studies on the mammalian reabsorptive mechanism have been reviewed extensively elsewhere (see, e.g., Bindels et al. 2012) and are beyond the scope of this volume. Only the major points are summarized here. Briefly, phosphate reabsorption along the proximal tubules of mammals is a transcellular process involving transport into the cells against an electrochemical gradient at the luminal membrane and movement from the cells to the peritubular fluid down an electrochemical gradient at the basolateral membrane (Bindels et al. 2012). The transport step at the luminal membrane is a secondary active, sodium-coupled process driven by the electrochemical gradient for sodium. It involves primarily two isoforms of a sodium-phosphate cotransporter (NaPi-IIa and NaPi-IIc). Sodium-phosphate cotransport by NaPi-IIa is electrogenic with a stoichiometry of 3 Na^+ ions to 1 (divalent) phosphate ion, whereas sodium-phosphate cotransport by NaPi-IIc is electroneutral with a stoichiometry of $2Na^+$ ions to 1 (divalent) phosphate ion (Bindels et al. 2012). The exit step across the basolateral membrane for phosphate undergoing transepithelial reabsorption, although down an electrochemical gradient, is poorly understood. It must be carrier mediated, but the nature of that carrier or whether transport involves some form of anion exchange is simply unknown at present (Bindels et al. 2012). A great deal of additional work is ongoing with regard to these transport processes and their intracellular regulation in mammals, but this is reviewed elsewhere (see, e.g., Bindels et al. 2012).

As noted above, among nonmammalian vertebrates, the most extensive studies on renal tubular phosphate transport involve teleost fishes. Studies with primary monolayer cultures of marine teleost renal proximal tubule cells (winter flounder, *P. americanus*) mounted in Ussing chambers by Renfro and his colleagues (Dickman and Renfro 1986; Gupta and Renfro 1989, 1991; Renfro 1995) indicate that both secretory and reabsorptive fluxes can be active and sensitive to extracellular phosphate concentration. Although both reabsorptive and secretory phosphate fluxes occur simultaneously in these studies, the reabsorptive flux predominates at the standard phosphate concentration (0.4 mM) used in the luminal and basolateral solutions (Gupta and Renfro 1989). When the medium phosphate concentration in both solutions is raised above 0.4 mM, net phosphate secretion occurs. Moreover, as the phosphate concentration is increased in small increments from 0.5 to 2.0 mM, the fish physiological range (Hickman and Trump 1969), both the secretory and reabsorptive fluxes increase but the secretory flux always increases more than the reabsorptive flux. Thus, in this preparation, net secretion predominates at normal plasma phosphate concentrations (Gupta and Renfro 1989; Renfro 1995), as seems appropriate for a marine teleost.

In teleosts, it now appears very likely that during net reabsorption, the entry step for phosphate against an electrochemical gradient at the luminal membrane occurs via a sodium-phosphate cotransporter on that membrane, as in mammals, and that during net secretion the entry step for phosphate against an electrochemical gradient at the basolateral membrane also occurs via a sodium-phosphate cotransporter. A sodium-phosphate cotransporter NaPi-IIb with substantial identity to the one in mammals (at least in the membrane spanning regions) has been cloned from a stenohaline marine teleost (winter flounder, *P. americanus*), a stenohaline freshwater teleost (zebrafish, *Danio rerio*), and a euryhaline (usually freshwater) teleost (rainbow trout, *O. mykiss*) (Graham et al. 2003; Kohl et al. 1996; Nalbant et al. 1999; Sugiura et al. 2003; Werner et al. 1994). In the marine flounder, where net renal tubular secretion is observed, this transporter is localized to the basolateral membrane of the second segment of the proximal tubule with apparently a lesser amount along the luminal membrane of the collecting tubule (Elger et al. 1998; Kohl et al. 1996). These localizations in this marine species support the concept of tubular phosphate secretion by the second segment of the proximal tubule (vide supra), but also suggest that the final excretion may be modified by some reabsorption in the collecting tubule.

Among the freshwater teleosts, where net renal tubular reabsorption is observed, there appear to be species differences in the renal localization of NaPi-IIb, possibly depending on whether or not they are adaptable to seawater. In stenohaline fresh-water zebra fish, two isoforms of NaPi-IIb (termed NaPi-IIb1 and NaPi-IIb2) have been identified (Graham et al. 2003; Nalbant 1999). NaPi-IIb1 is localized all along the luminal brush-border membrane of the second segment of the proximal tubule with some carryover to the collecting tubule, whereas NaPi-IIb2 is localized to the luminal brush-border membrane only at the distal end of the second segment of the proximal tubule and along the entire luminal membrane of the collecting tubule (Graham et al. 2003). Thus, in this freshwater species, it appears likely that phosphate reabsorption occurs throughout the length of the second segment of the proximal tubule and the collecting tubule.

However, the situation is quite different in euryhaline, freshwater-raised, and -adapted rainbow trout (Sugiura et al. 2003). In this species, NaPi-IIb (probably isoform NaPi-IIb2 as judged from antibody effectiveness) is localized only along the luminal brush-border membrane of the first segment of the proximal tubule and not in any other tubule segments (Sugiura et al. 2003), suggesting that phosphate reabsorption in these freshwater-adapted animals is limited to this one tubule segment. There are no specific studies on any changes that might occur when these animals are adapted to seawater. However, Renfro (Renfro 1999) suggests that, at the least, the animals cease to express NaPi-IIb at the luminal brush-border membrane.

Although many aspects of these NaPi-II cotransporters and their regulation are not yet clear in teleosts, they appear to be electrogenic with a stoichiometry of $3Na^+$ ions to 1 (divalent) phosphate ion regardless of whether they are located on the luminal or basolateral membrane of the renal tubules (Renfro 1999). The exit steps for phosphate in teleosts, whether across the basolateral membrane during reabsorption or across the luminal membrane during secretion, although down an electrochemical gradient are, as in mammals, poorly understood. However, studies with monolayer cultures of flounder proximal tubule cells mounted in Ussing chambers indicate that during secretion this luminal exit step, which is sodium-independent and voltage-dependent, transports only the monovalent (acid) form of phosphate (Lu et al. 1994a; Renfro 1999).

Very little is known about the mechanistic details of phosphate transport by the renal tubules of amphibians, reptiles, and birds. However, in the case of birds, studies on brush-border membranes from chicken kidneys indicate that there is a sodium-dependent uptake process for inorganic phosphate, which is PTH sensitive (vide infra), at the luminal side of the proximal tubule cells (Renfro and Clark 1984). Based on the results of the studies in teleost fishes and mammals, it appears likely that this process involves a sodium-phosphate cotransporter, possibly of the NaPi-II family. In addition, further studies on these brush-border membranes from chicken kidneys indicate that the exit step across the luminal membrane during PTH-stimulated secretion, as in flounders, is sodium-independent and voltage and potassium-dependent (Barber et al. 1993). Although avian renal brush-border membrane vesicles reflect the luminal membranes of proximal tubules, they almost certainly are derived from both reptilian-type and mammalian-type nephrons. However, because 70 % or more of the nephrons in birds are of the reptilian-type and they make up the bulk of the kidney mass, it appears that the majority of these vesicles come from this nephron type.

It seems likely that the cellular entry step against an electrochemical gradient at the basolateral membrane in birds during net secretion is a sodium-phosphate cotransporter similar to that apparently present in the luminal membrane, but this has yet to be studied. However, one puzzling study raises questions about the net inorganic phosphate secretion in birds. Wideman and Braun (Wideman and Braun 1981), taking advantage of the avian renal portal system to deliver phosphate-containing compounds directly to the basolateral side of the renal tubules without having them first pass through the glomerular circulation, demonstrated in domestic fowl that when net phosphate secretion occurs, the transepithelial peritubular-to-lumen flux of inorganic phosphate represents only a minor component of net tubular secretion. They were unable to identify the exact source of the phosphate secreted in these circumstances.

From the studies on mammals, fishes, and birds, it appears likely that the cellular entry step in the renal tubules of amphibians and reptiles, whether at the basolateral membrane during secretion or at the luminal membrane during reabsorption, is a sodium-coupled cotransport process. However, there are no studies on the renal phosphate transport processes in these animals. Clearly, much more work on the cellular and molecular mechanisms of renal phosphate transport in these other nonmammalian vertebrates is required.

4.6.3   Hormonal Regulation

4.6.3.1   Parathyroid Hormone

In mammals, PTH regulates phosphate excretion by inhibiting phosphate reabsorption along the entire proximal tubule. The mechanisms of inhibition are reviewed by Bindels et al. (2012). Briefly, this inhibition involves primarily NaPi-IIa and to a lesser extent NaPi-IIc. In the case of NaPi-IIa, PTH causes endocytosis of the transport protein from the brush-border membrane with trafficking to early endosomes for degradation. The steps involved in inhibition of NaPi-IIc are less well understood. Although PTH inhibits phosphate reabsorption primarily by its action on NaPi-IIa, it also inhibits Na-K-ATPase, thereby reducing the sodium gradient driving sodium-phosphate cotransport. Intracellular signaling pathways from receptors for the various PTH actions include at least the adenylate cyclase-protein kinase A (PKA) pathway, the phospholipase C-protein kinase C (PKC) pathway, the extracellular signal-regulated kinase (ERK) pathway, and phospholipase A_2.

In fishes, PTHrP stimulates the net active secretory flux of inorganic phosphate across primary monolayer cultures of winter flounder (*P. americanus*) proximal tubule cells mounted in Ussing chambers without affecting the reabsorptive flux (Guerreiro et al. 2007). It now seems most likely that this effect involves stimulation of NaPi-IIb located in the basolateral membrane of these proximal tubule cells, although data suggest that it might also involve the stimulatory action of PKC on the luminal exit step (Renfro 1999).

In birds and reptiles, as pointed out above, net tubular secretion of inorganic phosphate appears to require the presence of parathyroid hormone. However, whether the hormone solely inhibits a reabsorptive flux (as in mammals), directly stimulates a secretory flux, or does both in these animals is not yet known. In starlings (*S. vulgaris*), net secretion is converted to net reabsorption by parathyoidectomy (Clark and Wideman 1977; Wideman et al. 1980). Net secretion can be restored by the administration of PTH with or without phosphate loading but not by phosphate loading alone (Clark and Wideman 1977; Wideman et al. 1980). Similarly, in both freshwater snakes (*Nerodia* spp.) and freshwater turtles (*C. picta* and *C. picta elegans*), net secretion of phosphate changes to net reabsorption following parathyroidectomy and is restored by the administration of PTH (Clark and Dantzler 1972; Laverty and Clark 1981). Moreover, marked net tubular secretion in intact water snakes noted above can be greatly enhanced by the administration of PTH (Clark and Dantzler 1972).

As already noted, the presence of PTH-dependent tubular secretion of phosphate in birds and reptiles offers an opportunity to examine both the site and underlying mechanism involved. With regard to the site of this effect, preliminary micropuncture measurements on starlings indicate that in birds with intact parathyroid glands, the administration of exogenous PTH can induce net secretion by the proximal tubules of all superficial, reptilian-type nephrons (Laverty and Dantzler 1983). Moreover, if the change from net secretion to net reabsorption and vice versa in these superficial nephrons in the absence of exogenous PTH is time-dependent (vide supra), it may reflect fluctuations in the rate of secretion of endogenous PTH. If, instead, simultaneous differences in the direction of net transport occur among these nephrons (vide supra), it would raise intriguing questions about the distribution of endogenous hormone or the number and sensitivity of hormone receptors. Nothing is yet known about transport in other segments of the superficial reptilian-type nephrons or in any segment of the mammalian-type nephrons. The site or sites of hormonal effects on phosphate transport by the renal tubules of reptiles are not yet fully known. However, preliminary studies with isolated, perused proximal renal tubules from garter snakes (*Thamnophis* spp.) indicate that PTH can partially inhibit net reabsorption in these segments but cannot induce net secretion (Dantzler 1992).

For both net secretion and net reabsorption to occur in the same cells of a given nephron, both a unidirectional secretory flux and a unidirectional reabsorptive flux must exist, as demonstrated for fishes with primary monolayer cultures of winter flounder proximal tubule cells mounted in Ussing chambers (vide supra) (Guerreiro et al. 2007; Gupta and Renfro 1989, 1991; Renfro 1995). As noted above, in birds and reptiles, PTH could inhibit the reabsorptive flux alone, as in mammals, stimulate the secretory flux, or do both. Since the rate of PTH-stimulated net secretion of phosphate in vivo can be two or more times the rate of filtration of phosphate in reptiles and birds (Clark and Dantzler 1972; Laverty and Dantzler 1982, 1983), inhibition of reabsorption alone by the hormone would mean that very large unidirectional secretory and reabsorptive fluxes must always exist. Although this appears energetically wasteful, a precedent for the presence of relatively large unidirectional transepithelial fluxes in the same tubule segment exists in regard to the transport of some organic cations by reptilian proximal renal tubules (vide infra) (Dantzler and Brokl 1986).

Some data do exist on the possible cellular mechanism of the PTH effect on phosphate transport in the avian kidney. Sodium-dependent phosphate uptake is significantly stimulated in renal brush-border membrane vesicles from the kidneys of parathyroidectomized chickens and significantly inhibited in these vesicles from intact or parathyroidectomised chickens given exogenous PTH (Renfro and Clark 1984). These findings, which are similar to those for mammals (Bindels et al. 2012), apparently reflect hormone inhibition of the reabsorptive step at the luminal membrane of avian proximal tubules. In other studies with these brush-border membranes, PTH significantly stimulates the sodium-independent and voltage- and potassium-dependent movement of monovalent inorganic phosphate across this membrane (Barber et al. 1993). In intact tubules, this would enhance exit from the cells into the lumen and thus increase the secretory flux. However, as noted above, no studies have yet examined the phosphate transport step into the cells against an electrochemical gradient at the basolateral membrane involved in transepithelial secretion or the exit step down an electrochemical gradient at this membrane involved in transepithelial reabsorption, much less the effects of PTH on these steps.

However, specific receptors for PTH have been demonstrated in plasma membranes from the superficial regions of the chicken kidney (Nissenson et al. 1981). PTH also stimulates adenylate cyclase in this superficial avian renal tissue, such stimulation being modulated by guanidyl nucleotides (Nissenson et al. 1981). More detailed molecular studies with precise localization of the hormone action are required to explain the mechanism by which this hormone induces or enhances net tubular secretion of phosphate in birds and reptiles.

4.6.3.2 Stanniocalcin

Although stanniocalcin has no effect on renal tubular calcium transport in marine teleosts (vide supra), it appears to have a significant effect on renal tubular phosphate transport, at least in primary monolayer cultures of marine teleost (winter flounder, *P. americanus*) proximal tubule cells mounted in Ussing chambers (Lu et al. 1994b). Teleost stanniocalcin, applied in the physiological plasma range, significantly stimulates net phosphate reabsorption in a dose-dependent fashion in this in vitro preparation. This effect on net phosphate reabsorption results from a significant stimulation of the unidirectional phosphate reabsorptive flux without any significant effect on the unidirectional phosphate secretory flux. Thus, it is clearly the reabsorptive process that is stimulated. This stanniocalcin effect on phosphate transport is mimicked by forskolin, an activator of adenylate cyclase to increase the intracellular level of adenosine cyclic-3', 5'-monophosphate (cyclic AMP or cAMP), and it is inhibited significantly by H-89, a highly specific protein kinase A (PKA) inhibitor. These data suggest that stanniocalcin stimulates renal tubular phosphate reabsorption by an adenylate cyclase-cAMP-PKA pathway (Lu et al. 1994b).

This stimulation should involve the NaPi-IIb transport step at the luminal membrane, which is the driving step in reabsorption. However, it is difficult to reconcile these in vitro data with the immunocytochemical localization of this transporter in winter flounder nephrons (vide supra) (Elger et al. 1998). As noted above, these immunocytochemical studies indicate that the transporter is located at the basolateral membrane of the second segment of the proximal tubules and the luminal membrane of the collecting tubules, suggesting that only an active net secretory flux, not an active net reabsorptive flux, occurs in the proximal tubules. Of course, changes in localization could have occurred even in the primary cell cultures used in the in vitro transport studies. But, it appears even more likely that the immunocytochemical studies were not sensitive enough to demonstrate the additional presence of the NaPi-IIb at the luminal membrane. Clearly, further work needs to be done on the details of the stanniocalcin effect on renal tubular phosphate transport in teleosts.

A possible effect of stanniocalcin on renal tubular phosphate transport in mammals is controversial. Human stanniocalcin administered to rats reduces renal phosphate excretion without affecting the excretion of other ions (Wagner et al. 1997). Moreover, brush border membrane vesicles from stanniocalcin-treated animals show significantly enhanced sodium-phosphate cotransport compared to those from control animals (Wagner et al. 1997). However, Chang et al. (2005) found that knocking out the stanniocalcin 1 gene in mice had no detectable anatomical or physiological effects, including no effect on serum calcium and phosphate levels (Chang et al. 2005). These authors considered that these findings were the result of compensation by upregulation of stanniocalcin 2 expression, but could find no evidence of this. As they note, however, completely ruling out this possibility requires knocking out both the stanniocalcin 1 and 2 genes and this has yet to be done (Chang et al. 2005).

4.7   Magnesium

4.7.1   Direction, Magnitude, and Sites of Net Transport

Clearance studies reveal net reabsorption of magnesium by renal tubules of mammals and most nonmammalian tetrapods and net secretion by the renal tubules of all marine fishes. However, in some marine reptiles (e.g., sea snakes, A. laevis), either net secretion or net reabsorption by the renal tubules can occur (Benyajati et al. 1985). Net secretion of magnesium in these marine reptiles, as in marine teleosts, may play an important role in the elimination of magnesium ingested in seawater, particularly at low glomerular filtration rates.

Although net tubular secretion of magnesium occurs in all marine fishes examined (hagfishes, elasmobranchs, and teleosts), it is particularly striking among the pauciglomerular and aglomerular teleosts (Hickman and Trump 1969; Stolte et al. 1977a). Even among glomerular marine teleosts, almost all excreted magnesium derives from tubular secretion (Renfro 1980). However, net secretion ceases or is markedly reduced with adaptation of euryhaline teleosts to freshwater (Hickman and Trump 1969; Schmidt-Nielsen and Renfro 1975). When need to excrete magnesium exceeds need to excrete sulfate (vide infra), a relatively large fraction of filtered chloride also appears in the urine (Hickman and Trump 1969). In vivo micropuncture experiments on elasmobranchs (little skate, R. erinacea) (Stolte et al. 1977a) and in vivo tubule-perfusion experiments on marine teleosts (winter flounder, P. americanus) (Beyenbach 1982; Beyenbach et al. 1986; Cliff et al. 1986) suggest that the primary site of magnesium secretion is the second segment of the proximal tubule.

As noted above, clearance studies on sea snakes (A. laevis) reveal net secretion or net reabsorption of magnesium. The observation that net secretion of magnesium is most significant at low filtered loads, when almost all of the filtered magnesium should be reabsorbed, suggests that magnesium is reabsorbed at a proximal site by a transport system with a high capacity and then secreted at a more distal point (Benyajati et al. 1985). If magnesium were secreted at a relatively proximal location, most secreted magnesium should be reabsorbed distally and net secretion of magnesium should not be apparent with a low filtered load. However, these possible sites of magnesium transport need to be examined by more direct methods.

Although the amount of magnesium excreted by these sea snakes at low urine flow rates may exceed the filtered load by an order of magnitude, the estimated rate of tubular secretion, if it is assumed that all excreted magnesium is secreted, is much lower than that for marine teleosts (1 mmol kg^{-1} h^{-1} in sea snakes versus 17 mmol kg^{-1} h^{-1} in American eel, A. rostrata) (Benyajati et al. 1985; Foster

1975). Magnesium excretion has not yet been studied in other reptiles or in amphibians.

Although clearance studies on birds reveal only net tubular reabsorption of magnesium (Laverty and Dantzler 1982; Robinson and Portwood 1962), micropuncture measurements on starlings (*S. vulgaris*), made simultaneously with clearance measurements, indicate that both net reabsorption and net secretion can occur along the proximal tubules of the superficial, reptilian-type nephrons (Laverty and Dantzler 1982). The data suggest that net secretion occurs in the early portion of the tubules and net reabsorption in the late portion. If this is the case, then net reabsorption along the late proximal tubules or distal segments of these reptilian-type nephrons must more than compensate for the early proximal secretion. In addition, as in mammals (vide infra), reabsorption of magnesium in the loops of Henle of the mammalian-type nephrons may contribute significantly to the conservation of filtered magnesium by the whole kidney. However, no direct measurements of magnesium transport by these nephrons have yet been made.

In mammals, under normal conditions, about 5–10 % of the filtered magnesium appears in the final urine, the rest being reabsorbed by the renal tubules (Bindels et al. 2012; Quamme and Dirks 1985). However, with hypomagnesemia or severely reduced dietary intake, reabsorption can increase so that only about 1 % of the filtered load appears in the urine (Bindels et al. 2012). In contrast, with hypermagnesemia or very high dietary intake, reabsorption can decrease to the point that almost 100 % of the filtered load appears in the urine. In vivo micropuncture and microperfusion studies indicate that under normal conditions, only about 20 % of the filtered magnesium is reabsorbed along the proximal tubules, the bulk, about 70 %, being reabsorbed along the thick ascending limb of the loops of Henle (Bindels et al. 2012; Quamme and Dirks 1985). A final 5–10 % can be reabsorbed in the distal convoluted tubules, apparently adjusted in such a way as to define the final amount in the urine under most conditions (Bindels et al. 2012).

4.7.2 Mechanism of Transport

The mechanisms involved in magnesium transport by the renal tubules are not yet clear even for mammals. What is known about the mammalian transport process is reviewed in detail elsewhere (Bindels et al. 2012; Quamme and Dirks 1985). Briefly for mammals, the approximately 20 % of filtered magnesium that is reabsorbed in the proximal tubules is thought to move passively via a paracellular pathway, probably driven, like calcium, at least partially by the small lumen-positive transepithelial potential in the late distal tubule and solvent drag (vide supra). However, why only about 20 % of the filtered magnesium is reabsorbed in this tubule segment compared with about 65 % of the filtered calcium is not clear. Apparently, this involves permeability differences of the paracellular pathway, but the claudin that might be involved in magnesium permeability in this tubule region is not known at present. Similarly, the reabsorption of about 70 % of the filtered magnesium in the thick ascending limb of Henle's loop occurs via a paracellular route driven primarily by the lumen-positive transepithelial potential. As in the case of calcium, the permeability of this pathway to magnesium is determined by claudins 16 and 19 and can be regulated via the CaSR on the basolateral membrane and claudin 14 (vide supra) (Yu 2015). The reason why about 70 % of the filtered magnesium but only 20 % of the filtered calcium is reabsorbed in this nephron region is not clear, but probably has something to do with the claudins and the permeability of the paracellular pathway.

In mammals, the reabsorption of the final 5–10 % of the filtered magnesium in the distal convoluted tubules apparently occurs via an active process (Bindels et al. 2012). In this process, magnesium enters the cells at the luminal membrane via the TRPM6 magnesium channel driven by the electrical gradient. It is then transported out of the cells against an electrical gradient at the basolateral side by a secondary active process (probably a sodium-magnesium exchanger) or a primary active process (possibly some form of magnesium ATPase) or both. This has yet to be determined. Apparently, much of the regulation of this transepithelial reabsorptive process involves the TRPM6 magnesium channel at the luminal membrane (Bindels et al. 2012).

Very little information is available on the mechanism of renal tubular transport of magnesium for most nonmammalian vertebrates. However, a number studies on teleosts (e.g., winter flounder, *P. americanus*; killifish, *Fundulus heteroclitus*; tilapia, *Oreochromis mossambicus*; rainbow trout, *O. mykiss*) and elasmobranchs (dogfish shark, *Scyliorhinus caniculus*) have suggested possible mechanisms for transepithelial secretion of magnesium against its electrochemical gradient in the second segment of the proximal tubule (Beyenbach 2004; Bijvelds et al. 1998; Renfro 1999). All studies indicate that, during this process, magnesium enters the cells across the basolateral membrane down its electrical gradient via a magnesium channel. This channel is very likely to be a member of the transient receptor potential (TRP) family (Beyenbach 2004) like that found across the luminal membrane of the mammalian distal convoluted tubules (Bindels et al. 2012). It is even possible that it could be the same TRPM6 magnesium channel as that found in mammals, but this certainly has yet to be evaluated.

The transport of magnesium out of the cells at the luminal membrane during secretion in these fish proximal tubules is against an electrochemical gradient and must therefore be either a primary or secondary active process. One likely possibility for this transporter is a sodium-magnesium exchanger, which would extrude magnesium from the cells against its electrochemical gradient in exchange for sodium moving into the cells down its electrochemical gradient (Natochin and Gusev 1970; Renfro and Shustock 1985), but studies with brush border vesicles from a number of the teleost species noted above have provided equivocal evidence for this process (Bijvelds et al. 1997; Freire et al. 1996; Renfro and Shustock 1985). Beyenbach (2004) suggests that, in view of his immunohistochemical evidence for the V-type H^+-ATPase at the luminal membrane of killifish proximal tubules, a proton-magnesium exchanger at the luminal membrane could also fill the same role as the sodium-magnesium exchanger. Finally, electron microprobe and mass spectrometry studies on both teleost and elasmobranch proximal tubules indicate that magnesium might be in cytoplasmic vesicles that could move to the luminal membrane and deliver magnesium into the lumen by exocytosis (Beyenbach 2004; Chandra et al. 1997; Hentschel and Zierold 1994). Certainly a great deal more information is required to fully understand this process in fishes as well as renal tubular magnesium transport in other nonmammalian vertebrate classes.

No information is available on local or systemic control of magnesium transport in nonmammalian vertebrates. However, PTH, arginine vasopressin, calcitonin, glucagon, and aldosterone have all been implicated in regulating magnesium reabsorption in mammals (Bindels et al. 2012; Greger 1985). The details of these mammalian regulatory processes are far from clear and are reviewed elsewhere (Bindels et al. 2012).

4.8 Sulfate

4.8.1 Direction, Magnitude, and Sites of Net Transport

Clearance studies reveal net reabsorption of filtered sulfate by the renal tubules of mammals, birds, and amphibians (Mudge et al. 1973). This is probably the case for all tetrapods. However, as in the case of magnesium (vide supra), net tubular secretion occurs in all marine fishes (hagfishes, elasmobranchs, and teleosts) examined (Beyenbach 2004; Hickman and Trump 1969; Renfro 1999; Stolte et al. 1977a). Such net secretion is most obvious in pauciglomerular or aglomerular marine teleosts (Hickman and Trump 1969), but even in glomerular marine teleosts almost all the excreted sulfate derives from secretion by the renal tubules (Renfro and Dickman 1980). Again, as in the case of magnesium, net tubular secretion of sulfate ceases or is greatly reduced with adaptation of euryhaline teleosts to freshwater (Hickman and Trump 1969), but tubular secretion still occurs (Cliff and Beyenbach 1992). Micropuncture in vivo of the renal tubules of elasmobranchs (little skate, *R. erinacea*) (Stolte et al. 1977a) and perfusion in vitro of renal tubules from marine teleosts (winter flounder, *P. americanus*), freshwater teleosts (killifish, *F. heteroclitus*), and marine elasmobranchs (dogfish shark, *S. acanthias*) (Beyenbach 1982, 1986; Cliff and Beyenbach 1988a, 1992; Sawyer and Beyenbach 1985) suggest that the primary site of sulfate secretion is the second segment of the proximal tubule. In mammals, reabsorption of sulfate apparently occurs primarily along the proximal tubules (Mudge et al. 1973; Ullrich et al. 1980), but no information is available about the site of sulfate reabsorption in nonmammalian vertebrates.

4.8.2   Mechanism of Transport

In mammals, both saturable secretory and saturable reabsorptive sulfate fluxes exist in proximal convoluted tubules isolated and perfused in vitro, but the reabsorptive flux predominates (Brazy and Dennis 1981). Studies with intact tubules and membrane vesicles first demonstrated that the sulfate reabsorptive flux involves a sodium-sulfate cotransport step at the luminal membrane, sulfate being transported into the cells against its electrochemical gradient by being coupled to sodium moving into the cells down its electrochemical gradient (Lucke et al. 1979; Ullrich et al. 1980), and this transporter, NaSi-1, has been cloned from rat, mouse, and human kidneys (reviewed by Burckhardt and Burckhardt 2003). Movement of sulfate out of the cells across the basolateral membrane down its electrochemical gradient then involves a carrier-mediated anion exchange process (Brazy and Dennis 1981; Pritchard and Renfro 1983). This anion exchanger, sat-1, has been cloned from rat, mouse, and human kidneys and has been clearly localized to the basolateral membrane of proximal tubules in rats (Brzica et al. 2009; Burckhardt and Burckhardt 2003). It can exchange a number of anions (organic and inorganic for sulfate), but may work physiologically primarily as a sulfate-hydroxyl anion exchanger in the sulfate reabsorptive process (Burckhardt and Burckhardt 2003). This anion exchange process on the basolateral membrane may be involved in the secretory flux as well (Brazy and Dennis 1981; Pritchard and Renfro 1983).

All studies of the renal sulfate transport mechanism in nonmammalian vertebrates have involved only the secretory process in marine teleosts. These studies with isolated, nonperfused tubules and membrane vesicles from two species of flounder (winter flounder, *P. americanus*, and southern flounder, *P. lethostigma*) have supplied substantial information about the transport process and revealed similarities and differences from the process in mammals (Renfro and Dickman 1980; Renfro and Pritchard 1982, 1983). The vesicle studies suggest that the uptake process against an electrochemical gradient at the basolateral membrane is an electroneutral process driven by a pH gradient (inside of the vesicles alkaline) and probably involving a sulfate-hydroxyl anion exchanger (Renfro 1999; Renfro and Pritchard 1982, 1983). Indeed, this likely involves a homolog of the mammalian sat-1 exchanger discussed above, but this teleost exchanger has yet to be isolated.

The sulfate exit step at the luminal membrane during the secretory process also appears to involve electroneutral anion exchange, in this case for bicarbonate that is being reabsorbed (Renfro 1999; Renfro and Pritchard 1982, 1983). This luminal transporter involved in the teleost secretory process is quite different from the luminal sodium-sulfate cotransporter, NaSi-1, involved in the mammalian reabsorptive process. Both teleost transport steps are carbonic anhydrase dependent (Renfro et al. 1997). Apparently, catalyzed conversion of bicarbonate inside the cell to carbon dioxide and hydroxyl ions helps to maintain the chemical driving force for bicarbonate entry across the luminal membrane and, therefore, for the bicarbonate-sulfate exchange. It also helps maintain the hydroxyl ion concentration to drive the sulfate-hydroxyl exchange at the basolateral membrane.

References

Abdulnour-Nakhoul S, Boulpaep EL (1998) Transcellular chloride pathways in *Ambystoma* proximal tubule. J Membr Biol 166:15–35

Agus ZS, Goldfarb S (1985) Renal regulation of calcium balance. In: Seldin DW, Giebisch G (eds) The kidney: physiology and pathophysiology. Raven, New York, pp 1323–1335

Alt JM, Stolte H, Eisenbach GM, Walvig F (1980) Renal electrolyte and fluid excretion in the atlantic hagfish *Myxine glutinosa*. J Exp Biol 91:323–330

Ames E, Steven K, Skadhauge E (1971) Effects of arginine vasotocin on renal excretion of Na^+, K^+, Cl^-, and urea in the hydrated chicken. Am J Physiol 221:1223–1228

An KW, Kim NN, Choi CY (2008) Cloning and expression of aquaporin 1 and arginine vasotocin receptor mRNA from the black porgy, *Acanthopagrus schlegeli*: effect of freshwater acclimation. Fish Physiol Biochem 34:185–194

Ares GR, Caceres PS, Ortiz P (2011) Molecular regulation of NKCC2 in the thick ascending limb. Am J Physiol Renal Physiol 301:F1143–F1159

Babonis LS, Miller SN, Evans DH (2011) Renal responses to salinity change in snakes with and without salt glands. J Exp Biol 214:2140–2156

Barber LE, Coric V, Clark NB, Renfro JL (1993) PTH-sensitive K^+- and voltage-dependent P_i transport by chick renal brush-border membranes. Am J Physiol Regul Integr Comp Physiol 265:F822–F829

Benyajati S, Yokota SD, Dantzler WH (1985) Renal function in sea snakes. II. Sodium, potassium, and magnesium excretion. Am J Physiol Regul Integr Comp Physiol 249:R237–R245

Beyenbach KW (1982) Direct demonstration of fluid secretion by glomerular renal tubules in a marine teleost. Nature 299:54–56

Beyenbach KW (1984) Water-permeable and -impermeable barriers of snake distal tubules. Am J Physiol Renal Physiol 246:F290–F299

Beyenbach KW (1986) Secretory NaCl and volume flow in renal tubules. Am J Physiol Regul Integr Comp Physiol 250:R753–R763

Beyenbach KW (2004) Kidneys sans glomeruli. Am J Physiol Renal Physiol 286:F811–F827

Beyenbach KW, Dantzler WH (1978) Generation of transepithelial potentials by isolated perfused reptilian distal tubules. Am J Physiol Renal Physiol 234:F238–F246

Beyenbach KW, Fromter E (1985) Electrophysiological evidence for Cl secretion in shark renal proximal tubules. Am J Physiol Renal Physiol 248:F282–F295

Beyenbach KW, Koeppen BM, Dantzler WH, Helman SI (1980) Luminal Na concentration and the electrical properties of the snake distal tubule. Am J Physiol Renal Physiol 239(8):F412–F419

Beyenbach KW, Petzel DH, Cliff WH (1986) Renal proximal tubule of flounder. I. Physiological properties. Am J Physiol Regul Integr Comp Physiol 250:R608–R615

Biagi B, Kubota T, Sohtell M, Giebisch G (1981) Intracellular potentials in rabbit proximal tubules perfused in vitro. Am J Physiol Renal Physiol 240:F200–F210

Biemesderfer D, Payne JA, Lytle CY, Forbush B III (1996) Immunocytochemical studies of the Na-K-Cl cotransporter of shark kidney. Am J Physiol Renal Physiol 270:F927–F936

Bijvelds MJC, Kolar Z, Wendelaar Bonga SE, Flik G (1997) Mg^{2+} transport in plasma membrane vesicles of renal epithelium of Mozambique tilapia (*Oreochromis mossambicus*). J Cell Sci 110:1431–1440

Bijvelds MJC, Van der Velden JA, Kolar ZI, Flik G (1998) Magnesium transport in freshwater teleosts. J Exp Biol 201:1981–1990

Bindels RJM, Hoenderop JGJ, Biber J (2012) Transport of calcium, magnesium, and phosphate. In: Taal MW, Chertow GM, Marsden PA, Skorecki K, Yu ASL, Brenner BM (eds) Brenner and Rector's the kidney, 9th edn. Elsevier (Saunders), Philadelphia, pp 226–251

Blaine J, Chonchol M, Levi M (2015) Renal control of calcium, phosphate, and magnesium homeostasis. Clin J Am Soc Nephrol 10:1257–1272

Blazer-Yost B, Paunescu TC, Helman SI, Lee KD, Vlahos CJ (1999) Phosphoinositide 3-kinase is required for aldosterone-regulated sodium reabsorption. Am J Physiol Cell Physiol 277:C531–C536

Boron WF, Boulpaep EL (1983a) Intracellular pH regulation in the renal proximal tubule of the salamander: Basolateral HCO_3 transport. J Gen Physiol 81:53–94

Boron WF, Boulpaep EL (1983b) Intracellular pH regulation in the renal proximal tubule of the salamander: Na-H exchange. J Gen Physiol 81:29–52

Bott PA (1962) Micropuncture study of renal excretion of water, K, Na, and Cl in *Necturus*. Am J Physiol 203:662–666

Boulpaep EL (1972) Permeability changes of the proximal tubule of *Necturus* during saline loading. Am J Physiol 222:517–531

Boulpaep EL (1976) Electrical phenomena in the nephron. Kidney Int 9:88–102

Bradley EL, Holmes WN, Wright A (1971) The effects of neurohypophysectomy on the pattern of renal excretion in the duck (*Anas platyrhynchos*). J Endocrinol 51:57–65

Bradshaw FJ, Bradshaw SD (1996) Arginine vasotocin: locus of action along the nephron of the Ornate Dragon Lizard, *Ctenophorus ornatus*. Gen Comp Endocrinol 103:281–289

Bradshaw SD (1975) Osmoregulation and pituitary–adrenal function in desert reptiles. Gen Comp Endocrinol 25:230–248

Bradshaw SD, Rice GE (1981) The effects of pituitary and adrenal hormones on renal and postrenal reabsorption of water and electrolytes in the lizard, *Varanus gouldii* (gray). Gen Comp Endocrinol 44:82–93

Bradshaw SD, Shoemaker VH, Nagy KA (1972) The role of adrenal corticosteroids in the regulation of kidney function in the desert lizard *Dipsosaurus dorsalis*. Comp Biochem Physiol 43A:621–635

Braun EJ, Dantzler WH (1997) Vertebrate renal system. In: Dantzler WH (ed) Handbook of physiology: comparative physiology. Oxford University Press, New York, pp 481–576

Brazy PC, Dennis VW (1981) Sulfate transport in rabbit proximal convoluted tubules: presence of anion exchange. Am J Physiol Renal Physiol 241:F300–F307

Brett CL, Donowitz M, Rao R (2005) Evolutionary origins of eukaryotic sodium/proton exchangers. Am J Physiol Cell Physiol 288:C223–C229

Brewer KJ, Ensor DM (1980) Hormonal control of osmoregulation in the chelonia. I. the effects of prolactin and interrenal steroids in freshwater chelonians. Gen Comp Endocrinol 42:304–309

Brokl OH, Braun EJ, Dantzler WH (1994) Transport of PAH, urate, TEA, and fluid by isolated perfused and nonperfused avian renal proximal tubules. Am J Physiol Regul Integr Comp Physiol 266:R1085–R1094

Brokl OH, Martinez CL, Shuprisha A, Abbott DE, Dantzler WH (1998) Regulation of intracellular pH in proximal tubules of avian long-looped mammalian-type nephrons. Am J Physiol Regul Integr Comp Physiol 274:R1526–R1535

Brzica H, Breljak D, Krick W, Lovric M, Burckhardt G, Sabolic I (2009) The liver and kidney expression of sulfate anion transporter sat-1 in rats exhibits male-dominant gender differences. Pflugers Arch 457:1381–1392

Bulger RE, Trump BF (1968) Renal morphology of the English sole (*Parophrys vetulus*). Am J Anat 123:195–226

Burckhardt BC, Burckhardt G (2003) Transport of organic anions across the basolateral membrane of proximal tubule cells. Rev Physiol Biochem Pharmacol 146:95–158

Burg MB (1982) Thick ascending limb of Henle's loop. Kidney Int 22:454–464

Burg MB, Green N (1973) Function of thick ascending limb of Henle's loop. Am J Physiol 224:659–668

Butler DG (1972) Antidiuretic effect of arginine vasotocin in the Western painted turtle (*Chrysemys picta belli*). Gen Comp Endocrinol 18:121–125

Chan DKO, Chester Jones I (1969) Influences of the adrenal cortex and the corpuscles of stannius on osmoregulation in the European eel (*Anguilla anguilla* L.) adapted to freshwater. Gen Comp Endocrinol 2(Suppl):342–353

Chandra S, Morrison GH, Beyenbach KW (1997) Identification of Mg-transporting renal tubules and cells by ion microscopy imaging of stable isotopes. Am J Physiol Renal Physiol 273:F939–F948

Chang AC-M, Cha J, Koentgen F, Reddel RR (2005) The murine stanniocalcin 1 gene is not essential for growth and development. Mol Cell Biol 25:10604–10610

Chester Jones I, Chan DKO, Rankin JC (1969) Renal function in the european eel (*Anguilla anguilla* L.): changes in blood pressure and renal function of the freshwater eel transferred to sea-water. J Endocrinol 43:9–19

Cho KW, Kim SH, Koh GY, Seul KH (1988) Renal and hormonal responses to atrial natriuretic peptide and turtle atrial extract in the freshwater turtle, *Amyda japonica*. J Exp Zool 247:139–145

Churchill PC, Malvin RL, Churchill MC, McDonald FD (1979) Renal function in *Lophius piscatorius*: effects of angiotensin II. Am J Physiol Regul Integr Comp Physiol 236:R297–R301

Clark NB, Braun EJ, Wideman RF (1976) Parathyroid hormone and renal excretion of phosphate and calcium in normal starlings. Am J Physiol 231:1152–1158

Clark NB, Dantzler WH (1972) Renal tubular transport of calcium and phosphate in snakes: role of parathyroid hormone. Am J Physiol 223(6):1455–1464

Clark NB, Dantzler WH (1975) Renal tubular transport of calcium and phosphate in snakes: role of calcitonin. Gen Comp Endocrinol 26:321–326

Clark NB, Mok LLS (1986) Renal excretion in gull chicks: effect of parathyroid hormone and calcium loading. Am J Physiol Regul Intgr Comp Physiol 250:R41–R50

Clark NB, Sasayama Y (1981) The role of parathyroid hormone on renal excretion of calcium and phosphate in the Japanese quail. Gen Comp Endocrinol 45:234–241

Clark NB, Wideman RF Jr (1977) Renal excretion of phosphate and calcium in parathyroidecto-mized starlings. Am J Physiol Renal Physiol 233:F138–F144

Cliff WH, Beyenbach KW (1988) Fluid secretion in glomerular renal proximal tubules of freshwater-adapted fish. Am J Physiol Regul Integr Comp Physiol 254(1):R154–R158

Cliff WH, Beyenbach KW (1992) Secretory renal proximal tubules in seawater- and freshwater-adapted killifish. Am J Physiol Renal Physiol 262:F108–F116

Cliff WH, Sawyer DB, Beyenbach KW (1986) Renal proximal tubule of flounder II. Transepithelial Mg secretion. Am J Physiol Regul Integr Comp Physiol 250:R616–R624

Cobb CS, Brown JA (1992) Angiotensin II binding to tissues of the rainbow trout, *Oncorhyncus mykiss*, studied by auttoradiography. J Comp Physiol B 162:197–202

Cohen B, Giebisch G, Hansen LL, Teuscher U, Wiederholt M (1984) Relationship between peritubular membrane potential and net fluid reabsorption in the distal renal tubule of Amphiuma. J Physiol 348:115–134

Coulson RA, Hernandez T (1964) Biochemistry of the Alligator. Louisiana State University Press, Baton Rouge

Coulson RA, Hernandez T (1983) Alligator metabolism: studies on chemical reactions in vivo. Comp Biochem Physiol 74:1–182

Dantzler WH (1967) Glomerular and tubular effects of arginine vasotocin in water snakes (*Natrix sipedon*). Am J Physiol 212:83–91

Dantzler WH (1968) Effect of metabolic alkalosis and acidosis on tubular urate secretion in water snakes. Am J Physiol 215(3):747–751

Dantzler WH (1976) Renal function (with special emphasis on nitrogen excretion). In: Gans CG, Dawson WR (eds) Biology of Reptilia, vol 5, Physiology A. Academic, London, pp 447–503

Dantzler WH (1978) Urate excretion in nonmammalian vertebrates. In: Kelley WN, Weiner IM (eds) Uric acid, vol 51, Handbook of experimental pharmacology. Springer, Berlin, pp 185–210

Dantzler WH (1987) Comparative renal physiology. In: Gottschalk CW, Berliner RW, Giebisch G (eds) Renal physiology: people and ideas. American Physiological Society, Bethesda, MD, pp 437–481

Dantzler WH (1992) Comparative aspects of renal function. In: Seldin DW, Giebisch G (eds) The kidney: physiology and pathophysiology, 2nd edn. Raven, New York, pp 885–942

Dantzler WH, Bentley SK (1978) Fluid absorption with and without sodium in isolated perfused snake proximal tubules. Am J Physiol Renal Physiol 234:F68–F79

Dantzler WH, Bentley SK (1981) Effects of chloride substitutes on PAH transport by isolated perfused renal tubules. Am J Physiol Renal Physiol 241:F632–F644

Dantzler WH, Bradshaw SD (2009) Osmotic and ionic regulation in reptiles. In: Evans DH (ed) Osmotic and ionic regulation: cells and animals. CRC Press, Boca Raton, FL, pp 443–503

Dantzler WH, Brokl OH (1986) N^1-methylnicotinamide transport by isolated perfused snake proximal renal tubules. Am J Physiol Renal Physiol 250:F407–F418

De Falco M, Kaforgia V, Valiante S, Virgilio F, Varano L, DeLuca A (2002) Different patterns of expression of five neruopeptides in the adrenal gland and kidney of two species of frog. Histochem J 34:21–26

Deetjen P, Maren T (1974) The dissociation between renal HCO_3^- reabsorption and H^+ secretion in the skate, *Rajo erinacea*. Pflugers Arch 346:25–30

Delaney R, Stoner LC (1982) Miniature Ag-AgCl electrode for voltage clamping of the *Ambystoma* collecting duct. J Membr Biol 64:45–53

DiBona GF, Kopp UC (1997) Neural control of renal function. Physiol Rev 77:75–197

Dickman KG, Renfro JL (1986) Primary culture of flounder renal tubule cells: transepithelial transport. Am J Physiol 251:F424–F432

Dietl P, Oberleithner H (1987) Ca^{2+} transport in diluting segment of frog kidney. Pflugers Arch 410:63–68

Dietl P, Stanton BA (1993) The amphibian distal nephron. In: Brown JA, Balment RJ, Rankin JC (eds) New insights in vertebrate kidney function. Cambridge University Press, Cambridge, England, pp 115–134

Donald JA, Meier SK, Riddell SR (2002) Toad atrial natriuretic peptide: cDNA cloning and functional analysis in isolated perfused kidneys. Physiol Biochem Zool 75:617–626

Duff DW, Conklin DJ, Olson KR (1997) Effect of atrial natriuretic peptide on fluid volume and glomerular filtration in the rainbow trout. J Exp Zool 278:215–220

Elger M, Werner A, Herter P, Kohl B, Kinne RKH, Hentschel H (1998) Na-P$_i$ cotransport sites in proximal tubule and collecting tubule of winter flounder (*Pseudopleuronectes americanus*). Am J Physiol Renal Physiol 274:F374–F383

Elizondo RS, Lebrie SJ (1969) Adrenal-renal function in water snakes *Natrix cyclopion*. Am J Physiol 217:419–425

Fine LG, Trizna W (1977) Influence of prostaglandins on sodium transport of isolated medullary nephron segments. Am J Physiol Renal Physiol 232:F383–F390

Flik G, Klaren PHM, Schoenmakers TJM, Bijvelds MJC, Verbost PM, Wendelaar Bonga SE (1996) Cellular calcium transport in fish: unique and universal mechanisms. Physiol Zool 69:403–417

Foster RC (1975) Renal hydromineral balance in starry flounder, *Platichthys stellatus*. Comp Biochem Physiol A 55:135–140

Freire CA, Kinne RKH, Kinne-Saffran E, Beyenbach KW (1996) Electrodiffusive transport of Mg across renal membrane vesicles of the rainbow trout *Oncorhynchus mykiss*. Am J Physiol Renal Physiol 270:F739–F748

Friedman PA, Hebert SC (1990) Diluting segment in kidney of dogfish shark. I. Localization and characterization of chloride absorption. Am J Physiol Renal Physiol 258(2):R398–R408

Fujimoto M, Kotera K, Matsumara Y (1980) The direct measurement of K, Cl, Na, and H ions in bullfrog tubule cells. In: Boulpaep EL (ed) Current Topics in Membranes and Transport, vol 13, Cellular Mechanisms of Renal Tubular Ion Transport. Academic, New York, pp 49–61

Gallardo R, Pang PKT, Sawyer WH (1980) Neural influences on bullfrog renal functions. Proc Soc Exp Biol Med 165:233–240

Galli-Gallardo SM, Pang PKT (1978) Renal and vascular actions ofangiotensinsin amphibians. Physiologist 21:41 (abstract)

Garland HO, Henderson IW, Brown JA (1975) Micropuncture study of the renal responses of the urodele amphibian *Necturus maculosus* to injections of arginine vasotocin and an anti-aldosterone compound. J Exp Biol 63:249–264

Giebisch G (1956) Measurements of pH, chloride, and inulin concentrations in proximal tubule fluid of *Necturus*. Am J Physiol 185:171–174

Giebisch G (1961) Measurement of electrical potential difference on single nephrons of the perfused *Necturus* kidney. J Gen Physiol 44:659–768

Giebisch G (1975) Some reflections on the mechanism of renal tubular potassium transport. Yale J Biol Med 48:315–336

Giebisch G, Hunter M, Kawahara K (1990) Apical potassium channels in *Amphiuma* diluting segment effect of barium. J Physiol (Lond) 420:313–323

Giebisch G, Windhager EE (1964) Renal tubular transport of sodium, chloride, and potassium. Am J Med 36:643–669

Goldstein DL, Reddy V, Plaga K (1999) Second messenger production in avian medullary nephron segments in response to peptide hormones. Am J Physiol Regul Integr Comp Physiol 276: R847–R854

Gono O (1982) Uptake of ^{125}I-labelled prolactin by bullfrog tubules: an autoradiographic study. J Endocrinol 93:133–138

Graham C, Nalbant P, Schölermann B, Hentschel H, Kinne RKH, Werner A (2003) Characterization of a type IIb sodium-phosphate cotransporter from zebrafish (*Danio rerio*) kidney. Am J Physiol Renal Physiol 284:F727–F736

Grantham JJ, Wallace DP (2002) Return of the secretory kidney. Am J Physiol Renal Physiol 282: F1–F9

Gray DA (1993) Plasma atrial natriuretic factor concentrations and renal actions in the domestic fowl. J Comp Physiol B 163:519–523

Greger R (1985) Ion transport mechanisms in thick ascending limb of Henle's loop of mammalian nephron. Physiol Rev 65:760–797

Greger R, Lang FC, Knox FG, Lechene CP (1977) Absence of significant secretory flux of phosphate in the proximal convoluted tubule. Am J Physiol Renal Physiol 232:F235–F238

Greger R, Schlatter E (1983) Properties of the lumen membrane of the cortical thick ascending limb of Henle's loop of rabbit kidney. Pflugers Arch 396:315–324

Gregg CM, Wideman RF Jr (1986) Effects of atriopeptin and chicken heart extract in *Gallus domesticus*. Am J Physiol Regul Integr Comp Physiol 251:R543–R551

Gross JB, Imai M, Kokko JP (1975) A functional comparison of the cortical collecting tubule and the distal convoluted tubule. J Clin Invest 55:1284–1294

Guerreiro PM, Canario AVM, Power DM, Renfro JL (2004) Possible actions of PTHrP on calcium and phosphate transport by winter flounder (*Pseudopleuronectes americanus*) renal proximal tubule cells. In: Castaño JP, Malagón MM, Garcia Navarro S (eds) Avances en Endocrinologia Comparada. Universidad de Córdoba, Códoba, Spain, pp 113–117

Guerreiro PM, Renfro JL, Power DM, Canario AVM (2007) The parathyroid hormone family of peptides: structure, tissue distribution, regulation, and potential functional roles in calcium and phosphate balance in fish. Am J Physiol Regul Integr Comp Physiol 292:R679–R696

Guggino WB (1986) Functional heterogeneity in the early distal tubule of the *Amphiuma* kidney: evidence for two modes of Cl^- and K^+ transport across the basolateral cell membrane. Am J Physiol Renal Physiol 250:F430–F440

Guggino WB, Boulpaep EL, Giebisch G (1982a) Electrical properties of chloride transport across the *Necturus* proximal tubule. J Membr Biol 65:185–196

Guggino WB, London R, Boulpaep EL, Giebisch G (1983) Chloride transport across the basolateral cell membrane of the Necturus proximal tubule: dependence on bicarbonate and sodium. J Membr Biol 71:227–240

Guggino WB, Windhager EE, Boulpaep EL, Giebisch G (1982b) Cellular and paracellular resistances of the *Necturus* proximal tubule. J Membr Biol 67:143–154

Gupta A, Renfro JL (1989) Control of phosphate transport in flounder renal proximal tubule primary cultures. Am J Physiol Regul Integr Physiol 256:R850–R857

Gupta A, Renfro JL (1991) Effects of pH on phosphate tramsport by flounder renal tubule primary cultures. Am J Physiol Regul Integr Comp Physiol 260:R704–R711

Hall DA, Varney DM (1980) Effect of vasopressin on electrical potential difference and chloride transport in mouse medullary thick ascending limb of henle's loop. J Clin Invest 66:792–802

Harris PJ, Young JA (1977) Dose-dependent stimulation and inhibition of proximal tubular sodium reabsorption by angiotensin II in the rat kidney. Pflugers Arch 367:295–297

Harrison-Bernard LM, Navar LG, Ho MM, Vinson GP, el-Dahr S (1997) Immunohistochemical localization of ANG II AT_1 receptor in adult rat kidney using a monoclonal antibody. Am J Physiol Renal Physiol 273:F170–F177

Hass M, Forbush B III (1998) The Na-K-Cl cotransporters. J Bioenerg Biomembr 30:161–172

Hebert SC, Culpepper RM, Andreoli TE (1981a) NaCl transport in mouse medullary thick ascending limbs. I. Functional nephron heterogeneity and ADH-stimulated NaCl cotransport. Am J Physiol Renal Physiol 241:F412–F431

Hebert SC, Culpepper RM, Andreoli TE (1981b) NaCl transport in mouse medullary thick ascending limbs. II. ADH enhancement of transcellular NaCl cotransport; origin of transepithelial voltage. Am J Physiol Renal Physiol 241:F432–F442

Hebert SC, Friedman PA (1990) Diluting segment in kidney of dogfish shark. II. Electrophysiology of apical membranes and cellular resistances. Am J Physiol Regul Integr Comp Physiol 258:R409–R417

Henderson IW (1998) Endocrinology of the vertebrates. In: Dantzler WH (ed) Handbook of Physiology. Section 13: Comparative Physiology. American Physiological Society (Oxford University Press), New York p 623-749.

Henderson IW, Kime DE (1986) The adrenal cortical steroids. In: Pang PKT, Schreibman MP (eds) Vertebrate endocrinology: fundamental and biochemical implications. Academic, London, pp 121–142

Hentschel H, Elger M (1987) The distal nephron in the kidney of fishes. Adv Anat Embryol Cell Biol 108:1–151

Hentschel H, Zierold K (1994) Morphology and element distribution of magnesium-secreting epithelium: the proximal tubule segment PII of dogfish, *Scyliorhinus caniculus* (L.). Eur J Cell Biol 63:32–42

Hickman CP Jr, Trump BF (1969) The kidney. In: Hoar WS, Randall DJ (eds) Fish physiology, vol I, Excretion, ion regulation, and metabolism. Academic, New York, pp 91–239

Hillyard SD, Mobjerg N, Tanaka S, Larsen EH (2009) Osmotic and ion regulation in amphibians. In: Evans DH (ed) Osmotic and ionic regulation. Cells and animals. CRC Press, Boca Raton, FL, pp 367–441

Hodler J, Heinemann HO, Fishman AP, Smith HW (1955) Urine pH and carbonic anhydrase activity in marine dogfish. Am J Physiol 93:155–162

Holmes WN (1972) Regulation of electrolyte balance in marine birds with special reference to the role of the pituitary-adrenal axis in the duck (*Anas platyrhynchos*). Fed Proc 31:1587–1598

Holmes WN, Fletcher GL, Stewart DJ (1968) The patterns of renal electrolyte excretion in the duck (*Anas platyrhynchos*) maintained on freshwater and on hypertonic saline. J Exp Biol 48:487–508

Holmes WN, Phillips JG (1976) The adrenal cortex in birds. In: Chester Jones I, Henderson IW (eds) General, comparative and clinical endocrinology of the adrenal cortex, vol I. Academic, London, pp 293–420

Horisberger JD, Hunter M, Stanton BA, Giebisch G (1987) The collecting tubule of *Amphiuma*. II. Effects of potassium adaptation. Am J Physiol Renal Physiol 253:F1273–F1282

Horster M (1978) Loop of henle functional differentiation in vitro perfusion of the isolated thick ascending segment. Pflugers Arch 378:15–24

Hoshi T, Kuramochi G, Yoshitomi K (1983) Lumen-positive chloride transport in the early distal tubule of triturus kidney: its absolute dependence on the presence of Na^+ and K^+ in the luminal field. Jpn J Physiol 33:855–861

Hoshi T, Suzuki Y, Itoi K (1981) Differences in functional properties between the early and the late segments of the distal tubule of amphibian (*Triturus*) kidney. Jpn J Nephrol 23:889–896

Hunter M, Giebisch G (1988) Calcium-activated K-channels of *Amphiuma* early distal tubule: inhibition by ATP. Pflugers Arch 412:331–333

Hunter M, Horisberger JD, Stanton BA, Giebisch G (1987) The collecting tubule of *Amphiuma*. I. Electrophysiological characterization. Am J Physiol Renal Physiol 253:F1263–F1272

Hunter M, Oberleithner H, Henderson RM, Giebisch G (1988) Whole-cell potassium currents in single early distal tubule cells. Am J Physiol Renal Physiol 255:F699–F703

Hurst AM, Hunter M (1989) Apical K^+ channels of frog diluting segment: inhibition by acidification. Pflugers Arch 415:115–117

Hurst AM, Hunter M (1990) Acute changes in channel density of amphibian diluting segment. Am J Physiol Cell Physiol 259:C1005–C1009

Irish JM III, Dantzler WH (1976) PAH transport and fluid absorption by isolated perfused frog proximal renal tubules. Am J Physiol 230(6):1509–1516

Jard S, Morel F (1963) Actions of vasotocin and some of its analogues on salt and water excretion by the frog. Am J Physiol 204:222–226

Johns EJ, Kopp UC (2013) Neural control of renal function. In: Alpern RJ, Moe OW, Caplan M (eds) Seldin and Giebisch's the kidney: physiology and pathophysiology, 5th edn. Academic, San Diego, CA, pp 451–486

Johns EJ, Kopp UC, DiBona GF (2011) Neural control of renal function. Compr Physiol 1:731–767

Kashlan OB, Blobner BM, Zuzek Z, Tolino M, Kleyman TR (2015) Na^+ inhibits the epithelial Na^+ channel by binding to a site in an extracellular acidic cleft. J Biol Chem 290:568–576

Kato A, Muro T, Kimura Y, Li S, Islam Z, Ogoshi M, Doi H, Hirose S (2011) Differential expression of Na^+-Cl^- cotransporter and Na^+-K^+-Cl^- cotransporter 2 in the distal nephrons of euryhaline and seawater pufferfishes. Am J Physiol Regul Integr Comp Physiol 300:R284–R297

Kempton RT (1940) The site of acidification of urine within the renal tubule of the dogfish. Bull Mt Desert Island Biol Lab LAB 34–36

Kim YK, Brokl OH, Dantzler WH (1997) Regulation of intracellular pH in avian renal proximal tubules. Am J Physiol Regul Integr Comp Physiol 272:R341–R349

Kim YK, Dantzler WH (1995) Relation of membrane potential to basolateral TEA transport in isolated snake proximal renal tubules. Am J Physiol Regul Integr Comp Physiol 268:R1539–R1545

Kimura G, Spring KR (1978) Transcellular and paracellular tracer chloride fluxes in Necturus proximal tubule. Am J Physiol Renal Physiol 235:F617–F625

Kloas W, Hanke W (1992a) Atrial natriuretic factor (ANF) binding sites in frog kidney and adrenal. Peptides 13:297–303

Kloas W, Hanke W (1992b) Localisation and quantification of atrial natriuretic factor binding sites in the kidney of *Xenopus laevis*. Gen Comp Endocrinol 85:26–35

Kloas W, Hanke W (1993) Receptors for atrial natriuretic factor (ANF) in kidney and adrenal tissue of urodeles – lack of angiotensin II (AII) receptors in these tissues. Gen Comp Endocrinol 91:235–249

Knox FG, Haramati A (1985) Renal regulation of phosphate excretion. In: Seldin DW, Giebisch G (eds) The kidney: physiology and pathophysiology. Raven, New York, pp 1381–1396

Knox FG, Schneider EG, Willis LR, Strandhoy JW, Ott CE (1973) Site and control of phosphate reabsorption by the kidney. Kidney Int 3:347–353

Koeppen B, Giebisch G, Malnic G (1985) Mechanism and regulation of renal tubular acidification. In: Seldin DW, Giebisch G (eds) The kidney: physiology and pathophysiology. Raven, New York, pp 1491–1525

Koeppen BM, Biagi BA, Giebisch GH (1983) Intracellular microelectrode characterization of the rabbit cortical collecting duct. Am J Physiol Renal Physiol 244:F35–F47

Kohl B, Herter P, Hülseweh B, Elger M, Hentschel H, Kinne RKH, Werner A (1996) Na-P$_i$ cotransport in flounder: same transport system in kidney and intestine. Am J Physiol Renal Physiol 270:F937–F944

Kokko JP, Burg MB, Orloff J (1971) Characteristics of NaCl and water transport in the renal proximal tubule. J Clin Invest 50:69–76

Konno N, Hyodo S, Yamaguchi Y, Kaiya H, Miyazato M, Matsuda K, Uchiyama M (2009) African lungfish, *Protopterus annectens*, possess an arginine vasotocin receptor homologous to the tetrapod V2-type receptor. J Exp Biol 212:2183–2193

Lahlou B, Giordan A (1970) Le controle hormonal des echages et de la balance del'eau chez le teleostean d'eau douce *Carassius auratus*, intacte et hypophysectomise. Gen Comp Endocrinol 14:491–509

Lahlou B, Sawyer WH (1969) Electrolyte balance in hypophysectomized goldfish, *Carassius auratus* L. Gen Comp Endocrinol 12:370–377

Larsen EH, Deaton LE, Onken H, O'Donnell M, Grosell M, Dantzler WH, Weihrauch D (2014) Osmoregulation and excretion. Compr Physiol 4:405–573

Laverty G, Alberici M (1987) Micropuncture study of proximal tubule pH in avian kidney. Am J Physiol Renal Physiol 253:R587–R591

Laverty G, Clark NB (1981) Renal clearance of phosphate and calcium in the fresh-water turtle: effects of parathyroid hormone. J Comp Physiol 141:463–469

Laverty G, Dantzler WH (1982) Micropuncture of superficial nephrons in avian (*Sturnus vulgaris*) kidney. Am J Physiol Renal Physiol 243:F561–F569

Laverty G, Dantzler WH (1983) Effects of parathyroid hormone (PTH) on phosphate transport in superficial reptilian-type nephrons of the European starling. Fed Proc 42:478 (abstract)

Le Moellic C, Boulkroun S, Gonzalez-Nunez D, Dublineau I, Cluzeaud F, Fay M, Blot-Chabaud-M, Farman N (2005) Aldosterone and tight junctions: modulation of claudin-4 phosphorylation in renal collecting duct cells. Am J Physiol Cell Physiol 289:C1513–C1521

Lebrie SJ, Elizondo RS (1969) Saline loading and aldosterone in water snakes *Natrix cyclopion*. Am J Physiol 217:426–430

Lemieux G, Berkofsky J, Quenneville A, Lemieux C (1985) Net tubular secretion of bicarbonate by the alligator kidney. Antimammalian response to acetazolamide. Kidney Int 28:760–766

Levinsky NG, Davidson DG (1957) Renal action of parathyroid extract in the chicken. Am J Physiol 191:530–536

Li L, Wang YP, Capparelli AW, Jo OD, Yanagawa N (1994) Effect of luminal angiotesin II on proximal tubule fluid transport: role of apical phospholipase A2. Am J Physiol Renal Physiol 266:F202–F209

Logan AG, Moriarty RJ, Rankin JC (1980) A micropuncture study of kidney function in the river lamprey, *Lampetra fluviatilis*, adapted to fresh water. J Exp Biol 85:137–147

Long S, Giebisch G (1979) Comparative physiology of renal tubular transport mechanisms. Yale J Biol Med 52:525–544

Long WS (1973) Renal handling of urea in *Rana catesbeiana*. Am J Physiol 224:482–490

Lonnerholm G, Ridderstrale Y (1974) Distribution of carbonic anhydrase in the frog nephron. Acta Physiol Scand 90:764–778

Lu M, Barber LE, Renfro JL (1994a) Renal transepithelial phosphate secretion: luminal membrane voltage and Ca^{2+} dependence. Am J Physiol Regul Integr Comp Physiol 267:F624–F631

Lu M, Wagner GF, Renfro JL (1994b) Stanniocalcin stimulates phosphate reabsorption by flounder renal proximal tubule in primary culture. Am J Physiol Regul Integr Comp Physiol 267:R1356–R1362

Lucke H, Stange G, Murer H (1979) Sulfate ion/sodium co-transport by brush border membrane vesicles isolated from rat kidney cortex. Biochem J 182:223–229

Maren TH (1967) Carbonic anhydrase: chemistry, physiology, and inhibition. Physiol Rev 47:595–781

Martinez CL, Brokl OH, Shuprisha A, Abbott DE, Dantzler WH (1997) Regulation of intracellular pH in proximal tubules of avian loopless reptilian-type nephrons. Am J Physiol Regul Integr Comp Physiol 273:R1845–R1854

Meier SK, Toop T, Donald JA (1999) Distribution and characterisation of natriuretic peptide receptors in kidney of the toad, *Bufo marinus*. Gen Comp Endocrinol 115:244–253

Miwa T, Nishimura H (1986) Diluting segment in avian kidney. II. Water and chloride transport. Am J Physiol Regul Integr Comp Physiol 250:R341–R347

Mizgala CL, Quamme GA (1985) Renal handling of phosphate. Physiol Rev 65:431–466

Mobjerg N, Larsen EH, Jesperson A (1998) Morphology of the nephron in the mesonephros of *Bufo bufo* (Amphibia, Anura, Bufonidae). Acta Zool (Stockholm) 79:31–51

Mobjerg N, Larsen EH, Novak I (2002) K^+ transport in the mesonephric collecting duct system of the toad, *Bufo bufo*: microelectrode recordings from isolated and perfused tubules. J Exp Biol 205:897–904

Mok L (1978) The effect of parathyroid extract on the renal handling of ions in parathyroidectomized chickens. Am Zool 18:602 (abstract)

Montgomery H, Pierce JA (1937) The site of acidification of the urine within the renal tubule in amphibia. Am J Physiol 118:144–152

Moriarty RJ, Logan AG, Rankin JC (1978) Measurement of single nephron filtration rate in the kidney of the river lamprey, *Lampetra fluviatilis* L. J Exp Biol 77:57–69

Mount DB (2012) Transport of sodium, chloride, and potassium. In: Taal MW, Chertow GM, Marsden PA, Skorecki K, Yu ASL, Brenner BM (eds) Brenner and Rector's the kidney, 9th edn. Elsevier (Saunders), Philadelphia, pp 158–201

Mount DB (2014) Thick ascending limb of the loop of Henle. Clin J Am Soc Nephrol 9:1974–1986

Mudge GH, Berndt WO, Valtin H (1973) Tubular transport of urea, glucose, phosphate, uric acid, sulfate, and thiosulfate. In: Orloff J, Berliner RW (eds) Handbook of physiology: renal physiology. American Physiological Society, Washington, DC, pp 558–652

Mutig K, Paliege A, Kahl T, Jöns T, Müller-Esterl W, Bachmann S (2007) Vasopressin V_2 receptor expression along rat, mouse, and human renal epithelia with focus on TAL. Am J Physiol Renal Physiol 293:F1166–F1177

Nalbant P, Boehmer C, Dehmelt L, Wehner F, Werner A (1999) Functional characterization of a Na^+-phosphate cotransporter (NaP_i-II) from zebrafish and identification of related transcripts. J Physiol (Lond) 520(1):79–89

Natochin YV, Gusev GP (1970) The coupling of magnesium secretion and sodium reabsorption in the kidney of the teleost. Comp Biochem Physiol 37:107–111

Nielsen S, Maunsbach AB, Ecelbarger CA, Knepper MA (1998) Ultrastructural localization of Na-K-2Cl cotransporter in thick ascending limb and macula densa of rat kidney. Am J Physiol Renal Physiol 275:F885–F893

Nishimura H (1985) Endocrine control of renal handling of solutes and water in vertebrates. Renal Physiol (Basel) 8:279–300

Nishimura H (2001) Angiotensin receptors – evolutionary overview and perspectives. Comp Biochem Physiol A 128:11–30

Nishimura H, Imai M (1982) Control of renal function in freshwater and marine teleosts. Fed Proc 41:2355–2360

Nishimura H, Imai M, Ogawa M (1983a) Sodium chloride and water transport in the renal distal tubule of the rainbow trout. Am J Physiol Renal Physiol 244:F247–F254

Nishimura H, Imai M, Ogawa M (1983b) Transepithelial voltage in the reptilian- and mammalian-type nephrons from Japanese quail. Fed Proc 42:304 (Abstract)

Nishimura H, Imai M, Ogawa M (1986) Diluting segment in avian kidney. I. Characterization of transepithelial voltages. Am J Physiol Regul Integr Comp Physiol 250:R333–R340

Nishimura H, Miwa T, Bailey JR (1984) Renal handling of sodium chloride and its control in birds. J Exp Zool 232:697–705

Nishimura H, Sawyer WH (1976) Vasopressor, diuretic, and natriuretic responses to angiotensins by the american eel, *Anguilla rostrata*. Gen Comp Endocrinol 29:337–348

Nissenson RA, Nyiredy KO, Arnaud CD (1981) Guanyl nucleotide potentiation of parathyroid hormone-stimulated adenyl cyclase in chicken renal plasma membranes: a receptor independent effect. Endocrinology 108:1949–1953

O'Regan MG, Malnic G, Giebisch G (1982) Cell pH and luminal acidification in *Necturus* proximal tubule. J Membr Biol 69:99–106

Oberleithner H (1985) Intracellular pH in diluting segment of frog kidney. Pflugers Arch 404:244–251

Oberleithner H, Dietl P, Munich G, Weigt M, Schwab A (1985a) Relationship between luminal Na$^+$/H$^+$ exchange and luminal K$^+$ conductance in diluting segment of frog kidney. Pflugers Arch 405:S110–S114

Oberleithner H, Greger R, Neuman S, Lang F, Giebisch G, Deetjen P (1983a) Omission of luminal potassium reduces cellular chloride in early distal tubule of amphibian kidney. Pflugers Arch 398:18–22

Oberleithner H, Guggino W, Giebisch G (1982) Mechanism of distal tubular chloride transport in *Amphiuma* kidney. Am J Physiol Renal Physiol 242:F331–F339

Oberleithner H, Guggino W, Giebisch G (1983b) Potassium transport in the early distal tubule of the amphiuma kidney. Effects of potassium adaptation. Pflugers Arch 396:185–191

Oberleithner H, Guggino W, Giebisch G (1983c) The effect of furosemide on luminal sodium, chloride and potassium transport in the early distal of *Amphiuma* kidney. Pflugers Arch 396:27–33

Oberleithner H, Guggino WB, Giebisch G (1985b) Resistance properties of the diluting segment of *Amphiuma* kidney: influence of potassium adaptation. J Membr Biol 88:139–147

Oberleithner H, Kersting U, Hunter M (1988) Cytoplasmic pH determines K conductances in fused renal epithelial cells. Proc Natl Acad Sci U S A 85:8345–8349

Oberleithner H, Lang F, Greger R, Wang W, Giebisch G (1983d) Effect of luminal potassium on cellular sodium activity in the early distal tubule of *Amphiuma means*. Pflugers Arch 396:34–40

Oberleithner H, Lang F, Messner G, Wang W (1984) Mechanism of hydrogen ion transport in the diluting segment of frog kidney. Pflugers Arch 402:272–280

Oberleithner H, Weigt M, Westphale H-J, Wang W (1987) Aldosterone activates Na$^+$/H$^+$ exchange and raises cytoplasmic pH in target cells of the amphibian kidney. Proc Natl Acad Sci U S A 84:1464–1468

Oken DE, Solomon AK (1963) Single proximal tubules of necturus kidney. VI. Nature of potassium transport. Am J Physiol 204:377–380

Olsen HS, Rosen CA, Vozzolo BL, Jaworski EM, Wagner GF (1996) Human stanniocalcin, a possible hormonal regulator of mineral metabolism. Proc Natl Acad Sci U S A 93:1792–1796

Perrott MN, Sainsbury RJ, Balment RJ (1993) Peptide hormone-stimulated second messenger production in the teleostean nephron. Gen Comp Endocrinol 89:387–395

Pritchard JB, Renfro JL (1983) Renal sulfate transport at the basolateral membrane is mediated by anion exchange. Proc Natl Acad Sci U S A 80:2603–2607

Prunet P, Sturm A, Milla S (2006) Multiple corticosteroid receptors in fish: from old ideas to new concepts. Gen Comp Endocrinol 147:17–23

Quamme GA (1980) Effect of calcitonin on calcium and magnesium transport in rat nephron. Am J Physiol Endocrinol Metab 238:E573–E578

Quamme GA, Dirks JH (1985) Magnesium: cellular and renal exchanges. In: Seldin DW, Giebisch G (eds) The kidney: physiology and pathophysiology. Raven, New York, pp 1269–1280

Quan A, Baum M (1996) Endogenous production of angiotensin II modulates rat proximal tubule transport. J Clin Invest 97:2878–2882

Reinhart GA, Zehr JE (1994) Atrial natriuretic factor in the freshwater turtle *Pseudemys scripta*: a partial characterization. Gen Comp Endocrinol 96:259–269

Renfro JL (1978) Calcium transport across peritubular surface of the marine teleost renal tubule. Am J Physiol Renal Physiol 234:F522–F531

Renfro JL (1980) Relationship between renal fluid and Mg secretion in a glomerular marine teleost. Am J Physiol Renal Physiol 238:F92–F98

Renfro JL (1995) Solute transport by flounder renal cells in primary culture. In: Wood CM, Shutleworth TJ (eds) Cellular and molecular approaches to fish ionic regulation. Academic, New York, pp 147–171

Renfro JL (1999) Recent developments in teleost renal transport. J Exp Zool 283:653–661

Renfro JL, Clark NB (1984) Parathyroid hormone effect on chicken renal brush-border membrane phosphate transport. Am J Physiol Regul Integr Comp Physiol 247:R302–R307

Renfro JL, Dickman KG (1980) Sulfate transport across the peritubular surface of the marine teleost renal tubule. Am J Physiol Renal Physiol 239:F143–F148

Renfro JL, Dickman KG, Miller DS (1982) Effect of Na^+ and ATP on peritubular Ca transport by the marine teleost renal tubule. Am J Physiol Regul Integr Comp Physiol 243:R34–R41

Renfro JL, Maren TH, Patel C, Mills J, Swenson ER (1997) Sulfate secretion by flounder (*Pseudopleuronectes americanus*) renal epithelium is carbonic anhydrase dependent. Bull Mt Desert Biol Lab 36, 48–49 (Abstract)

Renfro JL, Pritchard JB (1982) H^+-dependent sulfate secretion in the marine teleost renal tubule. Am J Physiol Renal Physiol 243:F150–F159

Renfro JL, Pritchard JB (1983) Sulfate transport by flounder renal tubule brush border: presence of anion exchange. Am J Physiol Renal Physiol 244:F488–F498

Renfro JL, Shustock E (1985) Peritubular uptake and brush border transport of 28 Mg by flounder renal tubules. Am J Physiol Renal Physiol 249:F497–F506

Rice GE, Bradshaw SD, Prendergast FJ (1982) The effects of bilateral adrenalectomy on renal function in the lizard *Varanus gouldii* (gray). Gen Comp Endocrinol 47:182–189

Ridderstrale Y (1976) Intracellular localization of carbonic anhydrase in the frog nephron. Acta Physiol Scand 98:465–469

Roberts JR, Dantzler WH (1990) Micropuncture study of the avian kidney: infusion of mannitol or sodium chloride. Am J Physiol Regul Integr Comp Physiol 258:R869–R875

Roberts JR, Dantzler WH (1992) Micropuncture study of avian kidney: effect of prolactin. Am J Physiol Regul Integr Comp Physiol 262:R933–R937

Robinson RR, Portwood RM (1962) Mechanism of magnesium excretion by the chicken. Am J Physiol 202:309–312

Rocha AS, Kokko JP (1973) Sodium chloride and water transport in the medullary thick ascending limb of Henle. J Clin Invest 52:612–623

Rocha AS, Kudo LH (1982) Water, urea, sodium, chloride, and potassium transport in the in vitro isolated perfused papillary collecting duct. Kidney Int 22:485–491

Rosen S (1972) Localization of carbonic anhydrase activity in the vertebrate nephron. Histochem J 4:35–48

Roy A, Al-bataineh MM, Pastor-Soler NM (2015) Collecting duct intercalated cell function and regulation. Clin J Am Soc Nephrol 10:305–324

Sackin H, Boulpaep EL (1981a) Isolated perfused salamander proximal tubule. II. Monovalent ion replacement and rheogenic transport. Am J Physiol Renal Physiol 241:F540–F555

Sackin H, Boulpaep EL (1981b) Isolated perfused salamander proximal tubule: methods, electro-physiology, and transport. Am J Physiol Renal Physiol 241:F39–F52

Sackin H, Morgunov H, Boulpaep EL (1981) Intracellular microelectrode measurements in the diluting segment of the amphibian nephron. Fed Proc 40:395 (abstract)

Sakamoto T, McCormick SD (2006) Prolactin and growth hormone in fish osmoregulation. Gen Comp Endocrinol 147:24–30

Sandor T, Fazekas AG, Robinson BH (1976) The biosynthesis of corticosteroids throughout the vertebrates. In: Chester Jones I, Henderson I (eds) General, comparative and clinical endocri-nology. Academic, London, pp 25–142

Sasaki S, Imai M (1980) Effects of vasopressin on water and NaCl transport across the in vitro perfused medullary thick ascending limb of Henle's loop of mouse, rat, and rabbit kidneys. Pflugers Arch 383:215–221

Sawyer DB, Beyenbach KW (1985) Mechanism of fluid secretion in isolated shark renal proximal tubules. Am J Physiol Renal Physiol 249:F884–F890

Sawyer WH (1970) Vasopressor, diuretic, and natriuretic responses by lungfish to arginine vasotocin. Am J Physiol 218:1789–1794

Sawyer WH (1972) Neurohypophysial hormones and water and sodium excretion in the African lungfish. Gen Comp Endocrinol Suppl. 3:345–349

Schafer JA, Andreoli TE (1979) Perfusion of isolated mammalian renal tubules. In: Giebisch G (ed) Membrane transport in biology, vol IV, Transport organs. Springer, Berlin, pp 473–528

Schafer JA, Barfuss DW (1982) The study of pars recta function by the perfusion of isolated tubules. Kidney Int 22:434–448

Schafer JA, Patlak CS, Andreoli TE (1977) Fluid absorption and active and passive ion flows in the rabbit superficial pars recta. Am J Physiol Renal Physiol 233:F154–F167

Schmidt-Nielsen B, Renfro JL (1975) Kidney function of the american eel *Anguilla rostrata*. Am J Physiol 228:420–431

Schneider EG, Hanson RC, Childers JW, Fitzgerald EM, Gleason SD (1980) Is phosphate secreted by the kidney? In: Massry SG, Fleisch H (eds) Renal handling of phosphate. Plenum, New York, pp 59–78

Schütz H, Gray DA, Gerstberger R (1992) Modulation of kidney function in conscious Pekin duck by atrial natriuretic factor. Endocrinology 130:678–684

Seifter JL, Aronson PS (1984) Cl^- transport via anion exchange in *Necturus* renal microvillus membranes. Am J Physiol Renal Physiol 247:F888–F895

Shane MA, Nofziger C, Blazer-Yost B (2006) Hormonal regulation of the epithelial Na^+ channel: from amphibians to mammals. Gen Comp Endocrinol 147:85–92

Shareghi GR, Agus ZS (1982) Magnesium transport in the cortical thick ascending limb of Henle's loop of rabbit. J Clin Invest 69:759–769

Shareghi GR, Stoner LC (1978) Calcium transport across segments of the rabbit distal nephron in vitro. Am J Physiol Renal Physiol 235:F367–F375

Shoemaker VH (1967) Renal function in the mourning dove. Am Zool 7:736 (abstract)

Silbernagl S, Swenson ER, Maren TH (1986) Proximal tubule acidification in the isolated perfused kidney of the skate, *Raja erinacea*. Bull Mt Desert Isl Biol Lab 26:156–158

Skadhauge E (1973) Renal and cloacal salt and water transport in the fowl (*gallus domesticus*). Dan Med Bull 20:1–82

Skadhauge E, Schmidt-Nielsen B (1967) Renal function in domestic fowl. Am J Physiol 212:793–798

Smith WW (1939) The excretion of phosphate in the dogfish, *Squalus acanthias*. J Cell Comp Physiol 14:95–102

Spring KR, Kimura G (1978) Chloride reabsorption by renal proximal tubules of *Necturus*. J Membr Biol 38:233–254

Stallone JN, Nishimura H (1985) Angiotensin II (AH) – induced natriuresis in anesthetized domestic fowl. Fed Proc 44:1364 (abstract)

Stanton B, Biemesderfer D, Stetson D, Kashgarian M, Giebisch G (1984a) Cellular ultrastructure of *Amphiuma* distal nephron. Am J Physiol Cell Physiol 247:C204–C216

Stanton B, Biemesderfer D, Stetson D, Kashgarian M, Giebisch G (1984b) Cellular ultrastructure of *Amphiuma* distal nephron: effects of exposure to potassium. Am J Physiol Cell Physiol 247:C204–C216

Stanton B, Guggino W, Giebisch G (1982) Electrophysiology of isolated and perfused distal tubules of *Amphiuma*. Kidney Int 21:289 (abstract)

Stanton B, Omerovic A, Koeppen B, Giebisch G (1987a) Electroneutral H^+ secretion in distal tubule of *Amphiuma*. Am J Physiol Renal Physiol 252:F691–F699

Stanton BA (1988) Electroneutral NaCl transport by distal tubule: evidence for Na^+/H^+-$Cl^-/HCO3^-$ exchange. Am J Physiol Renal Physiol 254:F80–F86

Stanton BA (1990) Cellular actions of thiazide diuretics in the distal tubule. J Am Soc Nephrol 1:832–836

Stanton BA, Omerovic A, Koeppen BM, Giebisch G (1987b) Electroneutral H^+ secretion in distal tubule of *Amphiuma*. Am J Physiol Renal Physiol 247:F691–F699

Steinmetz PR (1967) Characteristics of hydrogen ion transport in urinary bladder of water turtle. J Clin Invest 46:1531–1540

Steinmetz PR (1974) Cellular mechanisms of urinary acidification. Physiol Rev 54:890–956

Steven K (1974) effect of peritubular infusion of angiotensin II on rat proximal nephron function. Kidney Int 6:73–80

Stiffler DF (1993) Amphibian calcium metabolism. J Exp Biol 184:47–61

Stokes JB (1979) Effect of prostaglandin E_2 on chloride transport across the rabbit thick ascending limb of henle. Selective inhibition of the medullary portion. J Clin Invest 64:495–502

Stokes JB (1981) Potassium secretion by cortical collecting tubule: relation to sodium absorption, luminal sodium concentration, and transepithelial voltage. Am J Physiol Renal Physiol 241: F395–F402

Stokes JB (1982) Na and K transport across the cortical and outer medullary collecting tubule of the rabbit: evidence for diffusion across the outer medullary portion. Am J Physiol Renal Physiol 242:F514–F520

Stokes JB, Tisher CC, Kokko JP (1978) Structural-functional heterogeneity along the rabbit collecting tubule. Kidney Int 14:585–593

Stolte H, Galaske RG, Eisenbach GM, Lechene C, Schmidt-Nielsen B, Boylan JW (1977a) Renal tubule ion transport and collecting duct function in the elasmobranch little skate, *Raja erinacea*. J Exp Zool 199:403–410

Stolte H, Schmidt-Nielsen B (1978) Comparative aspects of fluid and electrolyte regulation by cyclostome, elasmobranch, and lizard kidney. In: Barker Jorgenson C, Skadhauge E (eds) Osmotic and volume regulation. Alfred Benzon Symposium XI. Munksgaard, Copenhagen, pp 209–220

Stolte H, Schmidt-Nielsen B, Davis L (1977b) Single nephron function in the kidney of the lizard, *Sceloporus cyanogenys*. Zool Jb Physiol Bd 81:219–244

Stoner LC (1977) Isolated, perfused amphibian renal tubules: the diluting segment. Am J Physiol Renal Physiol 233:F438–F444

Stoner LC (1985) The movement of solutes and water across the vertebrate distal nephron. Renal Physiol 8:237–248

Subramanya AR, Ellison DH (2014) Distal convoluted tubule. Clin J Am Soc Nephrol 9:2147–2163

Sugiura SH, McDaniel NK, Ferraris RP (2003) In vivo fractional P_i absorption and NaP_i-II mRNA expression in rainbow trout are upregulated by dietary P restriction. Am J Physiol Renal Physiol 285:R770–R781

Sullivan LP, Welling DJ, Deeds DG, Simone JN (1977) Kinetic analysis of potassium transport in bullfrog kidney. Am J Physiol Renal Physiol 233:F464–F480

Sullivan WJ (1968) Electrical potential differences across distal renal tubules of amphiuma. Am J Physiol 214:1096–1103

Swenson ER, Maren TH (1986) Dissociation of CO_2 hydration and renal acid secretion in the dogfish, *Squalus acanthias*. Am J Physiol Renal Physiol 250:F288–F293

Takada M, Hokari S (2007) Prolactin increases Na^+ transport across bullfrog skin via stimulation of both ENaC and Na^+/K^+-pump. Gen Comp Endocrinol 151:325–331

Terreros DA, Tarr M, Grantham JJ (1981) Transmembrane electrical potential differences in cells of isolated renal tubules. Am J Physiol Renal Physiol 241:F61–F68

Tervonen V, Ruskoaho H, Lecklin T, Ilves M, Vuolteenaho O (2002) Salmon cardiac natriuretic peptide is a volume-regulating hormone. Am J Physiol Endocrinol Metab 283:E353–E361

Toop T, Donald JA (2004) Comparative aspects of natriuretic peptide physiology in non-mammalian vertebrates: a review. J Comp Physiol 174B:189–204

Tran U, Pickney LM, Özpolat D, Wessely O (2007) *Xenopus* Bicaudal-C is required for the differentiation of the amphibian pronephros. Dev Biol 307:152–164

Tseng D-Y, Chou M-Y, Tseng Y-C, Hsiao C-D, Huang C-J, Kaneko T, Hwang P-P (2009) Effects of stanniocalcin 1 on calcium uptake in zebrafish (*Danio rerio*) embryo. Am J Physiol Regul Integr Comp Physiol 296:R549–R557

Tsukada T, Takai Y (2006) Integrative approach to osmoregulatory action of atrial natriuretic peptide in seawater eels. Gen Comp Endocrinol 147:31–38

Uchiyama M, Konno N (2006) Hormonal regulation of ion and water transport in anuran amphibians. Gen Comp Endocrinol 147:54–61

Ullrich KJ, Rumrich G, Kloss S (1980) Active sulfate reabsorption in the proximal convolution of the rat kidney: specificity, Na$^+$ and HCO$_3^-$ dependence. Pflugers Arch 383:159–163

Vize PD (2003) The chloride conductance channel ClC-K is a specific marker for the *Xenopus* pronephric distal tubule and duct. Gene Expr Patterns 3:347–350

Wagner GF, Vozzolo BL, Jaworski E, Haddad M, Kline RL, Olsen HS, Rosen CA, Davidson MB, Renfro JL (1997) Human stanniocalcin inhibits renal phosphate excretion in the rat. J Bone Miner Res 12:165–171

Walker AM, Hudson CL (1937) The role of the tubule in the excretion of inorganic phosphates by the amphibian kidney. Am J Physiol 118:167–173

Walker AM, Hudson CL, Findley T, Richards AN (1937) The total molecular concentration and the chloride concentration of fluid from different segments of the renal tubule of Amphibia. The site of chloride reabsorption. Am J Physiol 118:121–129

Wallace DP, Rome LA, Sullivan LP, Grantham JJ (2001) cAMP-dependent fluid secretion in rat inner medullary collecting ducts. Am J Physiol Renal Physiol 280:F1019–F1029

Wang T, Agulian SK, Giebisch G, Aronson PS (1993) Effects of formate and oxalate on chloride absorption in rat distal tubule. Am J Physiol Renal Physiol 264:F730–F736

Wang T, Hropot M, Aronson PS, Giebisch G (2001) Role of NHE isoforms in mediating bicarbonate reabsorption along the nephron. Am J Physiol Renal Physiol 281:F1117–F1122

Wang W, Henderson RM, Geibel J, White S, Giebisch G (1989) Mechanism of aldosterone-induced increase of K$^+$ conductance in early distal renal tubule cells of the frog. J Membr Biol 111:277–289

Weigt M, Dietl P, Silbernagl S, Oberleithner H (1987) Activation of luminal Na$^+$/H$^+$ exchange in distal nephron of frog kidney. An early response to aldosterone. Pflugers Arch 408:609–614

Weinstein AM, Windhager EE (1985) Sodium transport along the proximal tubule. In: Seldin DW, Giebisch G (eds) The kidney: physiology and pathophysiology. Raven, New York, pp 1033–1062

Wendelaar Bonga SE (1997) The stress response in fish. Physiol Rev 77:591–625

Werner A, Murer H, Kinne RKH (1994) Cloning and expression of a renal Na-P$_i$ cotransport system from flounder. Am J Physiol Renal Physiol 267:F311–F317

White BA, Nicoll CS (1979) Prolactin receptors in *Rana catesbeiana* during development and metamorphosis. Science 204:851–853

Wideman RF, Braun EJ (1981) Stimulation of avian renal phosphate secretion by parathyroid hormone. Am J Physiol Renal Physiol 241:F263–F272

Wideman RF Jr, Clark NB, Braun EJ (1980) Effects of phosphate loading and parathyroid hormone on starling renal phosphate excretion. Am J Physiol Renal Physiol 239:F233–F243

Wiederholt M, Sullivan WJ, Giebisch G (1971) Potassium and sodium transport across single distal tubules of *Amphiuma*. J Gen Physiol 57:495–525

Wilkinson HL, Sullivan LP, Welling DJ, Welling LW (1983) Localization of tissue potassium pools in the amphibian kidney. Am J Physiol 245:F801–F812

Windhager EE, Whittembury G, Oken DE, Schatzmann HJ, Solomon AK (1959) Single proximal tubules of the necturus kidney. III. Dependence of H$_2$O movement on NaCl concentration. Am J Physiol 197:313–318

Wittner M, Mandon B, Roinel N, deRouffignac C, Di Stefano A (1993) Hormonal stimulation of Ca^{2+} and Mg^{2+} transport in the cortical thick ascending limb of Henle's loop of the mouse: evidence for a change in the paracellular pathway permeability. Pflugers Arch 423:387–396

Wright FS, Giebisch G (1985) Regulation of potassium excretion. In: Seldin DW, Giebisch G (eds) The kidney: physiology and pathophysiology. Raven, New York, pp 1223–1249

Yu ASL (2015) Claudins and the kidney. J Am Soc Nephrol 26:11–19

Zhou X, Vize PD (2004) Proximo-distal specialization of epithelial transport processes within the *Xenopus* pronephric kidney tubules. Dev Biol 271:322–338

Zucker A, Nishimura H (1981) Renal responses to vasoactive hormones in the aglomerular toadfish, opsanus tau. Gen Comp Endocrinol 43:1–9

Zuo Q, Claveau D, Hilal G, Leclerc M, Brunette MG (1997) Effect of calcitonin on calcium transport by the luminal and basolateral membranes of the rabbit nephrons. Kidney Int 51:1991–1999

Chapter 5
Transport of Fluid by Renal Tubules

Abstract This chapter initially considers net reabsorption of filtered fluid by the renal tubules. It describes the magnitude and tubule sites of net fluid reabsorption, indicating that only in mammals and birds are the renal tubules themselves capable of reabsorbing nearly all the filtered fluid. In other nonmammalian vertebrates and even in birds substantial fluid can be reabsorbed by structures distal to the kidneys (e.g., cloaca or bladder). Moreover, although more than half of the filtered fluid is reabsorbed along the proximal tubules of mammals and birds, much less is reabsorbed along the proximal tubules of other nonmammalian vertebrates. In some amphibian and reptile species, very large percentages of filtered fluid are reabsorbed along the late distal tubules, collecting tubules, and collecting ducts. The chapter then examines the mechanism involved in net fluid reabsorption, especially the mechanism by which apparently isosmotic fluid reabsorption can occur along proximal tubules. In general, water reabsorption depends on the reabsorption of sodium, but the detailed coupling mechanism is not completely understood, especially in the proximal tubule of nonmammalian vertebrates. Moreover, in at least one reptilian species fluid reabsorption in the proximal tubule can occur at control rates in the absence of sodium. Finally, the chapter examines the process of net fluid secretion by the renal tubules, most of which depends upon the net secretion of sodium and chloride described in Chap. 4.

Keywords Tubular fluid transport • Tubular fluid reabsorption • Fluid and ion transport coupling • Fluid transport sites • Tubular fluid secretion

5.1 Introduction

A large fraction of the fluid filtered by glomerular nephrons does not appear in the urine and, therefore, must be reabsorbed by the renal tubules. In addition, as already noted (vide supra; Chap. 3), the rate of fluid excretion in some nonmammalian vertebrates is determined, in part, by secretion by the renal tubules. The extent to which these processes have been studied or are understood in mammalian and nonmammalian vertebrates varies greatly, but some general patterns can be ascertained.

© The American Physiological Society 2016
W.H. Dantzler, *Comparative Physiology of the Vertebrate Kidney*,
DOI 10.1007/978-1-4939-3734-9_5

5.2 Fluid Reabsorption

5.2.1 Magnitude and Sites of Net Reabsorption

Although a large fraction of the filtered fluid is reabsorbed by the renal tubules of all tetrapods, only in mammalian and avian kidneys—kidneys capable of producing urine hyperosmotic to the plasma—are the tubules able to reabsorb nearly all the filtered fluid—more than 99 % during dehydration (Jamison and Kriz 1982; Skadhauge 1973). Moreover, among those vertebrates studied, only the proximal renal tubules of mammals and birds are capable of reabsorbing over half, 60–65 %, the filtered fluid (Table 5.1). About 15–45 % of the filtered fluid is reabsorbed along the proximal renal tubules of amphibians and reptiles (Table 5.1). However, a substantial fraction of the filtered fluid, as much as 25–45 %, can be reabsorbed along the late distal tubules and collecting ducts of these animals (Table 5.1). Significant fluid reabsorption, perhaps as much as 20 % of that filtered (Jamison and Kriz 1982; Skadhauge 1973), can also occur along the collecting ducts of mammals and birds, but the amount reabsorbed along these structures in mammals and birds and in some amphibians and reptiles depends on the requirements for water conservation and the presence of antidiuretic hormone (vide infra; Chap. 7). In addition, in amphibians, reptiles, and birds, substantial filtered fluid can be reabsorbed, depending on the requirements of the animals, by structures distal to the kidneys: colon, cloaca, or bladder.

Among fishes, fluid reabsorption by the renal tubules is even more variable and less well understood than among tetrapods. Approximately 40 % of the filtered fluid is reabsorbed by the nephrons of the primitive river lamprey (*L. fluviatilis*) in freshwater. However, only 10 % is reabsorbed along the proximal tubules, the remaining 30 % being reabsorbed along the collecting ducts (Table 5.1). An average of 40 % of the filtered fluid also is reabsorbed by the nephrons of freshwater teleosts, but this can vary from 25 % to 75 %; the sites of reabsorption along the renal tubules are unknown (Hickman and Trump 1969). It is difficult to determine the magnitude of net fluid reabsorption for marine elasmobranchs and teleosts and euryhaline teleosts adapted to seawater because of the net fluid secretion that also takes place (vide infra). However, the data suggest that the fraction of filtered water reabsorbed by the renal tubules can range from 9 % to 93 % (average, about 70–85 %) in marine elasmobranchs and can exceed 90 % in marine teleosts and euryhaline teleosts adapted to seawater (Hickman and Trump 1969). The major sites of reabsorption along the renal tubules in these marine fishes are unknown.

When the river lamprey is adapted to seawater to mimic its marine phase, almost 90 % of the filtered water is reabsorbed by the renal tubules (Logan et al. 1980a). As in the animals adapted to freshwater, most of this reabsorption appears to occur along the distal tubules and collecting ducts (Logan et al. 1980b). Finally, in the marine cyclostomes, e.g., the Atlantic hagfish (*M. glutinosa*), which conform to their environment, there is no net fluid reabsorption along the primitive archinephric ducts (Stolte and Schmidt-Nielsen 1978).

Table 5.1 Fluid transport by tubule segment

Tubule segment and species	J_v % filtered load	J_v nl min^{-1} mm^{-1}	Osmolal TF/P	References
PROXIMAL				
Fishes				
Petromyzonta				
River lamprey, *Lampetra fluviatilis* (freshwater)	10	0.2	1.0	Logan et al. (1980a), Moriarty et al. (1978)
Elasmobranchii				
Dogfish shark, *Squalus acanthias* (stimulated)		-27.6 ± 3.9 (21)		Beyenbach (1986), Sawyer and Beyenbach (1985)
Teleostei				
Winter flounder, *Pseudopleuronectes americanus*		-36.6 ± 4.2 (53)		Beyenbach (1986), Beyenbach et al. (1986)
Killifish, *Fundulus heteroclitus* (seawater)		-49.0 ± 8.7 (16)		Beyenbach (1986)
Amphibia				
Anura				
Bullfrog, *Rana catesbeiana*	17	0.34 ± 0.07 (29)	0.99 ± 0.02 (23)	Irish and Dantzler (1976), Long (1973)
Urodela				
Tiger salamander, *Ambystoma tigrinum*		0.26 ± 0.01 (14)		Sackin and Boulpaep (1981)
Mudpuppy, *Necturus maculosus*	29	0.39 ± 0.06 (47)	1.01 (22)	Boulpaep (1972), Garland et al. (1975)
Reptilia				
Squamata				
Ophidia				
Garter snake, *Thamnophis* spp.	45	0.87 ± 0.04 (127)	1.00 ± 0.01 (12)	Dantzler and Bentley (1978a)
Sauria				
Blue spiny lizard, *Sceloporus cyanogenys*	35	0.18	0.99	Stolte et al. (1977)
Aves				
Galliformes				
Domestic fowl, *Gallus gallus domesticus* Transitional nephrons, Proximal straight segment		2.01 ± 0.16 (26)		Brokl et al. (1994)
Passeriformes				
European starling, *Sturnus vulgaris* Reptilian-type nephrons	60	0.21		Laverty and Dantzler (1982)
Mammalia				
Rabbit, *Oryctolagus cuniculus*				
Convoluted segment	65	0.9–1.0	1.0	Jacobson (1982)
Straight segment	65	0.4–0.6	1.0	Schafer and Barfuss (1982)
EARLY DISTAL "DILUTING"				
Fishes				
Petromyzonta				
River lamprey, *Lampetra fluviatilis* (freshwater)	1	0.07	0.87	Logan et al. (1980a), Moriarty et al. (1978)

(continued)

Table 5.1 (continued)

Tubule segment and species	J_v % filtered load	J_v nl min^{-1} mm^{-1}	Osmolal TF/P	References
Amphibia				
Anura				
Leopard frog,		0.07 ± 0.09 (6)		Stoner (1977)
Rana pipiens				
Bullfrog,	1		0.75 ± 0.11 (11)	Long (1973)
Rana catesbeiana				
Urodela				
Mudpuppy,	0–1		0.48	Garland et al. (1975)
Necturus maculosus				
Congo eel,		0.07 ± 0.01 (12)		Oberleithner
Amphiuma means				et al. (1982)
Aves				
Galliformes				
Japanese quail,				
Coturnix coturnix				
Mammalian-type, TAL		-0.01 ± 0.04 (9)		Miwa and Nishimura (1986), Nishimura et al. (1984)
Mammalia				
Rabbit,				
Oryctolagus cuniculus				
Inner medullary ATL		~0		Imai and Kokko (1974)
Medullary TAL	0	~0		Rocha and Kokko (1973)
Cortical TAL	0–1	0.13 ± 0.02		Burg and Green (1973)
LATE DISTAL AND COLLECTING OR CONNECTING TUBULE				
Amphibia				
Urodela				
Tiger salamander,				
Ambystoma tigrinum				
(control)		0.01 ± 0.02 (7)		Stoner (1977)
(ADH)		0.02 ± 0.02 (7)		Stoner (1977)
Mudpuppy,	2–25		~0.25	Garland et al. (1975)
Necturus maculosus				
Congo eel,	25			Wiederholt et al. (1971)
Amphiuma means				
Reptilia				
Squamata				
Ophidia				
Garter snake,				
Thamnophis spp.				
(control)		0.07 ± 0.04 (14)		Beyenbach (1984)
(ADH)		0.08 ± 0.03 (14)		Beyenbach (1984)
Sauria				
Blue spiny lizard,	2.5	0.05	0.85	Stolte et al. (1977)
Sceloporus cyanogenys				
COLLECTING DUCT				
Fishes				
Petromyzonta				
River lamprey,	28.4	0.43	0.18	Logan et al. (1980a), Moriarty et al. (1978)
Lampetra fluviatilis				
(freshwater)				

(continued)

Table 5.1   (continued)

Tubule segment and species	J_v % filtered load	J_v nl min^{-1} mm^{-1}	Osmolal TF/P	References
Amphibia				
Anura				
Bullfrog, *Rana catesbeiana*	46.9			Long (1973)
Urodela				
Tiger salamander, *Ambystoma tigrinum*				
(control)		0.09 ± 0.09 (4)		Stoner (1977)
(ADH)		0.02 ± 0.05 (4)		Stoner (1977)
Reptilia				
Squamata				
Sauria				
Blue spiny lizard, *Sceloporus cyanogenys*	36.4			Stolte et al. (1977)
Mammalia				
Rabbit, *Oryctolagus cuniculus* Cortical				
No osmotic gradient		0.3		Grantham and Burg (1966)
200 mOsmol osmotic gradient				
−ADH		0.9		Grantham and Burg (1966)
+ADH		3.0		Grantham and Burg (1966)

Values are means, ranges, or means ± SE. They are taken directly or calculated from the cited references. Numbers in parentheses indicate number of determinations. J_v, net transepithelial movement of fluid. Minus sign in front of values indicates secretion rather than reabsorption. Osmolal TF/P, the ratio of osmolality in fluid collected from the tubule (TF) to osmolality in plasma (P). TAL, thick ascending limb of loop of Henle in mammalian nephrons or mammalian-type avian nephrons. ATL, ascending thin limb of loop of Henle in mammalian nephrons. ADH, antidiuretic hormone (arginine vasotocin in nonmammalian vertebrates and arginine vasopressin in mammals)

5.2.2   Mechanism and Control of Fluid Reabsorption

The process by which filtered fluid is reabsorbed by the proximal renal tubules, although now reasonably well understood for mammals (vide infra) (Schafer 2004), is still not well understood for nonmammalian vertebrates. Initially, micropuncture or microperfusion studies over a period of years on lampreys (Logan et al. 1980a), amphibians (Windhager et al. 1959), reptiles (Dantzler and Bentley 1978a; Stolte et al. 1977), birds (Laverty and Dantzler 1982), and mammals (Gottschalk and Mylle 1959; Walker et al. 1941) indicated that during proximal transepithelial fluid reabsorption the fluid in the tubule lumen remains isosmotic with the peritubular fluid. Thus, within the limits of the osmolality measurements, transepithelial fluid reabsorption appears to be an isosmotic process. Moreover, studies on urodele amphibians (*N. maculosus*) also demonstrated that fluid reabsorption, although apparently isosmotic, is secondary to and dependent upon sodium reabsorption (Windhager et al. 1959). How this coupling of solute and water movement occurs has been largely resolved for proximal tubules of mammals (Schafer 2004) but remains an unsolved problem for those of nonmammalian vertebrates.

First, in summary for mammals, the hydraulic conductivity of the proximal tubules is extremely high (Andreoli et al. 1978; Schafer et al. 1978). This is now known to be primarily due to the very high expression of the water channel aquaporin 1 (AQP1) in both the luminal and basolateral membranes (Agre et al. 1993). Although most of the reabsorbed water (at least 75 %) moves through the cells via the AQP1 channels, the pathway for the remaining amount is controversial. It appears likely that much of this water must move via a paracellular pathway, probably mediated by claudin 2, which is water permeable (Yu 2015). Second, this extremely high hydraulic conductivity indicates that a small transepithelial osmotic pressure difference (peritubular side > luminal side) of only 2–15 mOsmol/l is needed to account for proximal fluid reabsorption at the observed rates (Schafer 2004). Third, both in vivo and in vitro microperfusions of proximal tubules have now demonstrated that, despite the apparent isosmotic fluid reabsorption observed in earlier studies, the luminal fluid is actually diluted sufficiently by solute reabsorption to produce a transepithelial osmotic pressure gradient in the required range (Barfuss and Schafer 1984a, b; Green and Giebisch 1984; Schafer 1984, 2004). In the late mammalian proximal tubule, differences between the reflection coefficients for chloride and bicarbonate may provide sufficient force for fluid reabsorption.

As noted above, possible mechanisms for fluid reabsorption by the renal proximal tubules of nonmammalian vertebrates are not yet well defined. It seems likely that all proximal tubules in which fluid reabsorption appears to be isosmotic would have a very high hydraulic conductivity, but this has yet to be measured in the proximal tubules of nonmammalian vertebrates. Moreover, although homologues of various mammalian aquaporins have been cloned from fish, amphibians, and birds, including some homologues of AQP1, none has yet been localized to the luminal or basolateral membranes of the proximal tubules (Hillyard et al. 2009; Larsen et al. 2014; Nishimura and Fan 2003).

Nevertheless, there have been a few studies on amphibians and reptiles that provide some information on the proximal reabsorptive process in these animals. Analysis of the transport process in the proximal tubules of mudpuppies (*N. maculosus*) suggests that during net reabsorption, the reabsorbate, as in mammalian proximal tubules, may be slightly hyperosmotic even though the overall process is experimentally indistinguishable from isosmotic (Sackin and Boulpaep 1975). If this is the case, and it has yet to be demonstrated as directly as in mammals, it markedly increases the likelihood that these tubules would have a very high hydraulic conductivity. Certainly further studies on the possible dilution of the luminal fluid during proximal tubule reabsorption, the hydraulic conductivity of the proximal tubules, and the possible presence of amphibian aquaporins in the tubule membranes are warranted.

In additional studies, a quantitative structural evaluation of the proximal renal tubules of two urodele amphibian species (tiger salamander, *A. tigrinum*, and mudpuppy, *N. maculosus*) demonstrated that the basolateral membrane of the former has a more elaborate organization and much greater amplification than that of the latter (Maunsbach and Boulpaep 1984). Maunsbach and Boulpaep (1984) suggest that this elaboration of the basolateral membrane may form an additional compartment that plays some role in the reabsorptive process in this amphibian species. Furthermore, micropuncture studies on *N. maculosus* suggest that organic substrates such as lactate, alanine, glutamine, lysine, butyrate, and glucose in the proximal tubule lumen and all of them except glucose in the peritubular blood are important for fluid reabsorption (Forster et al. 1980). However, it is not clear which of these are most important or whether they act by enhancing luminal permeability for sodium, regulating metabolism, reducing passive paracellular backleak for sodium, or by some other process (Forster et al. 1980).

Among reptiles, in vivo micropuncture studies of lizard (*S. cyanogenys*) proximal tubules (Stolte et al. 1977) and in vitro microperfusion studies of garter snake (*Thamnophis* spp.) proximal tubules (Dantzler and Bentley 1978a) indicate that sodium and water can be reabsorbed at apparently osmotically equivalent rates. However, in contrast to studies on amphibians and mammals (vide supra), they do not prove that fluid reabsorption always must be dependent on sodium reabsorption. In fact, substitutions for sodium or chloride or both in solutions used for bathing and perfusing the snake proximal tubules in vitro suggest that neither one of these ions may be essential for normal fluid reabsorption (Dantzler and Bentley 1978a). When sodium in the perfusate is replaced with choline, net fluid reabsorption almost ceases (Fig. 5.1). However, when sodium in the bathing medium is also replaced with choline, so that both solutions are identical, net fluid reabsorption returns to the control rate (Fig. 5.1). The results are the same when sodium is replaced with tetramethylammonium, when sodium and the equivalent amount of chloride are replaced with sucrose, and when chloride alone is replaced with methyl sulfate (Fig. 5.1). However, net fluid reabsorption does not change from the control rate when lithium replaces sodium in the perfusate alone or in both the perfusate and bathing medium simultaneously (Fig. 5.1). Fluid reabsorption at the control rates, regardless of the composition of the perfusate and bathing medium, is isosmotic within the limits of the measurements and can be at least partly inhibited by cold and cyanide (Dantzler and Bentley 1978a). Thus, it appears that isosmotic fluid reabsorption can proceed at control rates when lithium replaces sodium or when some other substance replaces sodium or chloride or both in the perfusate and bathing medium simultaneously.

Even with sodium present, however, net fluid reabsorption by these snake tubules cannot be inhibited by ouabain or other cardiac glycosides or by the removal of potassium from the bathing medium (Dantzler and Bentley 1978a, b); it is also not dependent on the nature of the buffer (bicarbonate, phosphate, or Tris) used (Dantzler and Bentley 1978a). With sodium present, net fluid reabsorption is

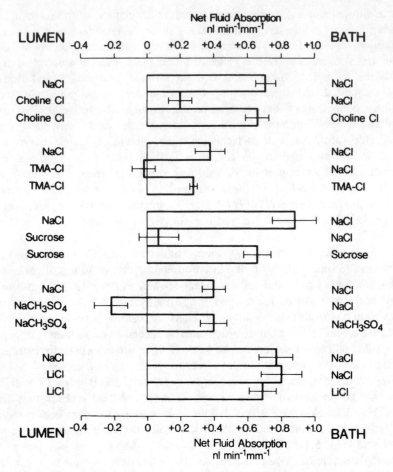

Fig. 5.1 Net fluid movement in isolated perfused snake proximal renal tubules. Composition of solution in lumen and bath in terms of sodium and chloride and of substitutes for them is shown at sides of figure for each experiment. Each bar represents mean net fluid movement with lumen and bath composition shown. Horizontal lines at end of each bar represent SE. From 6 to 13 tubules were used in each experiment (Dantzler 1978, with permission)

reduced to about 18–25 % by the removal of colloid from the peritubular fluid (Dantzler and Bentley 1978a).

Although these observations on snake proximal tubules suggest that isosmotic fluid reabsorption can occur in the absence of both sodium and chloride, they do not provide any information on the mechanism involved. However, quantitative structural studies on these isolated, perfused tubules (Dantzler et al. 1986) show that within a few minutes after substitution of choline for sodium in both the perfusate and bathing medium significant morphological changes take place (Fig. 5.2). Cells double in size and intercellular spaces nearly quintuple. At the same time, the areas of the lateral and apical cell membranes approximately double, but their surface

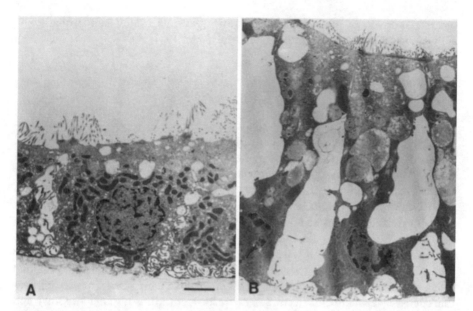

Fig. 5.2 Electron micrographs of cross section of isolated perfused proximal renal tubules of garter snake (*Thamnophis* sp.). (**a**) Proximal tubule perfused and bathed with medium containing sodium. (**b**) Proximal tubule perfused and bathed with medium in which sodium has been replaced with choline. Both tubules are shown at the same magnification. Note increase in cell height and widening of intercellular spaces in **b**, the tubule perfused and bathed with medium in which choline replaced sodium (Dantzler et al. 1986, with permission)

densities remain essentially constant. Therefore, the larger cells in the absence of sodium have proportionately larger surface areas but the volume-to-surface area ratio remains constant. The rapid increase in membrane area most likely involves incorporation of additional membrane from other areas, possibly from intracellular membrane vesicles, but the exact source is unknown. In any case, these changes are correlated with the maintenance of a control level of net fluid reabsorption and may permit a small, previously unimportant driving force (e.g., the osmotic pressure generated by colloid in the peritubular fluid) to produce a control level of net fluid reabsorption.

In general, the extent to which hydrostatic and colloid osmotic pressures in the peritubular capillaries or sinuses contribute to fluid reabsorption by the proximal tubules of nonmammalian or even mammalian vertebrates is unknown. However, these pressures, if they influence net fluid reabsorption significantly, could be particularly important in nonmammalian vertebrates because the renal venous portal system contributes to a variable extent to the peritubular blood supply. Alterations in the portal contribution could alter fluid reabsorption and, in this way, factors that regulate portal flow could also influence net fluid reabsorption by the proximal tubules. Net fluid reabsorption in the distal portions of the nephrons, although variable and regulated in many species by neurohypophysial peptides, clearly is driven by transepithelial osmotic gradients established primarily by sodium and chloride reabsorption (vide infra).

5.3 Fluid Secretion

As described in detail earlier (Chap. 3), net secretion of fluid by the proximal tubules has been demonstrated for numerous aglomerular and glomerular marine teleosts, marine elasmobranchs, euryhaline teleosts adapted to either seawater or freshwater, marine reptiles, and even mammals. The process is particularly important for marine aglomerular and glomerular teleosts. In vivo studies on marine teleosts originally suggested that such fluid secretion was dependent on magnesium and, to a lesser extent, sulfate and calcium secretion (Babiker and Rankin 1978; Berglund and Forster 1958; Bieter 1935; Hickman 1968; Hickman and Trump 1969; Renfro 1980). However, later studies with isolated, perfused proximal tubules from marine glomerular teleosts (winter flounder, *P. americanus*), euryhaline teleosts adapted to either seawater or freshwater (killifish, *F. heteroclitus*), and marine elasmobranchs (dogfish shark, *S. acanthias*) demonstrated that the dominant ions in the secreted fluid are sodium and chloride and that fluid secretion depends upon their secretion (Beyenbach 1982, 1986, 2004; Beyenbach et al. 1986; Cliff and Beyenbach 1988, 1992; Mandel 1986; Sawyer and Beyenbach 1985). The probable mechanism involved in the secretion of sodium and chloride by these tubule segments is described in detail above (Fig. 4.2). Proximal tubule secretion of magnesium and sulfate, the mechanism for which is also described above (Chap. 4), enhances fluid secretion, but the essential driving force remains the secretion of sodium and chloride (Beyenbach 1982, 2004; Cliff et al. 1986). The exact route for water movement across the proximal tubules during secretion in nonmammalian vertebrates is unknown, but is likely to be the same as the pathway during fluid reabsorption (vide supra).

Although small amounts of fluid can be secreted by mammalian proximal tubules, such secretion occurs secondary to the secretion of organic anions, physiologically probably hippurate, not inorganic cations, and is likely to be significant only during decreased glomerular function (Grantham and Wallace 2002). The mechanism involved in organic anion secretion is described in Chap. 6. The pathway for water movement during secretion in mammals is likely to be the same as that for water movement during reabsorption (vide supra).

Finally, as noted previously (vide supra; Chap. 4), small amounts of fluid can also be secreted by mammalian inner medullary collecting ducts. This fluid secretion, like fluid secretion by nonmammalian proximal tubules, depends upon the secretion of sodium and chloride for its driving force (Wallace et al. 2001). The mechanism for this sodium and chloride secretion in mammalian collecting ducts is also apparently the same as that found in proximal tubules of nonmammalian vertebrates (vide supra; Fig. 4.2).

References

Agre P, Preston GM, Smith BL, Jung JS, Raina S, Moon C, Guggino WB, Nielsen S (1993) Aquaporin CHIP: the archetypal molecular water channel. Am J Physiol Renal Physiol 265: F463–F476

Andreoli TE, Schafer JA, Troutman SL (1978) Perfusion rate-dependence of transepithelial osmosis in isolated proximal convoluted tubules: estimation of the hydraulic conductance. Kidney Int 14:263–269

Babiker MM, Rankin JC (1978) Neurohypophysial hormonal control of kidney function in the european eel (*Anguilla anguilla* L.) adapted to sea-water or fresh water. J Endocrinol 76:347–358

Barfuss DW, Schafer JA (1984a) Hyperosmolality of absorbate from isolated rabbit proximal tubules. Am J Physiol Renal Physiol 247:F130–F139

Barfuss DW, Schafer JA (1984b) Rate of formation and composition of absorbate from proximal nephron segments. Am J Physiol Renal Physiol 247:F117–F129

Berglund F, Forster RP (1958) Renal tubular transport of inorganic divalent ions by the aglomerular marine teleost, *Lophius americanus*. J Gen Physiol 41:249–440

Beyenbach KW (1982) Direct demonstration of fluid secretion by glomerular renal tubules in a marine teleost. Nature 299:54–56

Beyenbach KW (1984) Water-permeable and -impermeabile barriers of snake distal tubules. Am J Physiol Renal Physiol 246:F290–F299

Beyenbach KW (1986) Secretory NaCl and volume flow in renal tubules. Am J Physiol Regul Integr Comp Physiol 250:R753–R763

Beyenbach KW (2004) Kidneys sans glomeruli. Am J Physiol Renal Physiol 286:F811–F827

Beyenbach KW, Petzel DH, Cliff WH (1986) Renal proximal tubule of flounder. I. Physiological properties. Am J Physiol Regul Physiol 250:R608–R615

Bieter RM (1935) The action of diuretics injected into one kidney of the aglomerular toadfish. J Pharmacol Exp Ther 53:347–349

Boulpaep EL (1972) Permeability changes of the proximal tubule of *Necturus* during saline loading. Am J Physiol 222:517–531

Brokl OH, Braun EJ, Dantzler WH (1994) Transport of PAH, urate, TEA, and fluid by isolated perfused and nonperfused avian renal proximal tubules. Am J Physiol Regul Integr Comp Physiol 266:R1085–R1094

Burg MB, Green N (1973) Function of thick ascending limb of Henle's loop. Am J Physiol 224:659–668

Cliff WH, Beyenbach KW (1988) Fluid secretion in glomerular renal proximal tubules of freshwater-adapted fish. Am J Physiol Regul Integr Comp Physiol 254(1):R154–R158

Cliff WH, Beyenbach KW (1992) Secretory renal proximal tubules in seawater- and freshwater-adapted killifish. Am J Physiol Renal Physiol 262:F108–F116

Cliff WH, Sawyer DB, Beyenbach KW (1986) Renal proximal tubule of flounder. II. Transepithelial Mg secretion. Am J Physiol Regul Physiol 250:R616–R624

Dantzler WH (1978) Some renal glomerular and tubular mechanisms involved in osmotic and volume regulation in reptiles and birds. In: Barker Jorgenson C, Skadhauge E (eds) Osmotic and volume regulation, Alfred Benzon Symposium XI. Munksgaard, Copenhagen, pp 187–208

Dantzler WH, Bentley SK (1978a) Fluid absorption with and without sodium in isolated perfused snake proximal tubules. Am J Physiol Renal Physiol 234:F68–F79

Dantzler WH, Bentley SK (1978b) Lack of effect of potassium on fluid absorption in isolated, perfused snake proximal renal tubules. Renal Physiol 1:268–274

Dantzler WH, Brokl OH, Nagle RB, Welling DJ, Welling LW (1986) Morphological changes with Na^+-free fluid absorption in isolated perfused snake tubules. Am J Physiol Renal Physiol 251: F150–F155

Forster J, Steels PS, Boulpaep EL (1980) Organic substrate effects on and heterogeneity of necturus proximal tubule function. Kidney Int 17:479–490

Garland HO, Henderson IW, Brown JA (1975) Micropuncture study of the renal responses of the urodele amphibian *Necturus maculosus* to injections of arginine vasotocin and an anti-aldosterone compound. J Exp Biol 63:249–264

Gottschalk CW, Mylle M (1959) Micropuncture study of the mammalian urinary concentrating mechanism: evidence for the countercurrent hypothesis. Am J Physiol 196:927–936

Grantham JJ, Burg MB (1966) Effect of vasopressin and cyclic AMP on permeability of isolated collecting tubules. Am J Physiol 211:255–259

Grantham JJ, Wallace DP (2002) Return of the secretory kidney. Am J Physiol Renal Physiol 282: F1–F9

Green R, Giebisch G (1984) Luminal hypotonicity: a driving force for fluid absorption from the proximal tubule. Am J Physiol Renal Physiol 246:F167–F174

Hickman CP Jr (1968) Glomerular filtration and urine flow in the euryhaline southern flounder, *Paralichthys lethostigma*, in seawater. Can J Zool 46:427–437

Hickman CP, Trump BF (1969) The kidney. In: Hoar WS, Randall DJ (eds) Fish physiology, vol I, Excretion, ion regulation, and metabolism. Academic, New York, pp 91–239

Hillyard SD, Mobjerg N, Tanaka S, Larsen EH (2009) Osmotic and ion regulation in amphibians. In: Evans DH (ed) Osmotic and ionic regulation. Cells and animals. CRC Press, Boca Raton, FL, pp 367–441

Imai M, Kokko JP (1974) Sodium chloride, urea, and water transport in the thin ascending limb of henle. Generation of osmotic gradients by passive diffusion of solutes. J Clin Invest 53:393–402

Irish JM III, Dantzler WH (1976) PAH transport and fluid absorption by isolated perfused frog proximal renal tubules. Am J Physiol 230(6):1509–1516

Jacobson HR (1982) Transport characteristics of *in vitro* perfused proximal convoluted tubules. Kidney Int 22:425–433

Jamison RL, Kriz W (1982) Urinary concentrating mechanism. Oxford University Press, New York

Larsen EH, Deaton LE, Onken H, O'Donnell M, Grosell M, Dantzler WH, Weihrauch D (2014) Osmoregulation and excretion. Compr Physiol 4:405–573

Laverty G, Dantzler WH (1982) Micropuncture of superficial nephrons in avian (*Sturnus vulgaris*) kidney. Am J Physiol Renal Physiol 243:F561–F569

Logan AG, Moriarty RJ, Rankin JC (1980a) A micropuncture study of kidney function in the river lamprey, *Lampetra fluviatilis*, adapted to fresh water. J Exp Biol 85:137–147

Logan AG, Morris R, Rankin JC (1980b) A micropuncture study of kidney function in the river lamprey, *Lampetra fluviatilis*, adapted to sea water. J Exp Biol 88:239–247

Long WS (1973) Renal handling of urea in *Rana catesbeiana*. Am J Physiol 224:482–490

Mandel LJ (1986) Primary active sodium transport, oxygen consumption, and ATP: coupling and regulation. Kidney Int 29:3–9

Maunsbach AB, Boulpaep EL (1984) Quantitative ultrastructure and functional correlates in proximal tubule of *Ambystoma* and *Necturus*. Am J Physiol Renal Physiol 246:F710–F724

Miwa T, Nishimura H (1986) Diluting segment in avian kidney. II. Water and chloride transport. Am J Physiol Regul Integr Comp Physiol 250:R341–R347

Moriarty RJ, Logan AG, Rankin JC (1978) Measurement of single nephron filtration rate in the kidney of the river lamprey, *Lampetra fluviatilis* L. J Exp Biol 77:57–69

Nishimura H, Fan Z (2003) Regulation of water movement across vertebrate renal tubules. Comp Biochem Physiol A 136:479–498

Nishimura H, Miwa T, Bailey JR (1984) Renal handling of sodium chloride and its control in birds. J Exp Zool 232:697–705

Oberleithner H, Guggino W, Giebisch G (1982) Mechanism of distal tubular chloride transport in *Amphiuma* kidney. Am J Physiol Renal Physiol 242:F331–F339

Renfro JL (1980) Relationship between renal fluid and Mg secretion in a glomerular marine teleost. Am J Physiol Renal Physiol 238:F92–F98

Rocha AS, Kokko JP (1973) Sodium chloride and water transport in the medullary thick ascending limb of Henle. J Clin Invest 52:612–623

Sackin H, Boulpaep EL (1975) Models for coupling of salt and water transport. Proximal tubular reabsorption in Necturus kidney. J Gen Physiol 66:671–733

Sackin H, Boulpaep EL (1981) Isolated perfused salamander proximal tubule: methods, electrophysiology, and transport. Am J Physiol Renal Physiol 241:F39–F52

Sawyer DB, Beyenbach KW (1985) Mechanism of fluid secretion in isolated shark renal proximal tubules. Am J Physiol Renal Physiol 249:F884–F890

Schafer JA (1984) Mechanisms coupling the absorption of solutes and water in the proximal nephron. Kidney Int 25:708–716

Schafer JA (2004) Renal water reabsorption: a physiological retrospective in a molecular era. Kidney Int 91:S20–S27

Schafer JA, Barfuss DW (1982) The study of pars recta function by the perfusion of isolated tubules. Kidney Int 22:434–448

Schafer JA, Patlak CS, Troutman SL, Andreoli TE (1978) Volume absorption in the pars recta. II. Hydraulic conductivity coefficient. Am J Physiol Renal Physiol 234:F340–F348

Skadhauge E (1973) Renal and cloacal salt and water transport in the fowl (Gallus domesticus). Dan Med Bull 20:1–82

Stolte H, Schmidt-Nielsen B (1978) Comparative aspects of fluid and electrolyte regulation by cyclostome, elasmobranch, and lizard kidney. In: Barker Jorgenson C, Skadhauge E (eds) Osmotic and volume regulation. Alfred Benzon Symposium XI. Munksgaard, Copenhagen, pp 209–220

Stolte H, Schmidt-Nielsen B, Davis L (1977) Single nephron function in the kidney of the lizard, Sceloporus cyanogenys. Zool Jb Physiol Bd 81:219–244

Stoner LC (1977) Isolated, perfused amphibian renal tubules: the diluting segment. Am J Physiol Renal Physiol 233:F438–F444

Walker AM, Bott PA, Oliver J, MacDowell MC (1941) The collection and analysis of fluid from single nephrons of the mammalian kidney. Am J Physiol 134:580–595

Wallace DP, Rome LA, Sullivan LP, Grantham JJ (2001) cAMP-dependent fluid secretion in rat inner medullary collecting ducts. Am J Physiol Renal Physiol 280:F1019–F1029

Wiederholt M, Sullivan WJ, Giebisch G (1971) Potassium and sodium transport across single distal tubules of Amphiuma. J Gen Physiol 57:495–525

Windhager EE, Whittembury G, Oken DE, Schatzmann HJ, Solomon AK (1959) Single proximal tubules of the necturus kidney. III. Dependence of H_2O movement on NaCl concentration. Am J Physiol 197:313–318

Yu ASL (2015) Claudins and the kidney. J Am Soc Nephrol 26:11–19

Chapter 6
Transport of Organic Substances by Renal Tubules

Abstract This chapter considers tubular transport of glucose, bicarbonate, amino acids, urea, ammonia, organic acids (except amino acids, urate, and lactate), urate, lactate, and organic cations and bases. For each of these substances, it examines the direction or directions of transepithelial transport, the magnitude of such transport, the tubule sites of transport, and the cellular and molecular mechanisms of transport, so far as they are known for nonmammalian vertebrates. In so far as possible for each of these organic substances, the chapter describes and discusses the cellular and molecular characteristics of the transport steps at both the luminal and peritubular membranes of the renal tubule cells in the various tubule segments. The molecular characteristics of these steps are very poorly defined for the renal tubule cells of many nonmammalian vertebrates. However, they are defined for at least one of the cell membranes for a number of substances in a few nonmammalian vertebrates. Moreover, the physiological characteristic of the steps at both membranes of the renal tubule cells for a number substances are sufficiently well defined in many nonmammalian vertebrates to suggest that the molecular characteristics of the transporters are similar to those in mammals and to indicate avenues of future research.

Keywords Organic substances • Tubular transport • Tubular reabsorption • Tubular secretion • Tubule transport sites • Membrane transport • Molecular transporters

6.1 Introduction

The excretion of numerous organic substances is regulated by renal tubular transport. For a few of these substances, enough information is available on the transport processes in nonmammalian species that they can be discussed, compared, and contrasted. To a limited extent, comparisons and contrasts with transport processes in mammals are possible.

6.2   Glucose

6.2.1   Direction, Magnitude, and Sites of Transport

Glucose is freely filtered and almost completely reabsorbed by the glomerular nephrons of all those vertebrates studied (Table 6.1) (Dantzler 1981; Moe et al. 2012; Von Baeyer and Deetjen 1985). A low rate of net tubular secretion is observed in aglomerular teleosts (e.g., *Lophius americanus*) (Malvin et al. 1965). However, even aglomerular nephrons appear to have retained a reabsorptive process for glucose because the administration of phlorizin, an inhibitor of glucose reabsorption in glomerular nephrons, to these aglomerular teleosts produces an increase in the urine-to-plasma glucose concentration ratio (Malvin et al. 1965).

The proximal tubule has been identified as the primary site of net glucose reabsorption in amphibians by micropuncture (Walker and Hudson 1937a), in reptiles by perfusion of isolated renal tubules (Barfuss and Dantzler 1976), and in mammals by micropuncture (Walker et al. 1941) and by perfusion of isolated renal tubules (Tune and Burg 1971). Although some reabsorption may be possible in nephron segments distal to the proximal tubule in these vertebrates, this does not appear to occur under normal circumstances (Von Baeyer and Deetjen 1985). It is generally assumed that the proximal tubule is the primary site of glucose reabsorption in all other vertebrates. However, in elasmobranchs, distal nephron sites may also contribute to reabsorption (vide infra).

In garter snakes (*Thamnophis* spp.), the maximum rate of net tubular reabsorption is twice as great in the distal portion of the proximal tubule as in the proximal portion (Barfuss and Dantzler 1976), a pattern that is almost exactly the opposite of that in mammals (Moe et al. 2012; Tune and Burg 1971). However, no studies of intranephron heterogeneity have been made in other nonmammalian vertebrates to determine if the pattern of glucose transport is similar to that in either reptiles or mammals.

Finally, in birds, the overall renal tubular reabsorptive process for glucose is particularly interesting. Birds maintain a normal plasma glucose concentration over twice as great as mammals of comparable body mass (Braun and Sweazea 2008). Since birds and mammals of similar body mass have about equal whole-kidney glomerular filtration rates (Yokota et al. 1985), the filtered load of glucose must be twice as great in birds as in mammals of similar body mass. Nevertheless, essentially all this filtered load is reabsorbed by the renal tubules; none appears in the ureteral urine (Braun and Sweazea 2008). Thus, the avian renal tubules apparently have a very high capacity for reabsorbing glucose. Indeed, preliminary studies on domestic fowl indicate that the tubular maximum for glucose reabsorption in the avian kidney (calculated on the basis of renal mass) is about 4–5 times that in the human kidney (Morgan and Braun 2001).

Table 6.1 Normal direction of net tubular transport of some organic substances observed with whole-kidney clearance studies

Substance	Fishes		Amphibia		Reptilia		Aves		Mammalia	
	Net Reabsorption	Net Secretion	Net Reabsorption	Net Secretion	Net Reabsorption	Net Secretion	Net Reabsorption	Net Secretion	Net Reabsorption	Net Secretion
Glucose	+	+	+		+		+		+	
Bicarbonate	+		+		+		+		+	
Amino acids	+	+	+		+		+		+	
Urea	+					+		+?		+?
Ammonia		+		+		+		+		+
Organic acids and anions (except amino acids, urate, and lactate)		+	+	+		+		+		+
Urate				+	+	+		+	+	+
Lactate					+				+	
Organic bases and cations						+	+	+		+

6.2.2   Mechanism of Transport

6.2.2.1   Net Reabsorption

The mechanism for transepithelial glucose reabsorption by mammalian proximal tubules has now been well characterized and is summarized briefly here (Moe et al. 2012). Glucose enters the cells across the luminal membrane via secondary-active, carrier-mediated sodium-glucose cotransport in which glucose is moved against its chemical gradient by being coupled to sodium movement down its electrochemical gradient. Glucose then moves out of the cells at the basolateral membrane down its chemical gradient via carrier-mediated diffusion. The trans-porters involved in this process have now been identified and cloned. In the early proximal tubule, the sodium-glucose cotransporter at the luminal membrane is sodium-glucose-linked transporter 2 (SGLT2). This is a high-capacity, low-affinity transporter, well suited to transporting large quantities of glucose in this early part of the tubule. It has a sodium-to-glucose stoichiometry of 1:1. The transporter involved in carrier-mediated diffusion of glucose across the basolateral membrane in this tubule segment is a member of a large family of glucose transporters (GLUT2). This, too, is a low-affinity transporter.

In the late proximal tubule, the sodium-glucose cotransporter at the luminal membrane is sodium-glucose-linked transporter 1 (SGLT1). This is a high-affinity, low-capacity transporter, well suited to transporting the small quantities of glucose remaining in this region against a steep gradient. This is also reflected in its sodium-to-glucose stoichiometry of 2:1. The transporter involved in facilitated diffusion of glucose out of the cells at the basolateral membrane in this segment is GLUT1, a glucose transporter found in almost all cells. Now that these basic membrane transport steps are characterized, a great deal of work is being done on the detailed physical and molecular structure of SGLTs and its relationship to the kinetics of transport (see, e.g., Sala-Rabanal et al. 2012), but such detailed studies are beyond the scope of this discussion on mammals.

As in mammals, the transepithelial reabsorptive transport process in all nonmammalian vertebrate species studied is saturable and sodium-dependent (Barfuss and Dantzler 1976; Coulson and Hernandez 1964; Frazier and Vanatta 1972; Khuri et al. 1966; Sperber 1960; Vogel et al. 1965; Vogel and Stoeckert 1966). However, the details of the steps in transepithelial transport are much less well understood in nonmammalian vertebrates than in mammals.

Information is most complete for the reabsorptive transport step across the luminal membrane in fishes. Initially, studies with brush-border membrane vesicles (BBMV) from proximal tubules of cyclostomes (hagfish, *Myxine glutinosa*), elas-mobranchs (skate, *Raja erinacea*; dogfish shark, *Squalus acanthias*), aglomerular teleosts (toadfish, *Opsanus tau*), and glomerular teleosts (winter flounder, *P. americanus*; rainbow trout, *O. mykiss*) all indicated that, as in mammals, glucose crosses the luminal membrane against its concentration gradient by a secondary active, phlorizin-sensitive cotransport step with sodium (Fig. 6.1) (Eveloff

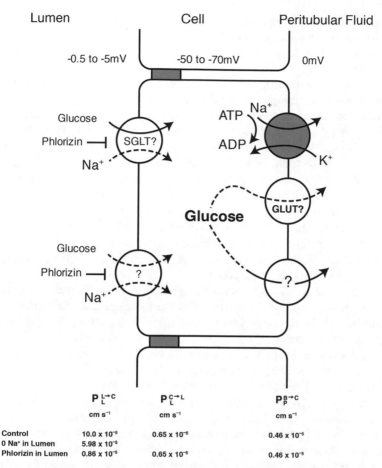

Fig. 6.1 Model for tubular reabsorption of glucose based on studies with fishes, amphibians, and reptiles. *Filled circle with solid arrows* and breakdown of ATP to ADP indicates primary active transport, in this case, Na-K-ATPase. *Open circles* indicate carrier-mediated diffusion or secondary active transport. *Broken arrows* indicate movement down an electrochemical gradient. *Solid arrows* indicate movement against an electrochemical gradient. *Question marks* indicate uncertainty about nature of transport steps. *Lines with bar* at end indicate inhibition. Apparent permeabilities of luminal membrane in lumen-to-cell ($P_L^{L \to C}$) and cell-to-lumen ($P_L^{C \to L}$) directions and of peritubular membrane in bath-to-cell ($P_P^{B \to C}$) direction are shown for various conditions for snake (*Thamnophis* spp.) renal tubules. See text for more detailed description of the processes diagrammed in the figure

et al. 1980; Flöge et al. 1984; Freire et al. 1995; Kipp et al. 1997; Wolff et al. 1987). The BBMV studies on elasmobranchs indicated that in skates, the transporter has a high affinity for glucose with an apparent coupling ratio of sodium to glucose of 2:1, whereas in dogfish sharks the transporter has a much lower affinity for glucose with a coupling ratio of about 1:1 (Kipp et al. 1997). In agreement with these BBMV studies on affinities and coupling ratios, molecular identification and cloning of the sodium-glucose cotransporters from the kidneys of these elasmo-branchs indicate that the only one in skates is a homologue of mammalian SGLT1 and the only one in dogfish sharks is a homologue of mammalian SGLT2 (Althoff et al. 2006, 2007).

In skates, the SGLT1 homologue is localized by immunohistochemistry to the second segment of the proximal tubule, somewhat analogous to its position in mammals, but also to an equal extent to the early distal tubule, collecting tubule, and collecting duct, in contrast to its position in mammals (Althoff et al. 2007). These sites suggest that the final, remaining glucose can still be reabsorbed at the end of the skate nephron. In dogfish sharks, the SGLT2 homologue is localized to both the first and second segments of the proximal tubule and to a lesser extent to the collecting tubule and collecting duct (Althoff et al. 2006). These sites differ to some extent from the sites in both mammals and skates. Again, the localization to the collecting tubule and collecting duct suggests that some glucose can be con-served in the terminal part of the nephrons in dogfish sharks. However, the low glucose affinity of this SGLT2 transporter certainly reduces its effectiveness in conserving small amounts of glucose.

It seems most likely that reabsorbed glucose in elasmobranchs exits the cells across the basolateral membrane down its chemical gradient via carrier-mediated diffusion (Fig. 6.1). By analogy to the findings with the transporters on the luminal membrane, these basolateral transporters are likely homologues of the mammalian GLUTs, but this has yet to be determined. In view of the BBMV studies, it appears likely that homologues of the transporters found in mammalian kidneys are also involved in glucose reabsorption in the kidneys of cyclostomes and teleosts, but no studies have yet been performed to clone the glucose transporters in these animals.

In amphibians (newt, *Triturus pyrrhogaster*), studies employing intracellular and transepithelial potential measurements on microperfused proximal tubules in situ indicate that, as in mammals and fishes, glucose is transported from the tubule lumen into the cells against its chemical gradient via a saturable, phlorizin-sensitive, electrogenic, secondary active cotransport step with sodium (Maruyama and Hoshi 1972) (Fig. 6.1). Although no molecular studies have yet been done with regard to this transport step in amphibians, it again seems likely that it involves a homologue of one of the mammalian SGLTs.

In reptiles (garter snakes, *Thamnophis* spp.), studies with BBMV (Benyajati and Dantzler 1988) and with isolated, perfused proximal tubules (Barfuss and Dantzler 1976) indicate that, as in mammals, fishes, and amphibians, glucose is transported into the cells across the luminal membrane by a saturable, phlorizin-sensitive, cotransport step with sodium. However, whereas the data from the BBMV studies indicate that, as in mammals, fishes, and amphibians, glucose is transported against its chemical gradient (Benyajati and Dantzler 1988), those from the perfused tubule studies indicate that glucose enters the cells down its chemical gradient (Barfuss and Dantzler 1976). Whether this difference between the isolated, perfused tubules and the BBMV from the same species reflects a purely technical difference between the types of studies or whether it reflects a difference between what can be demonstrated in an isolated membrane and what actually occurs in an intact asymmetrical epithelium containing cytosol and two membranes in series is not yet known. It may be, since the intracellular glucose concentration in the isolated, perfused snake tubules was measured only with sufficient glucose in the lumen to saturate the transport system, that normal transport against a concentration gradient when the luminal concentration is low was obscured (Barfuss and Dantzler 1976). In any case, it is not possible to decide at present whether the mediated entry step for glucose across the luminal membrane is always a form of secondary active transport or whether it may be a form of facilitated diffusion in some nonmammalian species (Fig. 6.1)

Moreover, glucose transport out of the cells at the peritubular side of these isolated, perfused snake tubules appears to be against a concentration gradient (Barfuss and Dantzler 1976). When glucose entry across the luminal membrane of these tubules is blocked with phlorizin, this gradient increases because the concentration of glucose within the cells decreases. In addition, the apparent passive permeability of the peritubular membrane of these snake tubules to glucose is too low to permit glucose reabsorption across this membrane, even if down a chemical gradient, at the observed rates by simple passive diffusion (Barfuss and Dantzler 1976). Therefore, glucose transport out of the cells at the peritubular membrane, although it may differ among vertebrate classes and perhaps species, at least always involves a carrier-mediated step (Fig. 6.1).

However, in view of the above BBMV data, not only for these reptiles but also for mammals and fishes and the intact tubule data for amphibians, it seems most likely that, as in other vertebrates studied, glucose enters the proximal tubule cells of most reptile species against its chemical gradient via a sodium-coupled secondary active transport system and leaves the cells at the peritubular side down its chemical gradient via a carrier-mediated process (Fig. 6.1). If this is the case, it also seems most likely that the luminal step involves a homologue of one of the mammalian SGLTs, and the basolateral step involves a homologue of one of the mammal GLUTs, but no molecular studies on this process have been performed on reptiles.

In view of the capacity of the renal tubular transport process in birds, discussed above, the steps in the process are particularly intriguing. However, there is as yet very little information available on this process. As in other vertebrates, BBMV studies indicate that glucose is transported across the luminal membrane of the proximal tubules by an electrogenic secondary active cotransport step with sodium (Villalobos and Braun 1995). Although these data suggest that this cotransporter is probably a homologue of the mammalian SGLTs, the actual transporter has yet to be identified and cloned. However, GLUTs 1, 2, and 3, which could be involved in glucose transport across the basolateral membrane, have been identified in avian kidneys (Kono et al. 2005; Sweazea and Braun 2006). Unfortunately, the cellular localization of these proteins has yet to be determined.

Clearly, much more work needs to be done to identify, clone, and determine the structure of the renal glucose transporters in all nonmammalian vertebrate classes for comparison with each other and with those of mammals. Eventually, the relationship of the kinetics of transport to the structure of the transporters relative to that now being explored in mammals will be of great importance.

6.2.2.2   Net Secretion

The small secretory flux observed in aglomerular teleosts could be explained by some form of passive process because the concentration in the urine is always far below that in the plasma even when reabsorption is blocked by phlorizin (Malvin et al. 1965). However, the steps in transport across the tubule epithelium have yet to be examined directly.

6.3   Bicarbonate

6.3.1   Direction, Magnitude, and Sites of Transport

Bicarbonate transport by mammalian nephrons, which has been studied extensively, is still the subject of intensive investigation. However, very little information is available on the transport of bicarbonate by the renal tubules of nonmammalian vertebrates or on the role of such transport in acid–base balance. As noted above (Chap. 4), in many nonmammalian vertebrates, extrarenal structures, such as the gills in teleost fishes and the bladder in chelonian reptiles, which have been intensively studied, are more important than the renal tubules in the removal of acid or alkali. For example, it is now well documented that the turtle bladder can secrete bicarbonate as well as hydrogen ions (Husted et al. 1979; Leslie et al. 1973). The tubular transport of bicarbonate and its regulation also may be related to the excretion of the end products of nitrogen metabolism and to the function of the extrarenal routes for ion excretion.

Most of the major aspects of bicarbonate transport have been covered in the discussion of hydrogen ion transport (Chap. 4), for there have been only a few studies in which bicarbonate excretion or the possible tubular transport of bicarbonate has actually been measured in nonmammalian vertebrates. As already noted (Chap. 4), mammals, birds, and amphibians are capable of producing ureteral urine substantially more acidic (ca. pH 4.5) than the initial filtered plasma (pH ca. 7.4) during an acid load, indicating indirectly that under these circumstances the filtered bicarbonate can be essentially completely reabsorbed by the renal tubules. Elasmobranch renal tubules also normally reabsorb essentially all the filtered bicarbonate (Hodler et al. 1955).

Although some reptiles (e.g., water snakes, N. sipedon) acidify ureteral urine to a lesser extent than mammals, birds, and amphibians during an acid load, the decrease in pH that does occur still indicates a substantial reabsorption of the filtered bicarbonate (Dantzler 1968, 1976b). Moreover, as already noted (Chap. 4), the alkaline urine resulting from the administration of carbonic anhydrase inhibitors to these water snakes supports the concept that bicarbonate reabsorption occurs and that it involves a process similar to that observed in mammals and amphibians (Dantzler 1968). Among crocodilian reptiles, however, as noted previously (Chap. 4), clearance studies have revealed net tubular secretion of bicarbonate that can be converted to net tubular reabsorption by the administration of carbonic anhydrase inhibitors (Lemieux et al. 1985).

The tubule sites of net bicarbonate transport, although well studied in mammals, are very poorly evaluated in nonmammalian vertebrates. In mammals, about 80 % of the filtered bicarbonate is normally reabsorbed by the proximal tubules, about 15 % by the thick ascending limbs of Henle's loops, and the remaining amount (about 5 %) by the collecting ducts (Weiner and Verlander 2012). In elasmobranchs, all filtered bicarbonate appears to be reabsorbed along the proximal tubules because the pH of the tubular fluid does not decrease beyond this point, and essentially no filtered bicarbonate appears in the final urine (Deetjen and Maren 1974; Hodler et al. 1955). In both anuran (*R. pipiens*) and urodele (*N. maculosus*) amphibians, however, micropuncture studies indicate that the pH of the tubular fluid does not fall below that of the plasma along the proximal tubules and that net bicarbonate reabsorption is proportional to net fluid reabsorption (Giebisch 1956; Montgomery and Pierce 1937; O'Regan et al. 1982). Similarly, micropuncture measurements on avian (*S. vulgaris*) proximal tubules indicating that sodium and chloride are reabsorbed at equivalent rates and that the pH of the tubule fluid does not fall below that of the peritubular plasma (Laverty and Alberici 1987; Laverty and Dantzler 1982) suggest that the bicarbonate concentration does not change and that bicarbonate and fluid reabsorption are proportional.

Additional bicarbonate reabsorption must occur in the distal portions of the nephrons in amphibians and birds to account for the low pH observed in the final ureteral urine during an acid load. The micropuncture experiments revealing a substantial fall in pH along the distal portions of amphibian nephrons and avian cortical collecting ducts suggest that this is the case (Giebisch 1956; Laverty and Alberici 1987; Montgomery and Pierce 1937). Studies involving direct measurements of bicarbonate transport with isolated, perfused renal tubules from urodele amphibians (*A. maculatum* and *A. tigrinum*) indicate that net bicarbonate reabsorption occurs to the greatest extent along the late distal tubules (Yucha and Stoner 1986). There is, of course, some net reabsorption along the proximal tubule. Although the study of Yucha and Stoner (1986) reveals no net bicarbonate reabsorption along the early distal tubules in urodeles, the electrical potential and bicarbonate concentration are still compatible with reabsorptive bicarbonate transport in this region. Moreover, net reabsorption of bicarbonate, as determined indirectly by hydrogen ion secretion, apparently can occur in the early distal tubules of anuran amphibians (*R. pipiens* and *R. esculenta*; vide supra; Chap. 4), but the relative importance of reabsorption in this region compared with more distal regions in these anurans is unknown. Some net bicarbonate reabsorption generally occurs along the initial collecting ducts of urodeles, but this is highly variable and net secretion may even occur along these segments in some animals (Yucha and Stoner 1986).

Nothing is known about the tubule site of net bicarbonate secretion in alligators. However, in view of the magnitude of the transport, which leads to net overall secretion for the entire kidney, it appears that the proximal tubule must be involved.

6.3.2   Mechanism of Transport

Most of the information available concerning the mechanism of bicarbonate reabsorption in the proximal tubules of nonmammalian vertebrates is derived from studies on urodele amphibians, which have been reviewed in connection with the discussion on hydrogen ion transport in Chap. 4. Briefly, these studies indicate that in the proximal tubules of amphibians, as in mammals, most bicarbonate reabsorption is driven by hydrogen ion secretion. Apparently, such reabsorption involves conversion of bicarbonate to carbon dioxide and water in the tubule lumen and then reconversion to bicarbonate within the tubule cells (Fig. 4.6). As also noted previously, carbonic anhydrase is presumed to play a role in this process within the tubule cells (Fig. 4.6), but its presence in the proximal tubule cells of amphibians is not clearly documented (Chap. 4). Nor is there any information on whether carbonic anhydrase is present in the proximal tubule lumen in urodele amphibians as it is in the proximal tubule lumen in mammals (Weiner and Verlander 2012).

The bicarbonate formed within the proximal tubule cells apparently exits the cells across the basolateral membrane via an electrogenic sodium-bicarbonate cotransporter (NBC) (Fig. 4.6), which has been cloned from amphibian (tiger salamander, *A. tigrinum*), rat, and human kidneys (Burnham et al. 1997; Romero et al. 1997, 1998). Its presence has also been suggested by studies on the regulation of intracellular pH in avian proximal tubules (Kim et al. 1997). This renal electrogenic NBC cotransporter has a sodium-to-bicarbonate coupling ratio of 1:3. This ratio could reflect either 3 HCO_3^- or 1 HCO_3^- and 1 CO_3^{-2}. In mammalian kidneys, immunocytochemical studies show this transporter strongly and exclusively expressed in the basolateral membrane of the proximal tubules, whereas in amphibian kidneys it is only weakly expressed in the basolateral membrane of the proximal tubules (Schmitt et al. 1999). However, in amphibian kidneys, it is strongly expressed in the basolateral membrane of late distal tubules, but not in any other tubule regions (Schmitt et al. 1999). The stronger distal than proximal signal in amphibian nephrons appears to reflect the relative importance of these regions in the reabsorption of filtered bicarbonate (vide supra).

As noted above (Chap. 4), no carbonic anhydrase is present in the kidneys of elasmobranchs; the titration studies of Deetjen and Maren (1974) on the proximal tubules of *R. erinacea* suggest that bicarbonate is reabsorbed directly without conversion to carbon dioxide and water. This reabsorptive process may involve the basolateral chloride-bicarbonate exchanger suggested by the studies of Silbernagl et al. (1986) (vide supra; Chap. 4). Although no direct information is available about the process of bicarbonate reabsorption in other nonmammalian vertebrates, the observation that the administration of carbonic anhydrase inhibitors to water snakes results in an alkaline urine, an increased excretion of sodium and potassium, and an unchanged excretion of chloride suggests that the mechanism for bicarbonate reabsorption in these reptiles is essentially the same as that in mammals and amphibians (Dantzler 1968, 1976b).

Few direct studies have been made on the mechanism of bicarbonate reabsorption in the distal portions of the nephrons of nonmammalian vertebrates. However, in the early distal tubules of amphibian nephrons, the reabsorptive process is probably driven by a sodium-hydrogen exchanger as it is in the comparable thick ascending limb of mammalian nephrons (Weiner and Verlander 2012) (vide supra; Chap. 4; hydrogen ion transport) (Fig. 4.3). Less complete studies on the late distal tubules of urodeles showing sodium and carbonic anhydrase dependence of bicarbonate reabsorption (Yucha and Stoner 1986) and involving pH measurements (Stanton et al. 1987) suggest that the reabsorptive process in this tubule region is also driven by a sodium-hydrogen exchanger and involves carbonic anhydrase (vide supra; Chap. 4; hydrogen ion transport). The localization of the sodium-bicarbonate (NBC) cotransporter to the basolateral membrane also supports the concept of strong bicarbonate reabsorption by this general process in this distal region. Perhaps a similar process occurs in the distal nephrons of other nonmammalian vertebrates. However, in no case have any molecular studies been performed to identify the sodium-hydrogen exchangers involved in this process in nonmammalian vertebrates.

In mammals, although most filtered bicarbonate is reabsorbed, varying amounts can be secreted by type B and type C intercalated cells in the connecting tubule (CNT), depending on the requirements of acid–base balance (Weiner and Verlander 2012). The secretory mechanism in these cells involves a chloride-bicarbonate exchanger (pendrin) in the luminal membrane to secrete bicarbonate into the lumen, carbonic anhydrase in the cytoplasm to facilitate bicarbonate and hydrogen ion production, and vacuolar-type H^+-ATPase to extrude hydrogen ions across the basolateral membranes (Weiner and Verlander 2012).

Except for the crocodilians, there are no data indicating that bicarbonate can be secreted by the renal tubules of nonmammalian vertebrates. However, as already noted (vide supra; Chap. 4), net renal tubular secretion of bicarbonate normally occurs in alligators (*A. mississippiensis*) (Coulson and Hernandez 1964, 1983; Lemieux et al. 1985). Moreover, the administration of carbonic anhydrase inhibitors to these animals results in a mildly acid urine and the conversion of net tubular secretion of bicarbonate to net reabsorption (vide supra; Chap. 4) (Coulson and Hernandez 1964; Lemieux et al. 1985). These observations suggest that carbonic anhydrase plays a role in the production of bicarbonate within the tubule cells and that hydrogen ion is normally secreted from the cells into the blood, accounting for the persistently low plasma pH in these animals. Bicarbonate is then secreted into the tubule lumen—in general the reverse of the process involved in bicarbonate reabsorption (Lemieux et al. 1985). However, this process has not been demonstrated directly, and absolutely nothing is known about the actual mechanism for hydrogen ion or bicarbonate movement across the cell membranes. Nevertheless, in view of what is now known about bicarbonate secretion by mammalian type B and type C intercalated cells, it seems very likely that alligator proximal tubule cells contain some type of chloride-bicarbonate exchanger in their luminal membranes and an H^+-ATPase in their basolateral membranes. Of course, the actual transporters may be unique to crocodilians. Clearly, molecular studies of the steps in bicarbonate secretion by alligator proximal tubules could be particularly informative.

6.4   Amino Acids

6.4.1   Direction, Magnitude, and Sites of Net Transport

Filtered amino acids are almost completely reabsorbed by the renal tubules of mammals and most nonmammalian vertebrates. However, clearance studies reveal that net tubular secretion of amino acids also can occur in some species of fishes and reptiles (Table 6.1). For example, the sulfonated amino acid, taurine, which plays an important role in cellular osmoregulation in marine fishes, can undergo net secretion by the renal tubules of euryhaline marine teleosts (winter flounder, *P. americanus*) and marine elasmobranchs (dogfish, *S. acanthias*; little skate, *R. erinacea*) (Schrock et al. 1982). Net secretion of taurine by the renal tubules of marine and terrestrial snakes (olive sea snake, *A. laevis*; garter snake, *T. sirtalis*) also can occur (Benyajati and Dantzler 1986a, b). Clearance studies reveal only net tubular secretion of taurine in flounders but either net tubular secretion or net tubular reabsorption in dogfish, skates, and snakes (Benyajati and Dantzler 1986a, b; Schrock et al. 1982). Moreover, either net tubular secretion or net tubular reabsorption of other endogenous β-amino acids (β-alanine and β-aminoisobutyric acid) and an endogenous analog of β-amino acids (L-cysteic acid) can occur in sea snakes and garter snakes (Benyajati and Dantzler 1986a, b). The variable direction of net tubular transport of taurine in some fishes may function to control the plasma levels, which are rapidly altered by osmotic stress and high protein meals. This is especially likely because taurine is not readily metabolized in these animals. However, the physiological function of tubular secretion of taurine and other β-amino acids in snakes is not at all clear. In addition, it is of interest that taurine is one of the few amino acids for which a significant fractional excretion, some 6–40 % of the filtered load, occurs in mammals (Silbernagl 1985), but the physiological significance of this high excretion rate in mammals is also not known.

In the snakes studied, the endogenous amino acids other than those discussed above are largely reabsorbed by the renal tubules (Benyajati and Dantzler 1986a, b). However, whereas in mammals and birds such reabsorption usually amounts to more than 99 % of the filtered load; in these reptiles it is often substantially less (Benyajati and Dantzler 1986a, b). For example, in both garter snakes and sea snakes, only about 80 % of the filtered histidine is reabsorbed. The reabsorption of filtered serine also is often this low, and in garter snakes adapted to cold the reabsorption of phenylalanine and glutamic acid also falls to about 60 % of the filtered load. In contrast, in the only other reptile species studied, the alligator (*A. mississippiensis*), reabsorption of all filtered amino acids appears to be more than 99 % complete (Hernandez and Coulson 1967). Because ureteral urine was collected from snakes and cloacal urine was collected from alligators, it appears possible that the apparent difference in tubular reabsorption results from the difference in the site of collection if amino acid reabsorption or, possibly even secretion, occurs in the cloaca as well as the renal tubules (Benyajati and Dantzler 1986a). This possibility is supported by preliminary data on sea snakes in which ureteral urine and cloacal urine samples were collected from the same animals (Benyajati and Dantzler 1986a), but a more direct and detailed evaluation is required before the idea can be accepted or discarded.

The sites of amino acid transport along the nephrons of nonmammalian vertebrates are only poorly understood. Much more information is available concerning the transport sites and pathways along the nephrons of mammals. However, the mammalian sites and pathways, which are multiple and complex, are beyond the scope of this volume (for detailed discussion, see Moe et al. 2012 and Silbernagl 1988). Nevertheless, it can clearly be stated that the primary site of amino acid reabsorption in the mammalian nephron is the proximal tubule (Moe et al. 2012; Silbernagl 1988). Among the nonmammalian vertebrates, studies with isolated tubules from flounders certainly indicate that secretion of taurine occurs along the proximal tubule (King et al. 1982), and it appears likely that the reabsorption of other filtered amino acids also occurs along this tubule segment. In snakes, net tubular secretion and reabsorption of taurine probably both occur long the proximal tubule, but analysis of the net tubular transport rates as a function of the filtered load suggests that net secretion occurs distal to the primary site of net reabsorption (Benyajati and Dantzler 1986b). Again, it appears likely that the reabsorption of other amino acids occurs along the proximal tubule in reptiles, but this has not been examined directly and no information is available about transport sites for amino acids beyond the proximal tubule. One micropuncture study on urodele amphibians (*N. maculosus*), for which only net reabsorption has been reported, suggests that this process occurs beyond the end of the proximal tubule (Oken and Weise 1978). For example, almost all filtered aspartic acid, but only 30 % of the filtered glutamic acid, is reabsorbed by the end of the proximal tubule in these animals. This pattern of fractional reabsorption is similar to that observed in the ureteral urine of snakes (Benyajati and Dantzler 1986a), but whether additional reabsorption of amino acids occurs in the more distal portions of nephrons of *N. maculosus*, in the cloaca, or not at all, has not been determined. As already noted, in birds, clearance studies indicate that more than 99 % of the total filtered amino acids are reabsorbed by the renal tubules (Boorman and Falconer 1972; Sykes 1971). The studies of Boorman (Boorman and Falconer 1972) suggest that there are at least two separate reabsorptive pathways in birds—one for basic amino acids and one for neutral and acidic amino acids—but no information is available on the sites of transport or the specificity for individual amino acids.

6.4.2 Mechanism of Transport

6.4.2.1 Net Tubular Reabsorption

In mammals, the process of amino acid reabsorption has been studied extensively, first with renal cortical brush-border membrane vesicles and perfusions of proximal tubules in vivo and in vitro (Silbernagl 1988) and then with molecular techniques

(Moe et al. 2012). The detailed studies are beyond the scope of this discussion. However, the initial studies all indicated that most amino acids are reabsorbed across the luminal membrane by a group of secondary active, electrogenic sodium cotransporters (Silbernagl 1988). The more recent molecular studies have led to the identification and cloning of most of these transporters as well as some of those involved in transport across the basolateral membrane. This work, which is ongoing, is well reviewed elsewhere (e.g., Moe et al. 2012) and is too extensive to be reviewed in detail here.

Among the nonmammalian vertebrates, the only studies on the possible mechanisms for amino acid reabsorption involve urodele amphibians, i.e., a newt, *T. pyrrhogaster* (Hoshi et al. 1976), and teleost fishes, i.e., mullet, *M. cephalus* (Lee and Pritchard 1983). The studies on newts utilized measurements of the potential difference across the whole epithelial wall and the basolateral membrane of the proximal tubule during the perfusion of the lumen with various amino acids in the presence and absence of sodium. These studies also involved the uptake of amino acids by isolated, nonperfused renal tissue in the presence and absence of sodium. The studies showed that neutral amino acids, e.g., alanine, in the lumen depolarize the peritubular membrane potential only in the presence of sodium; dibasic amino acids, e.g., lysine, depolarize it in the presence or absence of sodium; and acidic amino acids, e.g., aspartic acid, do not depolarize it at all. However, the uptake of acidic amino acids by the tubules, apparently from the luminal side, requires sodium, and some uptake of neutral amino acids and substantial uptake of dibasic amino acids can occur in the absence of sodium. When these uptake data are considered in light of the electrical data, they suggest that neutral amino acids cross the luminal membrane either in an electrogenic cotransport with excess sodium or in a non-electrogenic transport step without sodium (Fig. 6.2); dibasic amino acids cross the luminal membrane in a sodium-independent transport step that is electrogenic because of the positive charge of these acids at physiological pH (Fig. 6.2); and acidic amino acids apparently cross the luminal membrane in an electroneutral cotransport step with sodium (Fig. 6.2).

The studies on mullet were performed on brush-border membrane vesicles only (Lee and Pritchard 1983). The results, like those from the studies with mammalian vesicles and from the studies with newt tubules, support the concept that neutral amino acids, e.g., leucine, enter the cells across the luminal membrane via a secondary active transport process in which the transport of the amino acid is coupled to the electrochemical gradient for sodium (Fig. 6.2). They also support the concept that the transport of the dibasic amino acids, e.g., lysine, into the cells across the luminal membrane is independent of the sodium gradient (Fig. 6.2). In addition, they indicate that the transport of these basic amino acids may involve coupling to a proton gradient from the lumen into the cells and that this cotransport system is enhanced by the electrical gradient from the lumen to the cell interior (Fig. 6.2). Therefore, as suggested for the newt tubules, this entry step for the basic amino acids is probably electrogenic (Fig. 6.2). As yet, no attempts have been made to isolate and characterize molecularly the transporters involved in amino acid reabsorption in these nonmammalian vertebrates.

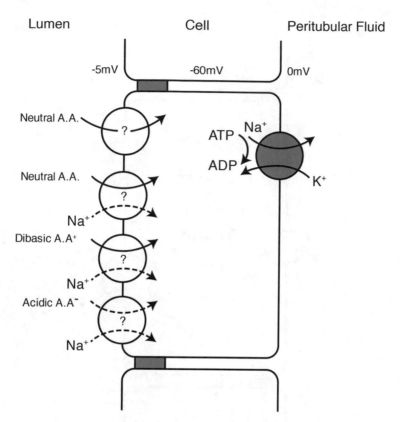

Fig. 6.2 Model for amino acid entry across luminal membrane during tubular reabsorption, based on studies on amphibians and fish. Symbols have same meaning as in Fig. 6.1. See text for more detailed description of the processes diagramed in the figure

6.4.2.2   Net Tubular Secretion

Net tubular secretion of taurine has been studied with kidney slices from elasmo-branchs (dogfish, *S. acanthias*) (Schrock et al. 1982), isolated perfused and nonperfused proximal tubules from teleosts (flounder, *P. americanus*; killifish, *F. heteroclitus*) (King et al. 1982; Wolff et al. 1986, 1987), and renal brush-border membrane vesicles from teleosts (flounder, *P. americanus*) (King et al. 1985). The primary energy-requiring step in the transepithelial secretory process appears to involve transport into the cells against a concentration gradient at the peritubular side (Fig. 6.3). This transport step across the basolateral membrane is apparently shared by other β-amino acids, e.g., β-alanine, but not by α-amino acids, e.g., α-aminoisobutyric acid, or other unrelated acids, e.g., ρ-aminohippurate. It is absolutely dependent on the presence of external sodium and is stimulated by the presence of external chloride. The kinetic studies on killifish suggest that, in the presence of external chloride, the stoichiometric relationship between sodium and

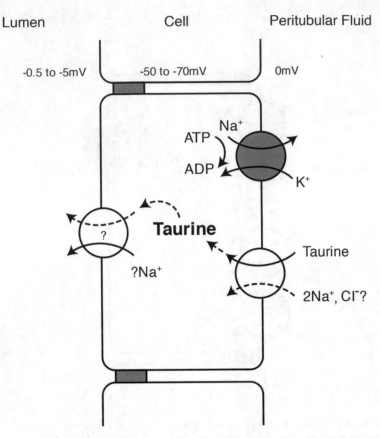

Fig. 6.3 Model for tubular taurine secretion, based on studies with teleost fishes. Symbols have same meaning as in Fig. 6.1. See text for more detailed description of the processes diagramed in the figure

taurine is 2:1 and that, in the absence of external chloride, it is 1:1. Wolff et al. (1986) suggest that chloride affects the binding of sodium to the carrier for taurine and that this transmembrane transport step may involve the cotransport of sodium, chloride, and taurine (Fig. 6.3).

Taurine that has been transported into the cells against a concentration gradient at the peritubular side moves down a concentration gradient from the cells to the lumen during the net secretory process (King et al. 1982; Wolff et al. 1986). However, the mechanism involved is less well understood than the transport step at the peritubular membrane. It probably cannot involve simple passive diffusion unless the permeability of the luminal membrane to taurine is much greater than would be expected; a carrier-mediated process appears more likely (King and Goldstein 1985). A sodium-dependent, electrogenic carrier-mediated transport step for taurine that is shared by other β-amino acids has been found in brush-border membrane vesicles from flounder proximal tubules (King et al. 1982). King et al. (1982) suggest that, under appropriate conditions in vivo, this transport process may facilitate taurine movement out of the cells into the lumen during the net secretory process (Fig. 6.3). However, it has yet to be demonstrated directly that the electrochemical gradients in vivo are appropriate to account for taurine and sodium movement into the lumen by this process. No attempts have yet been made to identify and isolate the actual transporters involved in taurine secretion.

6.5   Urea

6.5.1   Direction, Magnitude, and Sites of Net Transport

Urea is freely filtered and, in mammals and most other vertebrates, a variable amount is then reabsorbed passively by the renal tubules (Table 6.1) (Dantzler 1976b; Mudge et al. 1973). Filtration and net passive tubular reabsorption determine urea excretion even in most of those vertebrates in which it is the major excretory end product of nitrogen metabolism, i.e., all mammals; chelonian reptiles from mesic, semiaquatic, and aquatic habitats; adult urodele amphibians; and many anuran amphibians (Table 6.2) (Dantzler 1976b; Forster 1970; Mudge et al. 1973; Walker and Hudson 1937b). In mammals, some 50 % of the filtered urea is passively reabsorbed along the proximal tubule. An additional fraction is passively reabsorbed from the inner medullary collecting ducts, the amount depending on the presence or absence of antidiuretic hormone. The urea reabsorbed from the inner medullary collecting ducts is accumulated to a large extent in the inner medulla, playing a major role in the concentrating process (vide infra; Chap. 7). Some of it passively enters the descending and ascending thin limbs of Henle's loops and is recycled. Although some details of these movements of urea in the mammalian kidney and their relationship to the urine concentrating mechanism are considered in the discussion on urine concentration (vide infra; Chap. 7), most are beyond the scope of this volume. However, they are reviewed extensively elsewhere (Dantzler et al. 2011; Sands et al. 2012). In addition, there is some evidence for tubular secretion of urea under some circumstances in mammals, but this is still controversial (vide supra).

The site or sites of passive reabsorption of urea in the renal tubules of those nonmammalian vertebrates in which such reabsorption of filtered urea predominates are not yet clearly defined. However, in view of the large fraction reabsorbed, the proximal tubule must play an important role.

Although only net passive reabsorption of filtered urea occurs in many ureotelic anuran amphibians, net tubular secretion has been demonstrated by clearance and micropuncture studies in a number of species of primarily aquatic anuran amphibians of the genus *Rana* (*R. catesbeiana*, *R. pipiens*, and *R. clamitans*) (Long 1973; Mudge et al. 1973). In doubly perfused kidneys of the bullfrog (*R. catesbeiana*), there is a substantial unidirectional passive reabsorptive flux for urea as well as a

Table 6.2 Approximate percent of total urinary nitrogen as urates, urea, and ammonia

Environment and species	Percent of total urinary nitrogen as:			References
	Urates	Urea	Ammonia	
Amphibia				
Anura				
Mesic-xeric terrestrial				
African tree frog,	60–75	20–35	1–8	Loveridge (1970)
Chiromantis xerampelina				
South American tree	80–90	3–11	5	Shoemaker and
frog,				McClanahan (1975)
Phyllomedusa sauvagii				
Mexican tree frog,	2–7	90	5	Shoemaker and
Pachymedusa dacnicolor				McClanahan (1975)
Mesic terrestrial,				
Semiaquatic				
South American tree frog,	0	94	6	Shoemaker and
Hyla pulchella				McClanahan (1975)
Bullfrog,	0	84	12	Munro (1953)
Rana catesbeiana				
Aquatic (freshwater)				
South African clawed	0	20–28	72–80	McBean and Goldstein
toad				(1967), Munro (1953)
Xenopus laevis				
Reptilia				
Testudinea				
Wholly aquatic	5	20–25	20–25	Moyle (1949)
Semiaquatic	5	40–60	6–15	Baze and Horne (1970)
Freshwater turtle,	1–24	45–95	4–44	Dantzler and Schmidt-
Pseudemys scripta				Nielsen (1966)
Wholly terrestrial, mesic	7	30	6	Moyle (1949)
Wholly terrestrial, xeric	50–60	10–20	5	Baze and Horne (1970)
Desert tortoise,	20–50	15–50	3–8	Dantzler and Schmidt-
Gopherus agassizii				Nielsen (1966)
Crocodilia	70	0–5	25	Khalil and Haggag (1958)
Squamata				
Sauria	90	0–8	Insignificant	Dessauer (1952), Khalil
			to highly	(1951), Minnich (1972),
			significant	Perschmann (1956)
Ophidia	98	0–8	Insignificant	Khalil (1948a, b),
			to highly	Minnich (1972)
			significant	
Rhynchocephalia				
Tuatara,	65–80	10–28	3–4	Hill and Dawbin (1969)
Sphenodon punctatus				
Aves				
Galliformes				
Mesic terrestrial				
Domestic fowl[O],	55–72	2–11	11–21	McNabb and McNabb
Gallus gallus domesticus				(1975)
Semiaquatic				
Duck[O],	54	1.5	29	Stewart et al. (1969)
Anas platyrhynchos				

(continued)

Table 6.2 (continued)

Environment and species	Percent of total urinary nitrogen as:			References
	Urates	Urea	Ammonia	
Passeriformes				
Mesic terrestrial				
Anna's hummingbird[N],	35–65	8–13	17–53	Preest and Beuchat (1997)
Calypte anna				
Turkey vulture[C],	76–87	4	9–16	McNabb et al. (1980)
Cathartes aura				
Mammalia	1–2	80–90	2–8	Dantzler (1970)

Superscripts: O = omnivore; N = nectivore; C = carnivore

unidirectional secretory flux, but the net flux is always secretory (Love and Lifson 1958). Indeed, when net tubular secretion in these animals is abolished by the administration of dinitrophenol (DNP) (vide infra), passive net reabsorption is revealed (Forster 1970). In micropuncture studies on *R. catesbeiana*, Long (1973) observed that net secretion of urea occurs in both the proximal and distal tubules but not in the collecting ducts. Apparently, no net transtubular movement of urea occurs beyond the distal tubules (Long 1973).

Clearance studies have also revealed net tubular secretion of urea in some lizard species (*Lacerta viridans* and *S. cyanogenys*), in the only living representative of the rhynchocephalian reptiles (*Sphenodon punctatus*), and in ducks (*Anas platyrhynchos*) (Dantzler 1976b; Stewart et al. 1969). However, such net secretion is rarely observed except during extreme diuresis and, in ducks, may result from synthesis by the renal tubules (Dantzler 1976b; Stewart et al. 1969). Therefore, even in these animals, filtration and net passive reabsorption appear to play the primary roles in determining the excretion of urea.

Net tubular reabsorption of urea, which may have an active component (but see below for this and an alternative explanation), has been demonstrated by clearance and micropuncture studies in marine elasmobranchs (Forster 1970). In these animals, high concentrations of urea in the body fluids help to maintain osmotic equilibrium with the surrounding seawater, and such reabsorption may be of physiological significance in regulating these levels. Micropuncture studies on dogfish (*S. acanthias*) suggest that the site of such reabsorption is between the end of the proximal tubule and the beginning of the collecting duct, apparently in some portion of the distal nephron that is inaccessible to micropuncture (vide infra) (Schmidt-Nielsen 1972).

Stenohaline freshwater stingrays (*Potamotrygon* spp.), in contrast to marine elasmobranchs, have extremely low plasma levels of urea (Thorson 1970; Thorson et al. 1967). These low plasma levels reflect both a low rate of urea synthesis (ornithine-urea cycle enzymes in the liver are only one-half to one-twentieth of those found in marine rays) and a failure to reabsorb filtered urea (Goldstein and Forster 1971).

6.5.2 Mechanism of Transport

6.5.2.1 Net Tubular Reabsorption

As already noted, tubular reabsorption of urea appears to be a passive process in mammals and most other vertebrates. However, the very high polarity of urea molecules prevents them from readily crossing nonpolar lipid membranes. Therefore, passive reabsorption across the renal tubule epithelium must occur either between the cells (as in mammalian proximal tubules and possibly ascending thin limbs of Henle's loops) (Chou and Knepper 1993; Nawata et al. 2014) or via carriers or channels in the cell membranes (Knepper and Mindell 2009). In some regions of mammalian renal tubules, urea transport across cells is now known to be mediated by isoforms of the urea transporter (UT-A) (Lim et al. 2006). Although the UT proteins were originally considered to be urea uniporters, recent studies of crystal structures of bacterial and mammalian homologues have indicated that they are actually urea channels (Knepper and Mindell 2009; Levin et al. 2009, 2012). Isoforms UT-A1 and UT-A3 are found in the luminal and basolateral membranes of the principal cells of the middle and terminal inner medullary collecting ducts (IMCDs), and isoform UT-A2 is found in the membranes of the cells of descending thin limbs of Henle's loops in the outer medulla and upper inner medulla (Lim et al. 2006). These channels mediate the renal reabsorption and recycling involved in the urine concentrating mechanism as noted above (see Chap. 7 for more detail).

Although urea reabsorption generally appears to be a passive process in mammals, there is physiological, but not molecular, evidence for active transport under some circumstances. Early tissue-slice and micropuncture studies by Bodil Schmidt-Nielsen and her colleagues suggested that, in rats (and, possibly other mammals) maintained on a low protein diet, there is some form of active urea reabsorptive process in the IMCDs (Danielson and Schmidt-Nielsen 1972; Truniger and Schmidt-Nielsen 1964; Ullrich et al. 1967). Later studies with isolated segments of initial IMCDs from rats maintained on a similar low protein diet have provided significant evidence for the presence of a secondary active sodium-urea cotransporter mediating reabsorption in this tubule segment (Isozaki et al. 1994). Such a transporter, when activated, could be very important in conserving protein under extreme conditions, but, despite considerable effort by Sands and his colleagues (Jeff M. Sands, personal communication), the molecular identity of this transporter has yet to be determined.

No information is available on the mechanism involved in passive urea reabsorption by the renal tubules of reptiles and amphibians. Although a homologue of the mammalian UT proteins has been cloned from frog (*R. esculenta*) urinary bladder (Couriaud et al. 1999), there is no information about whether it or another homologue is present in the renal tubules.

Among the teleosts fishes, urea excretion occurs via gills, and generally little attention has been paid to urea transport by the kidney. However, in addition to a UT homologue identified in the gills of eels (*Anguilla japonica*) (eUT) (Mistry et al. 2001), a novel protein (UT-C), distantly related to known urea transporters, has been found in fugu (*Takifugu rubripes*) and eels and is restricted to the proximal renal tubules of eels (Mistry et al. 2005). As is the case for other UTs, these are probably channels, rather than uniporters (vide supra). However, both the gill eUT and the renal tubule eUT-C facilitate passive urea movement across the cell membrane when expressed in *Xenopus laevis* oocytes. Mistry et al. (2005) suggest that the eUT-C in the proximal tubule permits conservation of filtered urea by passive reabsorption for eventual excretion via eUT in the gills. They suggest that both transporters work in concert to eliminate urea efficiently in seawater conditions in which filtration and urine excretion are very low. This suggested physiological role for eUT-C is supported by the finding that this transporter is much more highly expressed in the proximal tubules of eels adapted to seawater than in the proximal tubules of animals adapted to freshwater.

As noted above, urea helps marine elasmobranchs and other cartilaginous fishes maintain osmotic equilibrium with the surrounding sea water and needs to be retained by the kidneys. The observations that the urea concentration in the urine and in the early collecting duct segments of marine elasmobranchs is below the concentration in the plasma and that this concentration difference cannot be explained by the addition of water to the tubule lumens suggest that urea is being removed from the lumen by some form of primary or secondary active transport system (Schmidt-Nielsen 1972). However, this transport process is not saturable or blocked by the usual metabolic inhibitors (Forster 1970). But it does appear to be shared by amide-containing compounds, e.g., acetamide, to be sodium-dependent, and to be suppressed by inhibitors of sodium transport and by phloretin (Roch-Ramel and Peters 1981; Schmidt-Nielsen 1972). Also, clearance studies on spiny dogfish (*S. acanthias*) indicated that tubular urea reabsorption is correlated with tubular sodium reabsorption at a ratio of 1.6:1 over a wide range of urine flows and quantities of urea reabsorbed (Schmidt-Nielsen et al. 1972). Except for the lack of saturation, these data suggest that tubular urea reabsorption could involve a secondary active transport step in which urea movement is coupled directly to the movement of sodium down its electrochemical gradient, although such a transport process has yet to be directly demonstrated physiologically, much less characterized molecularly.

Moreover, the complex anatomical arrangement of the elasmobranch nephrons (vide supra; Fig. 2.3) led Boylan (1972) to suggest that the apparent uphill movement of urea could be passive and indirectly coupled to the transport of sodium. He proposed that the proximal segments in some manner create an environment of low urea concentration around the terminal segment by reabsorbing sodium and water and that urea moves passively out of the terminal segment of the nephron into the surrounding space. More recently, Friedman and Hebert (1985, 1990) extended this idea involving the countercurrent arrangement of the elasmobranch nephron segments within the encapsulated bundle region (Fig. 2.3). They described a diluting segment with low water and urea permeabilities in the dogfish shark (*S. acanthias*). They then suggested that the countercurrent arrangement could concentrate the urea within the tubule lumen and if the terminal nephron segment within this encapsulated region had sufficiently high urea and low water permeabilities, the urea concentration could be maintained in this segment while it moved passively out of the lumen.

These general ideas are supported by a number of major findings. First, among the elasmobranchs, homologues of mammalian phloretin-sensitive facilitated UTs have been cloned from the kidneys of spiny dogfish (*S. acanthias*) (Smith and Wright 1999), little skate (*R. erinacea*) (Morgan et al. 2003), winter skate (*Leucoraja ocellata*) (Janech et al. 2008), Atlantic stingray (*Dasyatis Sabina*) (Janech et al. 2003, 2006), bluntnose stingray (*D. say*) (Janech et al. 2008), and the Japanese banded houndshark (*Triakis scyllium*) (Hyodo et al. 2004). The elasmobranch UTs have their highest sequence identity with mammalian UT-A2. In the houndshark and bullshark (*Carcharhinus leucas*), the UT is exclusively localized to the terminal segment of the nephron, the collecting tubule (Hyodo et al. 2004). Similarly, among the holocephalons, a UT that is orthologous to the elasmobranch UTs has been found only in the luminal and basolateral membranes of the collecting tubule of elephant fish (*Callorhinchus milii*) nephrons (Hyodo et al. 2014). These data suggest that the renal UTs in all cartilaginous fishes are found exclusively in the collecting tubules (Hyodo et al. 2014). This localization, which should account for a high urea permeability in the terminal nephron segment, agrees well with the models for urea reabsorption suggested by Boylan (1972) and Friedman and Hebert (1990). Second, the expression of message for UT in little skate (*R. erinacea*) kidneys decreases when the animals are adapted to 50 % seawater (Morgan et al. 2003). Third, as described previously (vide supra; Chap. 2), the kidneys of stenohaline freshwater elasmobranchs, which do not reabsorb large amounts of filtered urea, lack the bundle zone, the peritubular sheath, and nephron loops III and IV found in marine and euryhaline elasmobranchs (Fig. 2.3). These observations also support the importance of these structures in urea retention in marine cartilaginous fishes and perhaps a model like those proposed by Boylan (1972) and Friedman and Hebert (1990). However, as noted by Friedman (P. A. Friedman, personal communication), the current versions of the model are insufficient to explain a reduction in the concentration of urea in the tubule lumen below the concentration in the plasma. In any case, whether urea reabsorption along the nephrons of cartilaginous fishes involves a primary or secondary active process, a purely passive process, or a combination of these has not yet been determined.

6.5.2.2 Net Tubular Secretion

Schmidt-Nielsen (1955) first suggested from clearance studies on both laboratory rats and kangaroo rats (*Dipodomys merriami*) that urea might be actively secreted by the renal tubules of mammals, but this conclusion was not generally accepted at the time. Later clearance studies on dogs (Beyer and Gelarden 1988; Beyer et al. 1992) and humans (Beyer et al. 1990) also supported the concept of active urea secretion, and studies with isolated perfused terminal segments of IMCDs from rats have provided strong physiological evidence for secondary active secretion of urea via a sodium-urea antiporter in this nephron region (Kato and Sands 1998). Under normal conditions, such active urea secretion would not be apparent, but might contribute significantly to the excretion of urea during water diuresis (Kato and Sands 1998). However, as in the case of secondary active urea reabsorption, the molecular identity of this secondary active transporter for urea secretion has yet to be determined.

Studies of the secretory process have been performed only on freshwater frogs, primarily bullfrogs, *R. catesbeiana*. Secretory transport saturates in vivo, suggesting that it involves some form of mediated transport (Forster 1970). The urea concentration in renal tubule tissue obtained from animals that are secreting urea in vivo is greater than that in the plasma, suggesting that a primary or secondary active transport step for urea into the cells exists at the peritubular membrane (Fig. 6.4) (Long 1973; Schmidt-Nielsen and Shrauger 1963).The possibility of a mediated transport step at the peritubular side of the cells is supported by the observations that net secretion is inhibited by the urea analog, thiourea, (but not by methylurea or acetamide), by ρ-aminohippurate (PAH), and by probenecid, apparently in a competitive fashion (Forster 1970; Schmidt-Nielsen and Shrauger 1963). DNP also inhibits both net tubular secretion and tissue accumulation of urea and thiourea in vivo, further supporting the concept of an active transport step for these substances at the peritubular membrane (Fig. 6.4) (Forster 1970; Schmidt-Nielsen and Shrauger 1963). Studies by Vogel and Kurten (1967) with isolated, doubly perfused frog kidneys indicate that net urea secretion is dependent upon sodium reabsorption, suggesting that the transport step at the peritubular membrane is coupled in some fashion, direct or indirect, to sodium transport (Fig. 6.4). Possibly, this could involve a mechanism similar to that for the secretion of organic anions, such as PAH (vide infra), but there is no direct evidence for such a mechanism.

A linear correlation between the rate of net urea secretion and the urine flow rate in these frogs (Vogel and Kurten 1967) has been cited as indicating that urea moves passively from the tubule cells to the lumen during the net secretory process (Fig. 6.4), but, as noted above, there is no evidence for a urea transporter (or channel) that might facilitate such passive movement in frog kidneys. Furthermore, no direct measurements of the transport across the luminal membrane have been made.

Lastly, it is necessary to note that in vitro observations are not in agreement with the tentative model shown in Fig. 6.4. No transport into the cells against a concentration gradient occurs in kidney slices or teased proximal tubules from *R. catesbeiana*, although avid PAH transport into the tubule cells is observed under the same circumstances (Irish 1975; O'Dell and Schmidt-Nielsen 1961). Moreover, no net active secretion of urea is observed in isolated, perfused frog (*R. catesbeiana*) proximal tubules in which PAH secretion and net fluid reabsorption occur at rates comparable to those in vivo (Irish 1975). Some factor or factors necessary for urea transport may be missing from the in vitro preparations. However, Irish (1975) has also suggested that active transport occurs only in

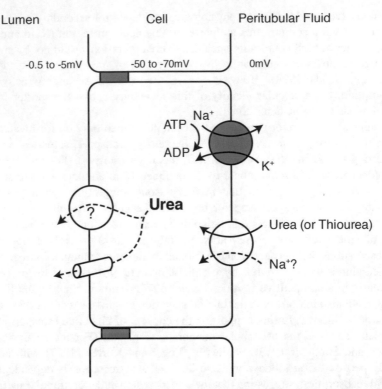

Fig. 6.4 Model for tubular secretion of urea, based on studies on amphibians. Tube crossing the cell membrane indicates channel for urea in the membrane. Other symbols have same meaning as in Fig. 6.1. See text for more detailed description of the processes diagramed in the figure

the distal tubule and, because of the convolutions of the frog nephron, recirculation may account for apparent active transport in the proximal tubule. Clearly, additional physiological and molecular studies are required to understand the mechanism involved in urea secretion by amphibian renal tubules.

6.6   Ammonia

6.6.1   Magnitude and Sites of Net Secretion

Ammonia is highly toxic and very little is normally present in the systemic blood of most vertebrate species; the ammonia excreted in the urine is produced by the renal tubule cells and secreted into the tubule lumen. In mammals, renal ammonia production occurs within the proximal tubule cells (Good and Knepper 1985; Weiner and Verlander 2012). Nothing is known about the exact site of production in the renal tubules of other vertebrates, but the proximal tubule is almost certainly involved.

The primary function of renal ammonia production and excretion in most vertebrates is to help maintain acid–base balance by enhancing the removal of excess acid or, more correctly, by leading to production of an equivalent amount of bicarbonate. Therefore, in most species, the magnitude of ammonia production and secretion by the renal tubules is dependent on acid–base status. Production and secretion increase with an acid load. This is certainly true for mammals, although it certainly varies among species, the capacity for enhanced production apparently being much greater in species such as rats and dogs, which generally excrete an acid urine, than in rabbits, which generally excrete an alkaline urine (Halperin et al. 1985; King and Goldstein 1985). An increase in ammonia production and excretion with an acid load is also true of those avian species (domestic fowl) and amphibian species studied (Craan et al. 1982; Long and Skadhauge 1983; Stetson 1978; Yoshimura et al. 1961).

Although acid–base regulation in fishes occurs primarily via the gills, there is substantial evidence of a role for the renal excretion of ammonia in acid-base balance even in some fish species, especially when they are exposed to an acid environment (King and Goldstein 1985;Wright et al. 2014). Apparently, at low environmental pH, acid–base regulation via the gills is impaired, and the animals actually appear to gain acid across the gills (McDonald and Wood 1981). This leaves the kidney as the only route for acid removal. There is now evidence for increased renal production and excretion of ammonia under these circumstances in freshwater teleosts (rainbow trout, *O. mykiss*; channel catfish, *Ictalurus punctatus*; goldfish, *Carassius auratus*) (Cameron and Kormanik 1982; King and Goldstein 1983a; McDonald and Wood 1981; Wheatly et al. 1984; Wood et al. 1999) and in marine elasmobranchs (dogfish shark, *S. acanthias*) (King and Goldstein 1983b). However, in at least one species of freshwater teleost (common carp, *Cyprinus carpio*), this does not occur (see below) (Wright et al. 2014).

Among amphibians, ammonia excretion via the kidney, although it increases with an acid load, is the primary means of nitrogen elimination in the aquatic clawed toad (*X. laevis*) (Table 6.2) (Balinski and Baldwin 1961). Among the reptiles, a clear ammoniagenic response to an acid load has not been demonstrated, although there is a suggestion that this might occur in freshwater snakes (*Nerodia* spp.) (Dantzler 1976b). Of course, few studies of the effects of an acid load on renal ammonia production and excretion have been made in reptiles. However, ammonia is a major end product of nitrogen metabolism in crocodilians (alligators, *A. mississippiensis*) (Coulson and Hernandez 1964) and in aquatic and semiaquatic chelonians (e.g., freshwater turtle, *Pseudemys scripta*) (Table 6.2) (Dantzler 1978). In alligators, urates account for about 75 % of the excreted nitrogen and ammonia for about 25 % (Dantzler 1978). The quantity of ammonia excreted by hydrated alligators on a standard meat diet (about 2.4 mEq kg^{-1} day^{-1}) is greater than that recorded for any other vertebrate species (Coulson and Hernandez 1970; King and Goldstein 1985). In semiaquatic and aquatic turtles, the excreted nitrogen is often about equally distributed between urea and ammonia, although urea tends to predominate (Table 6.2). In any case, the primary function of ammonia excretion in these reptiles is the elimination of nitrogen.

Finally, although ammonia production and excretion can increase with an acid load in birds, it also normally accounts for a significant fraction of the excreted nitrogen (average about 20 %) (Table 6.2). Moreover, although the largest fraction of excreted nitrogen is in the form of urates in birds, at least one nectivorous species (Anna's hummingbird, *Calypte anna*) can excrete over 50 % as ammonia at low temperatures when its intake of nectar is high (Preest and Beuchat 1997). Thus, this species and perhaps some other nectivore avian species are capable of changing from uricotelic to ammoniotelic under appropriate dietary and environmental circumstances (i.e., they are facultative ammoniouricoteles).

6.6.2   Process of Production and Secretion

In mammals, renal ammoniagenesis occurs in the proximal tubules by a process for which the steps and the control have been reviewed frequently and in great detail elsewhere (Goldstein 1976; Silbernagl and Scheller 1986; Tannen 1978; Weiner and Verlander 2012). Briefly, glutamine is taken up by the renal tubule cells and then enters the mitochondria. Here, it is deaminated by phosphate-dependent glutaminase I to form glutamate and NH_4^+. Glutamic acid is then deaminated via glutamate dehydrogenase to form α-ketoglutarate and NH_4^+. As noted by King and Goldstein (1985), the control of the increased production of ammonia in mammals during acidosis has been attributed to almost every step in the production pathway, and current information on this regulation is reviewed by Weiner and Verlander (2012). Weiner and Verlander (2012) also note that ammoniagenesis in mammals can be modified by a number of factors independent of acid–base balance, the most significant of which appears to be angiotensin II. Apparently, luminal angiotensin II stimulates and peritubular angiotensin II inhibits proximal tubule ammoniagenesis (Weiner and Verlander 2012).

Among birds, renal ammonia production has been investigated only for domestic chickens (Craan et al. 1982, 1983). The appropriate enzymes for the production of ammonia from glutamine or alanine are present in homogenates of the whole avian kidney and in tubules separated from the superficial regions of the kidney, apparently the reptilian-type nephrons. In the kidneys of acidotic chickens, as in those of mammals, the activities of glutaminase I, glutamine dehydrogenase, and alanine aminotransferase increase (Craan et al. 1982). Moreover, ammonia production is increased in superficial tubules from acidotic chickens compared with those from control chickens when the tubules are incubated in vitro with glutamine or alanine (Craan et al. 1982). Because only the uptake of glutamine, however, not alanine, is enhanced in tubules from acidotic animals, the former appears to be the preferred substrate for the increased ammonia production (Craan et al. 1983). Unfortunately, the exact tubule segment, or for that matter, the exact tubule population involved in ammonia production is not yet known.

Studies on the enzymes involved in ammonia production have also been performed on alligators (Lemieux et al. 1984). Glutaminase I is present in kidney mitochondria at suitable activities for ammonia production, but it is absent from the liver. However, glutamine synthetase is found only in the liver, suggesting that this organ may be a source of glutamine for ammonia production by the kidney. Glutamate dehydrogenase and alanine aminotransferase activities are also high in the kidney and low in liver and muscle. Moreover, isolated renal tubule fragments can produce ammonia from glutamine and alanine in vitro (Lemieux et al. 1984). Unfortunately, as noted above with regard to the avian kidney, there is no direct information available on the tubule segment involved in ammoniagenesis in the alligator kidney.

It should also be noted that during periods of dehydration in alligators, the renal production of ammonia as a major end product of nitrogen metabolism is reduced and even more nitrogen is excreted as urates (King and Goldstein 1985). It is possible that with low urine flow, accumulation of ammonia in the renal tissue could drive the reversible deamination reactions in the direction of amino acid formation (King and Goldstein 1985), but this process has not been examined directly. It also seems likely that a decrease in renal blood flow during dehydration would decrease the delivery of glutamine and alanine to the renal tissue, thereby reducing the amount of substrate for ammoniagenesis (King and Goldstein 1985).

Among amphibians, studies with renal tissue from X. laevis and R. temporaria indicate that glutamine and alanine, among the amino acids, are most readily deaminated in vitro (Balinski and Baldwin 1961). Moreover, acidosis leads to a 30 % increase in renal glutaminase I activity in X. laevis (Stetson 1978). The data suggest that, in X. laevis at least, glutamine is an important source for the increased ammonia excretion observed during acidosis and that regulation of renal ammoniagenesis may be similar to that in mammals. The very high concentrations of ammonia in the urine of X. laevis compared to the concentrations in the blood also suggest that, although ammonia is the major excretory end product of nitrogen metabolism in these animals, it is produced almost entirely by the kidney (King and Goldstein 1985). As in crocodilian reptiles, renal ammonia production as a major excretory end product of nitrogen metabolism in these anuran amphibians is reduced during dehydration, when more nitrogen is excreted as urea (King and Goldstein 1985). The mechanisms involved in this reduction in ammonia production may be the same as those suggested above for alligators (King and Goldstein 1985), but they have yet to be examined directly.

King and Goldstein (1983a, b) explored some aspects of ammoniagenesis in a species of marine elasmobranch (dogfish shark, S. acanthias) and a species of freshwater teleost (goldfish, C. auratus). As noted above, acidosis or an acid environment increases renal ammonia production in both species. In vitro, the dogfish kidney has the capacity to synthesize ammonia from a number of amino acids, but the greatest production apparently comes from glutamine. The activities of glutaminase I and glutamine synthetase, localized to the mitochondria, suggest that the production of ammonia by renal tissue in these animals may be regulated by the activities of these enzymes working antagonistically in a cycle of substrates between glutamine and glutamate plus NH_4^+ (King and Goldstein 1983b, 1985). In vitro studies also indicate that the goldfish kidney has the capacity to synthesize ammonia from a number of amino acids. In addition to glutamine, aspartate and alanine appear to be important precursors of ammonia production. In fact, measurements of enzyme activities and of ammonia production both suggest that aspartate has the greatest potential of any substrate analyzed as a precursor of renal ammonia synthesis, at least in vitro (King and Goldstein 1983a). Moreover, ammonia production from aspartate in these teleost kidneys occurs via transamination rather than via the purine nucleotide cycle as in mammalian kidneys. In contrast to goldfish, increased renal ammonia production during metabolic acidosis in another freshwater teleost species, rainbow trout (*O. mykiss*), occurs primarily from glutamine and alanine and involves increased renal activity of glutaminase, glutamate dehydrogenase, and alanine aminotransferase (Wood et al. 1999).

It was long assumed that the transport of ammonia from the renal tubule cells, where it is produced, into the tubule lumen (net secretion) in mammals and other vertebrates involved nonionic diffusion of free ammonia (NH_3) across the luminal membrane and trapping in the more acidic lumen as ammonium ion (NH_4^+). However, it is now known to be more complicated than this and to involve a number of transporters, including the Rhesus (Rh) glycoprotein ammonia (NH_3) transporters (Rhbg, Rhcg) (Marini et al. 2000; Weiner and Verlander 2012). In mammals, the transporters in different tubule segments have been at least partially identified and reviewed well elsewhere (Weiner and Verlander 2012). Briefly, in the proximal tubule, where newly produced ammonia must be transported into the lumen to be excreted or recycled, transport apparently involves primarily the substitution of ammonium ions for hydrogen ions on the NHE3 sodium-hydrogen exchanger on the luminal membrane (Good and Knepper 1985; Weiner and Verlander 2012). Nonionic diffusion and diffusion trapping probably account for only a small portion of ammonia secretion in the proximal tubule because the membrane permeability to free ammonia is not high enough to permit sufficient simple diffusion, no Rh glycoprotein is present to permit carrier-mediated diffusion, and the pH gradient from cells to lumen is insufficient to permit the observed secretion (Good and Knepper 1985; Weiner and Verlander 2012). Most of the ammonia in the tubule lumen is then reabsorbed in the thick ascending limb of Henle's loop as ammonium ions, apparently by replacing potassium ions on the Na-K-2Cl cotransporter (NKCC2) (Weiner and Verlander 2012). Some of this reabsorbed ammonia enters the descending thin limbs of Henle's loops (probably as free ammonia, NH_3) and is recycled between loops in the inner medulla, producing an increasing interstitial accumulation toward the papilla tip (Good and Knepper 1985; Weiner and Verlander 2012). About 80 % of the ammonia finally excreted in mammalian urine appears to be secreted by intercalated cells in the collecting ducts. This process apparently involves the diffusion of free ammonia (NH_3) from the interstitium into these cells via Rhcg and Rhbg transporters in the basolateral membrane, followed by diffusion from the cells to the lumen via Rhcg transporters in the luminal membrane. Active secretion of hydrogen ions via H-ATPase and H-K-ATPase in the luminal membrane rapidly converts the luminal ammonia (NH_3) to ammonium (NH_4^+), trapping it in the lumen and maintaining the gradient for ammonia (NH_3) diffusion (Weiner and Verlander

2012). The expression of Rhbg and Rhcg glycoproteins in the kidneys increases during acidosis (Weiner and Verlander 2012).

The processes involved in renal ammonia excretion are much less well understood in nonmammalian vertebrates. However, Rh glycoproteins are found in kidneys of both teleost (e.g., rainbow trout, *O. mykiss*; puffer fish, *Takifugu rubripes*; common carp, *C. carpio*; zebra fish, *Danio rerio*; mangrove rivulus, *Kryptolebias marmoratus*) and elasmobranch (e.g., little skate, *L. erinacea*; banded hound shark, *T. scyllium*) fishes (Cooper et al. 2013; Larsen et al. 2014; Nakada et al. 2007; Wright et al. 2014). Rhcg1 is localized to the luminal membrane of the distal tubule cells or both distal tubule and collecting duct cells in freshwater carp and zebrafish and the euryhaline mangrove rivulus (Cooper et al. 2013; Nakada et al. 2007; Wright et al. 2014). These data suggest that secretion of ammonia is likely to occur in the distal nephrons in these teleosts with Rhcg1 facilitating its movement from the cells into the tubule lumen. In the case of the carp, the expression of Rhcg1 increases with metabolic acidosis but the ammonia excretion does not, apparently because urine flow decreases (Wright et al. 2014). Wright et al. (2014) suggest that in these freshwater teleosts during metabolic acidosis, the gills continue to get rid of ammonia while urine flow decreases to conserve inorganic ions. In contrast, as noted above, in freshwater rainbow trout during a similar metabolic acidosis, branchial ammonia excretion decreases while renal ammonia production and excretion increase (McDonald and Wood 1981). Unfortunately, there is no information on changes in expression of Rh glycoproteins during metabolic acidosis in these animals.

The transport of ammonia as NH_4^+ appears to be a very likely possibility for net secretion by alligator nephrons, which, as noted above and in Chap. 4, always produce alkaline urine containing large amounts of bicarbonate. This transport may involve the substitution of ammonium ions for hydrogen ions on a sodium-hydrogen exchanger (probably a homologue of a mammalian NHE) at the brush-border membrane of proximal tubule cells (vide supra; Fig. 4.1). It is also possible, since alligators maintain a rather low blood pH (about 7.1), that the initial filtrate has a pH below that of the cells and that some ammonia may enter the lumen by nonionic diffusion in the first part of the proximal tubule (Lemieux et al. 1985). Because the pH gradient would be quite small, this appears unlikely, but, if it does occur, it would certainly require carrier-mediated diffusion, probably via Rh glycoproteins. However, there is, as yet, no information on the presence of these ammonia transporters in reptiles.

Neither mode of ammonia secretion into the tubule lumen would explain the lack of diffusion from the lumen back into the cells further along the alligator nephrons as the tubule fluid becomes highly alkaline. If secretion is indeed a carrier-mediated process in the proximal tubules, the luminal membrane of the cells in later tubule regions may be quite impermeable to ammonia, as is the case with the proximal straight tubule in mammals (Garvin et al. 1987). The details of this intriguing problem of large amounts of ammonia secretion in alkaline urine have yet to be examined in detail. Moreover, nothing further is known about the mechanism involved in ammonia secretion in those other nonmammalian vertebrates that produce acid urine and in which renal ammonia production increases with acidosis.

6.7   Organic Anions and Acids (Except Amino Acids, Urate, and Lactate)

6.7.1   Direction and Sites of Net Transport

A structurally diverse group of relatively small (molecular weight generally < 500 Da) organic anions (or organic acids that exist as anions at physiological pH; collectively OAs), such as p-aminohippurate (PAH), iodopyracet (Diodrast), and phenolsulfonphthalein (PSP, phenol red), undergo net secretion by a common process in the renal tubules of representatives of all five vertebrate classes (Table 6.1) (Dantzler and Wright 1997). Indeed, such net secretion has been demonstrated in every vertebrate species ever studied except a species of urodele amphibian, *N. maculosus*, and a species of hagfish, *M. glutinosa* (Table 6.1) (Dantzler 1992; Weiner 1973). Both net secretion and net reabsorption of PAH and iodopyracet by the proximal tubules of *N. maculosus* have been described, but net reabsorption is usually observed (Tanner 1967; Tanner and Kinter 1966). Moreover, no net tubular transport of phenol red is observed in these animals (Tanner et al. 1979). The renal tubules of *M. glutinosa* also do not transport phenol red or, it is generally assumed, any other organic anion (Fange and Krog 1963; Rall and Burger 1967). Among nonmammalian vertebrates, renal tubular secretion of OAs has been the most widely studied of all renal tubular secretory processes.

The proximal tubule was first designated as the site of net OA secretion in the classical experiments of Marshall and Grafflin (1928) showing net secretion of phenol red by the kidneys of the aglomerular goosefish, *Lophius americanus*, whose kidneys consist almost entirely of proximal tubules. Because these animals lack glomeruli, this study was also the first to convincingly demonstrate that net secretion by the renal tubules could play a role in urine formation. The exact proximal sites of secretion have since been determined in fish (flounders, *P. americanus*), amphibians (bullfrogs, *R. catesbeiana*), reptiles (garter snakes, *Thamnophis* spp.), birds (chickens, *Gallus gallus domesticus*), and mammals (rabbits, *Oryctolagus cuniculus*) by perfusion of isolated renal tubules (Brokl et al. 1994; Burg and Weller 1969; Dantzler 1974a; Irish and Dantzler 1976; Woodhall et al. 1978). In flounders, net secretion from bath to lumen against a concentration gradient apparently occurs in all segments of the proximal tubule (Burg and Weller 1969). In frogs, however, it is limited to the proximal and intermediate segments of the proximal tubule (Irish and Dantzler 1976). In snakes, such net secretion occurs only in the distal portion of the proximal tubule (Dantzler 1974a). In chickens, net secretion occurs in the straight portion of the proximal tubule of transitional nephrons (Brokl et al. 1994). It has not yet been possible to isolate and study proximal segments of small reptilian-type or large mammalian-type nephrons from birds. In rabbits, and probably other mammals, net OA secretion occurs primarily in the S2 segment of the proximal tubule, the central segment that includes approximately the second half of the convoluted portion and the first half of the straight portion (Woodhall et al. 1978).

6.7.2   Mechanism of Transport

6.7.2.1   Net Reabsorption

As pointed out above, net tubular reabsorption does not occur in most vertebrate species and the passive backflux that does occur is usually small and apparently crosses the cells (vide infra). However, in the proximal tubules of *N. maculosus*, net reabsorption of PAH and iodopyracet occurs against a transepithelial concentration gradient (Tanner 1967). Arterial injections of fatty acids, such as octanoate, inhibit net reabsorption (Tanner 1967), but no other information is available about this transport process.

6.7.2.2   Net Secretion

Transepithelial Transport Saturation of the net transepithelial secretory system has been demonstrated by clearance techniques in many species (Weiner 1973) and by perfusion of isolated proximal tubules from flounders, frogs, snakes, and rabbits (Burg and Weller 1969; Dantzler 1974a; Irish and Dantzler 1976; Shimomura et al. 1981). Saturation of the transepithelial transport system occurs at about the same bath concentration with isolated tubules from flounders, frogs, and snakes, but at about five times that concentration with isolated tubules from rabbits. In the case of the frogs, this in vitro concentration corresponds closely to the plasma concentration at which the secretory system saturates in vivo (Irish and Dantzler 1976; Schmidt-Nielsen and Forster 1954).

The apparent K_t for net PAH secretion, determined from saturation studies with isolated tubules, is about the same for frog and snake tubules but about 15–20 times higher for rabbit tubules (Table 6.3) (Dantzler 1976a; Irish and Dantzler 1976; Shimomura et al. 1981). However, J_{max} for net PAH secretion is about 10–15 times greater in rabbit tubules than in frog and snake tubules (Table 6.3) (Dantzler 1976a; Irish and Dantzler 1976; Shimomura et al. 1981). Thus, the ratio of J_{max} to K_t, which reflects the transport efficiency for net PAH secretion, is about the same for

Table 6.3 PAH transport by isolated, perfused proximal tubules

Species	Tubule segment	K_t (µM)	J_{max} (fmol min^{-1} mm^{-1})	References
Bullfrog, *Rana catesbeiana*	Proximal	15	659	Irish and Dantzler (1976)
Garter snake, *Thamnophis* spp.	Distal-proximal	10	325	Dantzler (1974b, 1976a)
Rabbit, *Oryctolagus cuniculus*	Proximal, S2 segment	195	7430	Shimomura et al. (1981)

K_t indicates substrate (PAH) concentration at which rate of transport is one-half the maximum rate. J_{max} indicates the maximum rate of transport for PAH

these amphibian, reptilian, and mammalian proximal tubules. The K_t and J_{max} values for the rabbit tubules are about the same whether determined for transepithelial transport or for transport into the cells at the basolateral membrane only, thereby indicating that the basolateral entry step is the rate-limiting step for transepithelial transport (Dantzler et al. 1995; Shimomura et al. 1981).

During net transepithelial secretory transport of PAH or iodopyracet by isolated, perfused renal proximal tubules from flounders, frogs, snakes, chickens, and rabbits, the steady-state concentration of the OA in the tubule lumen is greater than that in the peritubular bathing medium, and the concentration in the cells is greater than that in either the lumen or the peritubular bathing medium (Brokl et al. 1994; Burg and Weller 1969; Dantzler 1974b; Irish and Dantzler 1976; Tune et al. 1969). Because the inside of the tubule cells is negative compared to both the lumen and the bath, because these compounds are transported as anions at physiological pH, and because these anions do not appear to be bound inside the cells, the above observations are compatible with transport into the cells against an electrochemical gradient at the basolateral membrane and movement from the cells to the lumen down an electrochemical gradient (Fig. 6.5). This is the accepted process for transepithelial secretion in all species.

The apparent permeabilities of the luminal and basolateral membranes to PAH have been determined from the fluxes of PAH from the cells across these individual membranes in isolated proximal tubules from frogs, snakes, and rabbits (Table 6.4) (Fig. 6.5) (Dantzler 1974a, b; Irish and Dantzler 1976; Tune et al. 1969). In each species, the apparent permeability of the luminal membrane is always substantially greater than that of the basolateral membrane (Table 6.4). These findings are appropriate and in agreement with the general model if PAH transported into the cells across the basolateral membrane is to move into the lumen and not back into the bathing medium. However, the luminal membrane permeability certainly is an "apparent" permeability, for it must involve some sort of carrier-mediated process.

For tubules from those species studied, the backflux from lumen to peritubular bathing medium is small and, for frogs, snakes, and rabbits at least, can be predicted almost perfectly from the independently measured luminal and basolateral membrane permeabilities shown in Table 6.4 (Dantzler 1974b; Irish and Dantzler 1976; Tune et al. 1969). These data indicate that this small backflux occurs across the cells. Backflux appears to occur across the cells in avian renal tubules as well, but initial data suggest that it may also involve an uphill transport step at the luminal membrane (Brokl et al. 1994). However, this needs to be confirmed. In any case, net secretion of PAH in frog, snake, and rabbit tubules is not influenced by the very small backflux and thus is independent of perfusion rate (Dantzler 1974a; Irish and Dantzler 1976;Tune et al. 1969).

Basolateral Transport Many studies over the years have concentrated on the transport steps at the basolateral and luminal membranes of the proximal tubule cells. Although it could easily be shown, starting with the earliest studies, that transport of OAs into the cells across the basolateral membrane of renal tubules from both mammalian and nonmammalian vertebrates required energy (Chambers

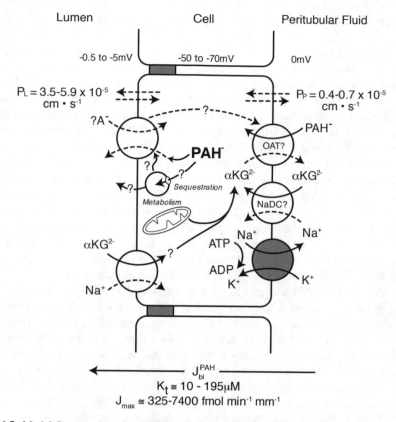

Fig. 6.5 Model for organic anion (PAH) secretion by proximal renal tubules, based on studies with fishes, amphibians, reptiles, birds, and mammals. A⁻ indicates anion of unknown type. αKG^{2-} represents α-ketoglutarate. *Broken arrow with question mark* indicates some form of possible feedback from transport step across luminal membrane to transport step across basolateral membrane. Other symbols have same meaning as in Fig. 6.1. Values for K_t and J_{max} for net transepithelial transport and for apparent permeabilities of luminal (P_L) and peritubular (P_P) membranes are from those given in Tables 6.3 and 6.4. See text for more detailed description of processes diagrammed in the figure

Table 6.4 Apparent permeabilities of tubule cell membranes to PAH

Species	Luminal membrane (cm s^{-1} × 10^{-5})	Basolateral membrane (cm s^{-1} × 10^{-5})	References
Bullfrog *Rana catesbeiana*	3.8	0.7	Irish and Dantzler (1976)
Garter snake *Thamnophis* spp.	3.5	0.5	Dantzler (1974a, b)
Rabbit *Oryctolagus cuniculus*	5.9	0.4	Tune et al. (1969)

et al. 1935; Cross and Taggart 1950), it could never be shown that it was a primary active process, i.e., that it was directly coupled to the hydrolysis of ATP (Maxild 1978; Ross and Weiner 1972). In addition, although this basolateral transport step could be shown to be sodium-dependent, it could never be shown to be directly coupled to the sodium gradient (Berner and Kinne 1976; Eveloff et al. 1979; Kasher et al. 1983; Kinsella et al. 1979; Ullrich et al. 1987a). However, studies showing (1) inhibition of basolateral PAH uptake by the anion exchange inhibitor SITS (Dantzler and Bentley 1980); (2) basolateral PAH/PAH exchange (Berner and Kinne 1976; Kasher et al. 1983; Kinsella et al. 1979); (3) inhibition or stimulation of basolateral PAH uptake by a number of anionic metabolites, whose entry across the basolateral membrane was directly coupled to the sodium gradient; and (4) production of a brief uptake of PAH above equilibrium in basolateral membrane vesicles by a combination of an inwardly directed sodium-gradient and an outwardly directed PAH or hydroxyl gradient (Eveloff 1987; Kasher et al. 1983) suggested to a number of investigators that PAH uptake across the basolateral membrane might normally involve exchange for an anionic metabolite (Berner and Kinne 1976; Sheikh and Moller 1983; Ullrich et al. 1987b, c).

In 1987 and 1988, Burckhardt and coworkers and Pritchard (Pritchard 1987, 1988; Shimada et al. 1987), using rat basolateral membrane vesicles, independently extended this general idea to produce the model shown in Fig. 6.5. In this model, transport of PAH (or other OAs that share this system) into the cells across the basolateral membrane is a tertiary active process, the final step in which is the countertransport of PAH against its electrochemical gradient in exchange for an intracellular metabolite (physiologically, α-ketoglutarate, α-KG) moving down its electrochemical gradient (Fig. 6.5). The outwardly directed gradient for αKG is maintained in turn by metabolism and by transport into the cells across the basolateral and luminal membranes by the Na^+-dicarboxylate cotransport system (Fig. 6.5). Finally, the inwardly directed sodium gradient essential to this process is maintained by the primary energy-requiring transport step in the whole process, the transport of sodium out of the cells by Na-K-ATPase at the basolateral membrane (Fig. 6.5).

As noted above, this model was initially developed and demonstrated with studies involving basolateral membrane vesicles from proximal tubules of rat kidneys. It was supported less directly by studies of OA uptake in rat kidney slices (Pritchard 1990) and masses of southern flounder (*Paralichthys lethostigma*) and killifish (*F. heteroclitus*) renal tubules (Miller and Pritchard 1991; Miller et al. 1996;). It was then shown in more detail to hold for OA transport in intact isolated, perfused proximal tubules from snake (*Thamnophis* spp.), chicken (*G. gallus domesticus*), and rabbit (*O. caniculus*) kidneys (Brokl et al. 1994; Chatsudthipong and Dantzler 1991, 1992; Dantzler et al. 1995). When tubules from all three species are preloaded with αKG from the bathing medium, both uptake of PAH into the cells and net transepithelial secretion of PAH increase to a comparable extent. This *trans*-stimulatory effect is inhibited by the addition of SITS to the bathing medium along with PAH, after the tubules are preloaded with αKG, supporting the concept that such stimulation involves PAH/αKG exchange.

Moreover, this stimulatory effect is eliminated if the preloading of the tubules with αKG via the basolateral Na^+-dicarboxylate cotransporter is prevented by the addition of LiCl (a specific inhibitor of the Na^+-dicarboxylate cotransporter) to the bath or the removal of sodium from the bath (replacement with N-methyl-D-glucamine) (Brokl et al. 1994; Chatsudthipong and Dantzler 1991, 1992). In isolated, perfused rabbit tubules, uptake of αKG by the Na^+-dicarboxylate cotransporter in the luminal membrane produces a similar effect on the uptake and transport of OAs placed in the bathing medium (Dantzler and Evans 1996; Shuprisha et al. 1999). It seems likely that this is the case in teleostean, reptilian, and avian proximal tubules as well, but this has yet to be examined. Studies of the kinetics of PAH uptake at the basolateral membrane of isolated rabbit tubules demonstrated that the increase in uptake with αKG preloading results from an increase in J_{max} with little change in K_t (Dantzler et al. 1995). Again, it seems likely that this is also the case for teleostean, reptilian, and avian proximal tubules, but it has yet to be determined.

   In addition, an analysis with isolated, perfused rabbit proximal tubules (Dantzler and Evans 1996; Shuprisha et al. 1999; Welborn et al. 1998) indicated that during uptake of OAs at the basolateral membrane, about 40 % of the αKG exchanged for the organic anions by the basolateral $OA/\alpha KG$ exchanger comes from metabolism (Fig. 6.5). The remaining 60 % of the αKG that is exchanged for OAs comes from uptake by the basolateral Na^+-dicarboxylate cotransporter, about half of it being recycled intracellular αKG and about half of it being new extracellular αKG (Fig. 6.5). Although no such analysis of the source of αKG involved in the $OA/\alpha KG$ exchange process has been made for renal tubules from nonmammalian vertebrates, it would be of particular interest to examine this problem under varying metabolic conditions in the isolated, perfused chicken proximal tubules (an example of a nonmammalian homeotherm) and snake proximal tubules (an example of a nonmammalian poikilotherm).

   In mammalian proximal tubules, the molecular identity of the transporters involved in basolateral OA transport process has been determined. The basolateral $OA/\alpha KG$ exchange process involves organic anion transporters 1 (OAT1) and 3 (OAT 3) (Pelis and Wright 2011; Wright and Dantzler 2004). A homologue of mammalian OAT1 has also been identified in winter flounder (*P. americanus*) tubules (Wolff et al. 1997), and mRNA for an OAT-like organic anion transporter has been found in chicken proximal tubule epithelium (Dudas et al. 2005). OATs have not been isolated or cloned from the kidneys of other nonmammalian vertebrates, but given the physiological similarity of basolateral $OA/\alpha KG$ exchange in fish, reptiles, birds, and mammals (Fig. 6.5), it appears highly likely that homologues of OATs are present in the proximal tubules of species from all nonmammalian vertebrate classes. The molecular identification of these specific proteins in nonmammalian vertebrates certainly deserves future work.

   Also, in mammals, NaDC1 and NaDC3 are the Na^+-dicarboxylate cotransporters in the proximal tubule luminal and basolateral membranes, respectively (Pajor 2000; Pelis and Wright 2011). Again, the molecular nature of these transporters is probably similar in the renal tubules of fish, amphibians, reptiles, and birds, but they have yet to be identified and cloned.

Luminal Transport As pointed out above, during the net tubular secretion of OAs, such as PAH, the movement from the cells to the lumen is apparently down an electrochemical gradient (Fig. 6.5). Although an exit could involve some passive diffusion, this cannot be the sole process given the usual low permeability of cell membranes to charged molecules. Indeed, the initial studies with isolated, perfused proximal renal tubules from frogs, snakes, and rabbits that revealed an apparent PAH permeability of the luminal membrane much higher than that of the basolateral membrane (Table 6.4) suggested that luminal transport was mediated in some way (Fig. 6.5) (Dantzler 1974a, b; Irish and Dantzler 1976; Tune et al. 1969). Moreover, the luminal exit step for fluorescein measured in real time with isolated, perfused rabbit proximal tubules saturates and can be described using Michaelis–Menten kinetics (Shuprisha et al. 2000). The concept for mediated luminal transport is also strongly supported by numerous studies with brush-border membrane vesicles from flounder, snake, rabbit, and rat tubules (Benyajati and Dantzler 1988; Blomstedt and Aronson 1980; Eveloff et al. 1979; Kinsella et al. 1979) and by additional studies with isolated, perfused snake and rabbit renal tubules (Chatsudthipong et al. 1999; Dantzler and Bentley 1979, 1980, 1981). However, this OA efflux process is much less well understood than the basolateral uptake process and, indeed, may vary among vertebrate classes and species within vertebrate classes. Also, in marked contrast to the basolateral OA/αKG exchanger, which can transport an enormous array of OAs of widely differing structures, the efflux process may involve separate transporters for different chemical types of OAs or even individual OAs.

In the case of mammals, mechanisms proposed for luminal efflux of OAs have included electrogenic carrier-mediated diffusion, anion exchange, and ATP-dependent efflux, and a number of OA transporters representing each of these modes of transport have been cloned and localized to the luminal membrane of the proximal tubules (Pelis and Wright 2011). These studies on mammals are beyond the scope of this volume and are reviewed well elsewhere (Pelis and Wright 2011; Wright and Dantzler 2004).

In the case of nonmammalian vertebrates, the luminal exit step is even less clearly defined than in mammals; there are no specific luminal transporter proteins for OAs yet identified in the proximal tubules of any nonmammalian vertebrate species. In isolated, perfused snake proximal renal tubules, the movement of PAH from cells to lumen is inhibited by SITS (Dantzler and Bentley 1980), suggesting that it involves anion exchange (Fig. 6.5). This inhibition of PAH transport by SITS also occurs with renal brush-border membrane vesicles from these snakes (Benyajati and Dantzler 1988). However, in contrast to what would be expected in the case of anion exchange, the luminal efflux of radiolabeled PAH from isolated, perfused snake proximal tubules is blocked, rather than stimulated, by unlabeled PAH or phenol red in the lumen (Dantzler and Bentley 1979). Of course, some anion exchange could still be occurring under these circumstances but with the loaded carrier translocating more slowly than the unloaded carrier. The movement of PAH into the lumen of these isolated perfused snake tubules is not influenced by the presence or absence of sodium, chloride, or potassium in the lumen (Dantzler

1974b; Dantzler and Bentley 1976, 1981). Although the movement of PAH into the lumen is not dependent on the presence of chloride per se, substitutions for chloride suggest that it may still be dependent on an anion in the lumen to which the membrane is highly permeable (Dantzler and Bentley 1981). This could be further evidence of anion exchange (Fig. 6.5) at the luminal membrane or it could be evidence of an effect on membrane potential. In any case, with these isolated, perfused snake proximal tubules, inhibition, by any means, of PAH movement from cells to lumen during net PAH secretion appears secondarily to reduce transport of PAH into the cells at the basolateral membrane (Dantzler and Bentley 1979, 1980, 1981; Dantzler and Brokl 1984a, b). These data suggest that in snake proximal tubules, there may be some type of feedback coupling between the transport systems on the luminal and basolateral membranes (Fig. 6.5).

Intracellular Transport It has generally been assumed that OAs, such as FL or PAH, transported into the renal tubule cells at the basolateral membrane diffuse through the cells to the luminal membrane during the process of net transepithelial transport. However, over the years, questions have been raised about this process. In fact, fairly early studies suggested that intracellular binding of some transported OAs, such as probenecid and phenol red, might occur (Berndt 1967; Eveloff et al. 1976), but this could not be confirmed by further studies (Pegg and Hook 1977). In more recent studies, Miller and Pritchard and their colleagues (Miller and Pritchard 1994; Miller et al. 1993;), using epifluorescence video imaging and confocal microscopy to study transepithelial fluorescein transport in flounder proximal tubules, obtained evidence for accumulation of this OA in intracellular punctate compartments. They considered these to be vesicles in which the OA could move across the cells and possibly empty into the lumen by exocytosis (Fig. 6.5). This may be the case for flounder tubules. However, preliminary studies with similar techniques provided little evidence of such accumulation in isolated snake and chicken proximal renal tubules (S. Shpun and W. H. Dantzler, unpublished observations). Accumulation of fluorescein in punctate compartments does occur in S2 segments of rabbit tubules but only when these tubules are in a depressed metabolic state and, under these circumstances, little fluorescein enters the tubule lumens (Shpun and W. H. Dantzler, unpublished observations). When these tubules are maintained in an optimal metabolic state, fluorescein is secreted into the lumen and is no longer accumulated in punctate compartments (S. Shpun and W. H. Dantzler, unpublished observations). Thus, accumulation of OA in vesicles for transport across the cells, if it occurs, apparently only occurs in flounder tubules under appropriate metabolic conditions. This concept of intracellular accumulation in vesicles for transepithelial transport in renal tubules of nonmammalian vertebrates certainly merits further study.

6.8   Urate[1]

6.8.1   Direction, Magnitude, and Sites of Net transport

Urate forms the major excretory end product of nitrogen metabolism in birds; in most reptiles, except chelonians from mesic, semiaquatic, and aquatic habitats; and in some amphibians (South American tree frogs of the genus *Phyllomedusa* and African tree frogs of the genus *Chiromantis*) (Table 6.2) (Dantzler 1978). Urate is always a very minor end product of nitrogen metabolism in mammals (Table 6.2), but in those mammals lacking uricase to convert urate to allantoin—humans, great apes, and Dalmatian coach hounds—it is the major excretory end product of purine metabolism (Pelis and Wright 2011; Weiner 1985). Urate appears to be freely filtered in mammals, reptiles, and amphibians, and probably in birds, although a small amount of binding to plasma proteins may occur in the last (Dantzler 1978; Weiner 1985). Clearance measurements on mammals generally reveal that some 96–99 % of the filtered urate is reabsorbed by the renal tubules (Weiner 1985). However, net tubular secretion is observed normally in pigs and frequently in rabbits (Weiner 1985). Thus, either net tubular reabsorption or net tubular secretion can be observed in mammals, depending on the species studied (Table 6.1). In all those birds, reptiles, and uricotelic amphibians studied, clearance measurements reveal only net secretion of urate by the renal tubules (Table 6.1) (Dantzler 1978). This is true even for those chelonian reptiles in which urate is not the primary excretory end product of nitrogen metabolism (Dantzler 1978). However, in chickens, infusions of the diuretics ethacrinic acid and furosemide (blockers of the Na-K-2Cl cotransporter) via the renal portal system can increase net urate secretion, suggesting that these drugs may be inhibiting a reabsorptive flux (Shideman et al. 1981), although this has yet to be confirmed. Moreover, urate synthesized by the renal tubule cells contributes about 3 % of that excreted in fasted chickens, about 20 % in normally fed chickens, and up to 50 % in chickens infused systemically with hypoxanthine (Chin and Quebbemann 1978). It contributes about 17 % of the urate excreted in normally fed alligators (Lemieux et al. 1985). No information is available on the synthesis of urate by the tubules of other species.

The site or sites of net urate transport in the renal tubules have been studied by micropuncture and perfusion of isolated renal tubules in mammals; by perfusion of isolated renal tubules in reptiles (garter snakes, *Thamnophis* spp.); and by micropuncture and perfusion of isolated renal tubules in birds (starlings, *Sturnus vulgaris*). In mammals, in which both reabsorption and secretion have been found in the proximal tubules, it is now well accepted that urate excretion is the sum of filtration, tubular secretion, and tubular reabsorption (Weiner 1985). In fact, in humans, a model of reabsorption, followed by secretion, followed by reabsorption, all in the proximal tubule, has generally been accepted on the basis of inhibitor

---

[1] The term "urate" in this volume refers to all forms that contain the urate anion (uric acid, uric acid dehydrate, and monobasic urate salts).

studies, but it has also been questioned because of the absence of direct evidence (Moe et al. 2012) (Von Baeyer and Deetjen 1985). In pigs, in which clearance studies reveal net secretion, the secretory process appears to involve the proximal convoluted tubule and the proximal straight tubule, but the proximal convoluted tubule appears to predominate (Roch-Ramel et al. 1980; Schali and Roch-Ramel 1981). In rabbits, the primary secretory flux, which can result in overall net secretion, appears to occur in the S2 segment of the proximal tubules (Weiner 1985). In snakes, net secretion from bath to lumen against a concentration gradient occurs throughout the proximal tubule but not in the distal tubule (Dantzler 1973, 1976a). There is no evidence of net reabsorption in these animals, but a passive unidirectional reabsorptive flux can also occur throughout the proximal tubule (Dantzler 1973). In starlings and chickens, net secretion occurs along the proximal tubules of the loopless superficial reptilian-type nephrons and transitional looped nephrons, but it is not clear whether any additional secretion can occur in more distal segments of these nephrons (Brokl et al. 1994; Laverty and Dantzler 1983). There is no micropuncture evidence of net reabsorption in these birds (Laverty and Dantzler 1983). No data are yet available on the site or direction of net urate transport by avian mammalian-type nephrons.

6.8.2 Mechanism of Transport

6.8.2.1 Net Reabsorption

Studies of the mechanism involved in the net reabsorptive process have all been made on mammals in which urate is not a major excretory end product of nitrogen metabolism. These have involved older physiological studies and more recent molecular studies. The details of the physiological studies involving brush-border membrane vesicles from kidneys of dogs, rats, and rabbits are beyond the scope of the current discussion but are reviewed in detail elsewhere (Kahn and Weinman 1985). Briefly, however, the studies with renal brush-border vesicles of dog and rat indicate that urate transport, presumably the initial step in the saturable net transepithelial reabsorptive process, involves an anion exchange system with an affinity, not only for urate but also for PAH, hydroxyl ions, and bicarbonate (Blomstedt and Aronson 1980; Kahn et al. 1983; Kahn and Aronson 1983; Kahn and Weinman 1985). However, a slightly later study revealed a urate-anion exchanger in brush-border membrane vesicles from human kidneys that did not function with PAH or hydroxyl ions (Roch-Ramel et al. 1994).

For rabbits, in which net tubular secretion frequently predominates, studies with renal brush-border vesicles suggest that urate transport across this luminal membrane is predominantly nonmediated (Boumendil-Podevin et al. 1979; Kahn and Weinman 1985; Kippen et al. 1979). Similarly, studies with renal brush-border membrane vesicles from snakes, animals in which there is no evidence for net tubular urate reabsorption, suggest that urate movement across this membrane is largely nonmediated (Benyajati and Dantzler 1988).

More recent molecular studies have identified a number of transporters that may be involved in the urate reabsorptive process in the luminal and basolateral membranes of mammalian proximal tubules. But the involvement of any of these in the actual transport process in vivo is by no means completely clear. These concerns and other details about these possible mammalian urate transporters are reviewed in detail by others (Mandal and Mount 2015; Pelis and Wright 2011; Von Baeyer and Deetjen 1985). However, of most significance in light of the earlier brush-border vesicle studies noted above is the identification of URAT1, a urate-organic anion exchanger on the luminal membrane. Although the intracellular anion or anions that *trans*-stimulate urate uptake by the URAT1 under physiological conditions are unknown, Pelis and Wright (2011) suggest that they could involve a number of endogenous monocarboxylates, such as lactate, nicotinate, acetoacetate, and pyrazinoate that have been shown to be countertransported for urate in human brush-border membrane vesicles (Roch-Ramel et al. 1994). Pelis and Wright (2011) further suggest that, like PAH uptake at the basolateral membrane, urate uptake at the luminal membrane may be a tertiary active process. In this model process, high intracellular concentrations of these monocarboxylates would be maintained by uptake across the luminal membrane via the secondary active sodium-monocarboxylate cotransporters 1 and 2 (SMCT1 and SMCT2). The sodium gradient necessary for sodium-monocarboxlyate cotransport would, of course, be maintained by the primary, energy-requiring Na-K-ATPase at the basolateral membrane. However, the situation in mammals is probably more complicated than this because other transporters, such as OAT4 and OAT10, that could exchange urate for monocarboxylates or dicarboxylates are also present in the luminal membrane and, as noted above, these problems for mammals are extensively reviewed elsewhere (Mandal and Mount 2015; Moe et al. 2012; Pelis and Wright 2011).

6.8.2.2   Net Secretion

Among mammals, as noted above, overall net secretion of urate by the renal tubules occurs in pigs and often in rabbits, although some degree of urate secretion by the proximal tubules can occur in all mammalian species studied. In isolated, perfused rabbit proximal tubules, both urate and PAH are secreted primarily by the S2 segment and appear to share the same transport system, at least at the basolateral membrane (see Fig. 6.5) (Grantham 1982). This is thought to be the case for any urate secretion by mammalian proximal tubules and is considered to involve OAT1, OAT3, and possibly OAT2, which is not involved in the standard transport process for PAH and similar organic anions (vide supra) (Pelis and Wright 2011). Studies with brush-border membrane vesicles from dogs, rats, rabbits, pigs, and humans indicate that the exit step at the luminal during secretion may involve both electrogenic facilitated diffusion and anion exchange depending on the species (Roch-Ramel et al. 1994). Numerous molecular candidates have been suggested for this

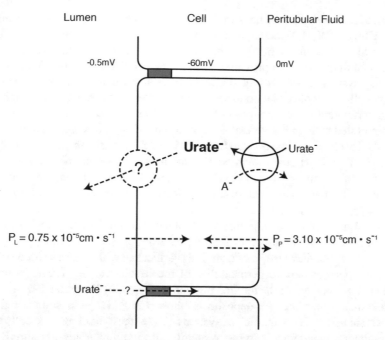

Lumen   Cell   Peritubular Fluid

-0.5mV   -60mV   0mV

Urate⁻

Urate⁻

A⁻

$P_L = 0.75 \times 10^{-5} \mathrm{cm} \cdot \mathrm{s}^{-1}$ $P_P = 3.10 \times 10^{-5} \mathrm{cm} \cdot \mathrm{s}^{-1}$

Urate⁻ --- ? --

Fig. 6.6 Model for tubular secretion of urate in reptiles, based on studies with snake (*Thamnophis* spp.) proximal renal tubules. Symbols have same meaning as in Figs. 6.1 and 6.5. Apparent permeabilities of luminal (P_L) and peritubular (P_P) are shown. Apparent K_t and J_{max} for net transepithelial transport are shown at bottom of figure. See text for more detailed description of processes diagramed in the figure

transport step, but the actual ones involved are not known with certainty. The discussion of these for mammals are beyond the scope of the current work and are well reviewed elsewhere (Moe et al. 2012; Pelis and Wright 2011).

Net secretion of urate by the renal tubules, as already noted, is the dominant process in urate excretion by nonmammalian vertebrates and is especially important in uricotelic species. Among reptiles, studies with isolated, perfused proximal renal tubules of snakes (*Thamnophis* spp.) have provided most of the detailed information on tubule secretion (Dantzler 1978, 1992). As in the case of the transport of other organic anions discussed above, these studies on urate transport indicate that net secretion involves transport into the cells against an electrochemical gradient at the peritubular side followed by movement from the cells into the lumen down an electrochemical gradient (Fig. 6.6) (Dantzler 1973, 1978). The effects of SITS on transport into the cells across the peritubular membrane in these tubules suggest that this transport step involves anion exchange (Fig. 6.6) (Mukherjee and Dantzler 1985). However, this transport step has no dependence on sodium (Randle and Dantzler 1973), indicating that the anion exchange cannot involve a dicarboxylate, such as αKG, that is transported into the cells by the sodium-dicarboxylate cotransporters. Also, preloading the isolated snake tubules with numerous mono-, di-, and tricarboxylates does not stimulate urate uptake at the basolateral membrane (Y. K. Kim and W. H. Dantzler, unpublished observations) the way αKG stimulates PAH uptake (vide supra). In addition, urate is secreted to a similar extent throughout the length of the snake proximal tubule, whereas PAH and organic anions that share that system are only secreted along the distal segment of the proximal tubule (Dantzler 1973, 1974b). Net urate secretion by snake renal tubules is not inhibited by high basolateral concentrations of PAH in vivo or in vitro, and net PAH secretion is not inhibited by high basolateral concentrations of urate in vitro (Dantzler 1978). Thus, the basolateral transport step for urate is independent of that for other organic anions, such as PAH, and the anion that might drive basolateral urate uptake by countertransport remains unknown (Fig. 6.6). Finally, there is no information on the possible molecular identity of this basolateral urate countertransport system in reptilian proximal renal tubules.

Urate transport by these reptilian renal tubules has a number of other distinctive features, which not only further differentiate it from the secretory transport of other organic anions but may also be of adaptive significance with regard to fluid balance. First, the apparent passive permeability of the basolateral membrane to urate is much greater than that of the luminal membrane (Fig. 6.6) (Dantzler 1976a). These observations, which are the opposite of those for PAH (vide supra; Fig. 6.5), suggest that this is a very inefficient system if urate transported into the cells across the basolateral membrane is to move readily into the lumen and not simply back into the peritubular bathing medium. Thus, the basolateral transport step is always working against a large backleak. Second, in these isolated, perfused snake renal tubules, the transport step into the cells across the basolateral membrane appears to be dependent, in part, on the presence of an artificial perfusate or, in vivo, glomerular filtrate flowing along the tubule lumen (Dantzler 1973). Third, net urate secretion by these tubules varies directly with the perfusion rate, suggesting that there is significant transepithelial backdiffusion from lumen to bath at low perfusion rates (Dantzler 1973). A flux from lumen to bath that appears to be passive and that varies with perfusion rate has been demonstrated with these tubules (Dantzler 1973). Moreover, the transepithelial permeability determined from this lumen-to-bath flux (about 2.4×10^{-5} cm s^{-1}) is four times that (0.60×10^{-5} cm s^{-1}) calculated from the independently measured luminal and basolateral membrane permeabilities, suggesting that much of this backflux must occur between the cells (Fig. 6.6) (Dantzler 1976a). Fourth, the kinetic data for net urate secretion in isolated, perfused snake renal tubules differ from those for net PAH secretion in the same preparation. The net secretory system for urate saturates at a much higher bath concentration than that for the net PAH secretory system (Dantzler 1973, 1974b). The apparent K_t for net urate secretion (about 150 μM), determined from the saturation data, is approximately 15 times that obtained for PAH under similar conditions (Table 6.3) (Dantzler 1982). Moreover, the J_{max} for net urate secretion (about 150 fmol min^{-1} mm^{-1}) in these tubules is only about one-half that for net PAH secretion (Table 6.3) (Dantzler 1973, 1974b). Even though the K_t for urate transport is substantially higher than the K_t for PAH transport, it is still well below the normal plasma urate level (400–500 μM) in these animals. This observation suggests that the urate secretory mechanism in snakes is normally saturated and that changes in plasma urate levels do not greatly alter the net urate secretion. Instead, the rate of flow through the lumen and, thus, the back-diffusion described above may be particularly important in determining net secretion and final net excretion.

As discussed previously (vide supra; Chap. 3), nephrons filter intermittently in reptiles, the number filtering increasing with increasing hydration and decreasing with dehydration. It appears likely that the relatively high passive permeability of the basolateral membrane, the low apparent passive permeability of the luminal membrane, the apparent dependence of basolateral transport into the cells on filtrate flowing through the lumen, and the apparent large paracellular lumen-to-bath backleak all function to reduce the accumulation of urate in the cells or lumens of nephrons that are not filtering.

Of additional significance with regard to net transepithelial urate secretion by reptilian renal tubules is the lack of convincing evidence for mediated urate transport across the luminal membrane. Neither unlabeled urate, nor probenecid, nor SITS has any effect on the movement of radioactively labeled urate from the cells to the lumen or on the apparent permeability of the luminal membrane in isolated, perfused snake proximal renal tubules (Dantzler and Bentley 1979; Mukherjee and Dantzler 1985). Studies with brush-border membrane vesicles from these snake renal tubules also provide no evidence of carrier-mediated transport of urate across this membrane (Benyajati and Dantzler 1988). Although these negative findings are compatible with simple passive diffusion of urate across the luminal membrane (Fig. 6.6), they do not prove it. Moreover, given the relatively large flux of urate across the luminal membrane during net secretion, simple passive diffusion appears unlikely.

In birds, studies ranging from the in vivo renal clearance level to the in vitro cellular and membrane level suggest that net renal tubular urate secretion involves primarily the pathway for PAH and other organic anions discussed above and, to a lesser extent, a separate urate-specific pathway (Fig. 6.7). Early in vivo studies on gouty chickens showed that in these animals, net urate secretion is impaired, apparently from impairment of the transport step into the cells at the basolateral membrane, but net PAH secretion is only impaired when the plasma concentration is near the level at which transport is maximum (Austic and Cole 1972; Zmuda and Quebbemann 1975). Other early in vivo data, this time on normal chickens, indicated that even levels of PAH sufficient to saturate the secretory system only partially inhibit net urate secretion and that, at this time, urate secretion can be further inhibited by adenine without influencing PAH secretion (Cacini and Quebbemann 1978). Studies with isolated, perfused, and nonperfused chicken proximal tubules indicate that the net secretion involves uptake into the cells at the basolateral membrane against an electrochemical gradient and movement from the cells into the lumen down an electrochemical gradient (Fig. 6.7) (Brokl et al. 1994). These studies and other studies with chicken proximal tubule epithelial cells in primary culture (Dudas et al. 2005) and turkey proximal tubule basolateral membrane vesicles (Grassl 2002) indicate that much, but not all, of the urate transport into the cells at the basolateral membrane involves the same pathway as

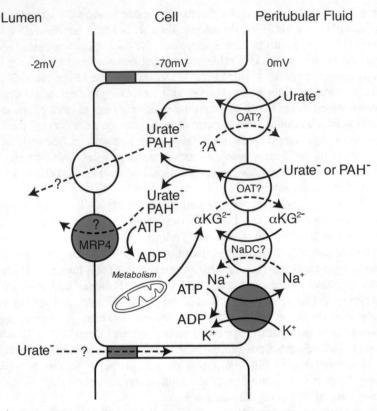

Fig. 6.7 Model for tubular secretion of urate and PAH in avian proximal tubules, based on studies with chickens and turkeys. Symbols have the same meaning as in Figs. 6.1 and 6.5. See text for more detailed description of processes diagramed in the figure

PAH with countertransport for αKG (Fig. 6.7). Since mRNA for OAT-like organic anion transporters has been found in cultured chicken proximal tubule epithelium (Dudas et al. 2005), it seems very likely that an OAT homolog is involved in countertransport of PAH and urate (and other organic anions) for αKG (Fig. 6.7) (vide supra; Fig. 6.5). However, inhibitor studies with isolated individual tubules and cultured tubule cells suggest that at least a portion of the urate transported into the cells at the basolateral membrane involves some other pathway with probable countertransport for some other unknown anion (Fig. 6.7) (Brokl et al. 1994; Dudas et al. 2005).

Although the transport step for both PAH and urate from the cells to the lumen in avian proximal tubules is down an electrochemical gradient, little is known about it. This step may be different for urate and PAH or other organic anions (Fig. 6.7). Indeed, as noted above with regard to PAH and other organic anions, there may be separate pathways across the luminal membrane for a number of organic anions of differing structures. However, studies with brush border membrane vesicles from turkey proximal tubules suggest that, in these tubules at least, urate and PAH move by the same facilitated diffusion pathway (Grassl 2002). In addition, mRNA for a multidrug resistance peptide 4 (MRP4)-like transporter has been found in epithelium of chicken proximal tubules, and inhibitor studies suggest that such a transporter may be involved in the movement of both urate and PAH across the luminal membrane (Fig. 6.7) (Dudas et al. 2005). Of course, if an MRP4 homolog is involved in urate and PAH transport across the luminal membrane, this movement would not be completely passive because this is an ATP-dependent transporter. Thus, both electrogenic carrier-mediated diffusion and ATP-dependent carrier-mediated flux may be involved in urate and PAH movement across the luminal membrane during net secretion in avian proximal renal tubules (Fig. 6.7).

Some transepithelial passive backflux of urate from lumen to peritubular fluid occurs during net secretion with isolated, perfused avian proximal tubules (Brokl et al. 1994). Because no urate accumulates in the tubule cells during this process, it seems likely that this backflux occurs by a paracellular route.

The concentration of urate in reptilian and avian plasma is well within the solubility limits of the predominant urate salts (Dantzler and Bradshaw 2009; Laverty and Dantzler 1983). However, as urate is secreted, the concentrations developed in the lumens of the renal tubules rapidly exceed even the solubility for potassium urate (12.06 mmol L^{-1}) and are well above the solubility for sodium urate (6.76 mmol L^{-1}) or the ammonium urate (3.21 mmol L^{-1}) found in alligator tubule fluid (Coulson and Hernandez 1964, 1983; Dantzler and Bradshaw 2009; Gudzent 1908; Laverty and Dantzler 1983). Nevertheless, no precipitates are observed in tubule lumens until the collecting ducts are reached (W. H. Dantzler, unpublished observations) (Laverty and Dantzler 1983). Precipitation in the proximal and distal tubules of avian nephrons is prevented by binding to serum albumen, which, as noted above (Chap. 3), is filtered to a significant extent by avian glomeruli (Braun 2009; Casotti and Braun 1996). A similar binding to proteins also appears to occur in reptilian renal tubules, but the proteins have not been defined and protein filtration has not been studied (Minnich 1972, 1976; Porter 1963). In birds, urate and its associated albumin form small spheres beginning in the early part of the proximal tubules and these remain in colloidal suspension until the urine becomes more concentrated in the collecting ducts (Casotti and Braun 2004; Laverty and Dantzler 1983). Apparently, this same situation occurs in reptilian renal tubules because the precipitates in collecting ducts also consist of small spheres, but this has not been examined as closely as in birds (Minnich and Piehl 1972). These spheres, even when precipitated, move smoothly along the collecting ducts without damaging the epithelium (W. H. Dantzler, unpublished observations) (Minnich and Piehl 1972).

Finally, with regard to birds and uricotelic reptiles, the relationship of inorganic cations to urate excretion may have consequences for other tubular transport systems. Sodium and potassium and, sometimes, calcium or magnesium may be found with the urate precipitates being excreted in the ureteral urine of many birds and uricotelic reptiles and amphibians (Dantzler 1978, 1992). The predominant cation may be determined by the diet and ionic requirements of each species. The chemical structure of the urate precipitates and the manner in which the cations are combined with them (not necessarily simple salts) have yet to be clearly defined (Dantzler 1978). Regardless of the nature of this chemical combination, the inorganic cations held in urate precipitates are excreted without contributing to the osmotic pressure of the urine. Therefore, the limited ability of the avian kidney and the lack of any ability of the reptilian and amphibian kidneys to concentrate solutes in the urine (vide infra; Chap. 7) may not be true indications of the ability of these kidneys in uricotelic species to excrete inorganic cations.

Some data suggest that when large amounts of sodium are combined with urate precipitates, the fraction of filtered water reabsorbed by the renal tubules may exceed the fraction of filtered sodium reabsorbed (Dantzler 1980). If this is the case, then isosmotic fluid reabsorption without sodium of the type discussed above may be required (see Chap. 5). The combination of sodium with urate precipitates also may function in reptilian distal nephrons to keep the concentration of free sodium low enough to permit continued reabsorption and maximum dilution by the sodium reabsorptive mechanism discussed above (see Chap. 4). In addition, the complexing of calcium with urate precipitates in the tubular fluid of some crocodiles (*Crocodylus porosus*) adapted to seawater (G. Grigg, personal communication) may facilitate its excretion in these animals. A portion of these inorganic cations combined with urate precipitates in the ureteral urine of reptiles and birds may be reclaimed—depending on the requirements of the animals—by reabsorption in the colon, coprodeum, cloaca, or bladder where less complex urate precipitates may be formed and urate salts may be converted to uric acid (Dantzler 1978, 1980; Larsen et al. 2014).

6.9   Lactate

Lactate, an endogenous monocarboxylic organic acid, which may be important in the energy metabolism of renal tubules, is freely filtered and then reabsorbed by the nephrons of all those vertebrates in which its renal handling has been studied (Table 6.1) (Dantzler 1992; Kinne and Kinne-Saffran 1985). Among nonmammalian vertebrates, this has involved only reptiles. However, studies with isolated proximal tubules from garter snakes (*Thamnophis* spp.) have been particularly revealing with regard to the transport processes by which lactate might be conserved and utilized by the renal tubules (Brand and Stansbury 1980a, b, 1981). Net transepithelial reabsorption occurs throughout the length of these snake proximal tubules via an energy-requiring, sodium-dependent process (Brand and Stansbury 1980a, b). It appears likely that the primary reabsorptive transport step in these reptilian tubules, as demonstrated for mammalian nephrons with rat brush-border membrane vesicles (Barac-Nieto et al. 1980, 1982), involves an electrogenic, sodium-coupled secondary active transport process at the luminal membrane. In mammals, this process has been identified as involving sodium-coupled monocarboxylate transporters 1 and 2 (SMCT1 and SMCT2) (Pelis and Wright

2011), and it seems likely that the molecular identity of the transporters is similar in reptiles. However, this remains to be examined directly.

Of particular interest is the observation that lactate is transported by perfused snake tubules from lumen to the bathing medium without being metabolized, whereas lactate that is transported into the cells of nonperfused tubules from the peritubular side is metabolized to a significant extent (Brand and Stansbury 1980a, b). In mammalian tubules, neither the transport step at the basolateral membrane nor the relationship of this step to the metabolism of the cells is well understood (Kinne and Kinne-Saffran 1985). However, these data on snake tubules suggest, although they do not demonstrate conclusively, that reabsorbed lactate and metabolized lactate are taken up by the tubule cells at opposite membranes and that the pools of reabsorbed and metabolized lactate are separate. For reptiles, at least, these data support the idea first put forward by Cohen (1964) that renal reabsorptive transport conserves substrate for the entire organism, whereas basolateral accumulation provides nutrients for the renal tubule cells (Brand and Stansbury 1980a).

6.10 Organic Cations and Bases

6.10.1 Direction and Sites of Net Transport

Many toxic, or potentially toxic, compounds in the natural environment are organic cations (or organic bases that exist as cations at physiological pH; collectively OCs) that, if ingested in any manner, must be eliminated by the kidneys. This includes about 40 % of prescribed drugs in humans. In addition, the systemic concentrations of a number of essential endogenous organic cations, e.g., choline and catecholamines, are regulated, at least in part, by the kidneys. Therefore, understanding the renal transport of these compounds in both mammalian and nonmammalian vertebrates is of considerable importance.

Because of the high toxicity of many of these compounds in vivo, early studies involved renal portal infusions in nonmammalian vertebrates, primarily chickens (*Gallus domesticus*), thereby permitting examination of tubular transport without the production of high systemic plasma levels (Rennick 1981a, b). These studies indicate that a number of endogenous organic cations, e.g., N^1-methylnicotinamide and catecholamines, and exogenous organic cations, e.g., tetraethylammonium (TEA) and morphine, undergo net tubular secretion (Table 6.1) (Rennick 1981b). Other endogenous organic cations, most notably choline, were shown to undergo net tubular reabsorption at normal plasma levels and net tubular secretion at elevated plasma levels in these studies on birds (Table 6.1) (Rennick 1981b). Additional early studies involving direct infusions through the renal arteries as well as standard clearance techniques for nontoxic organic cations indicate that the same pattern holds for mammals (Table 6.1) (Rennick 1981b; Weiner 1985).

A number of studies indicate that transport of organic cations, like transport of organic anions occurs along the proximal tubule, but there are differences in the quantitative importance of proximal regions. For example, studies with isolated, nonperfused proximal tubules from teleost fishes (Southern flounder, *Paralichthys lethostigma*; killifish, *Fundulus heteroclitus*) showed that transport of organic cations occurs in the proximal tubules of these animals (Smith et al. 1988). Work with isolated, perfused renal tubules of snakes (*Thamnophis* spp.) indicate that net tubular transport, secretion or reabsorption, of organic cations occurs throughout the proximal tubule of reptiles with no difference in magnitude between the proximal–proximal and distal–proximal segments (Dantzler, unpublished observations) (Dantzler and Brokl 1986; Hawk and Dantzler 1984), in contrast to secretion of organic anions, which occurs only in the distal-proximal segment (vide supra) (Dantzler 1974a). Studies with isolated perfused and nonperfused renal tubules from chickens also demonstrated that organic cation secretion occurs along the proximal tubule of birds (Brokl et al. 1994). Similarly, stop-flow studies on dogs, micropuncture studies on rats, and perfusions of isolated rabbit tubules indicate that secretion and reabsorption in mammals are functions of the proximal tubules (Schali et al. 1983; Weiner 1985). Moreover, studies with isolated, perfused rabbit proximal tubules clearly show that net secretion of the model exogenous organic cation, TEA, is greatest in the S1 segment of the proximal tubule and decreases along the tubule in the segment order S1 > S2 > S3, in contrast to net secretion of organic anions, which occurs primarily in the S2 segment (vide supra) (Schali et al. 1983; Woodhall et al. 1978). A similar pattern may be found in other mammalian species as well.

6.10.2 Mechanism of Transport

6.10.2.1 Net Secretion

Most renal tubular transport of organic cations in vertebrates involves net secretion, primarily of ingested xenobiotics. The vast majority of these are relatively small (molecular weight generally < 400 Da), monovalent, and structurally diverse compounds, frequently referred to as type I organic cations (Pelis and Wright 2011). Type II organic cations are larger (generally > 500 Da), frequently polyvalent, and hydrophobic. In mammals, at least, these type II organic cations are mainly secreted by the liver into the bile, but a few are also secreted by the renal tubules (Pelis and Wright 2011). The possible mechanism for their secretion is discussed briefly below, but most of the current discussion is concentrated on the mechanism for secretion of the dominant type 1 organic cations. It now appears that transepithelial secretion of these type 1 organic cations in proximal renal tubules occurs by a similar process in all vertebrates studied (Fig. 6.8).

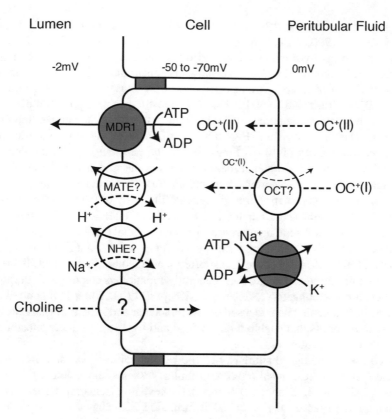

Fig. 6.8 Model for tubular secretion of organic cations and possible tubular reabsorption of choline, based on studies in fishes, reptiles, birds, and mammals. $OC^+(I)$ indicates Type I organic cations; $OC^+(II)$ indicates Type II organic cations. Other symbols have the same meaning as in Figs 6.1 and 6.5. See text for more detailed description of processes diagramed in the figure

Type I Organic Cations

Basolateral Organic Cation Entry Some studies of transepithelial organic cation secretion with isolated, perfused rabbit and snake renal tubules originally suggested that transport into the cells across the basolateral membrane might be against an electrochemical gradient (Hawk and Dantzler 1984; Schali et al. 1983), but the conclusions in these studies were based on measured intracellular concentrations that were above those expected from passive entry based on the membrane potential. It now appears likely that those high concentrations in these isolated, perfused tubules were the result of binding to intracellular proteins (Berndt 1981) or pH-dependent uptake into intracellular vesicles (Martinez-Guerrero et al. 2016; Pritchard et al. 1994). Most studies with basolateral membrane vesicles (BLMV) and isolated perfused and nonperfused renal tubules, including all recent studies, indicate that such transport is down an electrochemical gradient in fishes, reptiles, birds, and mammals (Fig. 6.8) (Dantzler, unpublished observations) (Brokl et al. 1994; Dantzler et al. 1991b; Pelis and Wright 2011; Smith et al. 1988; Sokol and McKinney 1990; Wright and Dantzler 2004). These same studies indicate that this entry step involves either facilitated diffusion or electroneutral exchange of organic cations and that these two mechanisms appear most likely to be alternative modes of action of a single transporter (Fig. 6.8) (Busch et al. 1996; Pelis and Wright 2011; Wright and Dantzler 2004). The inside-negative potential difference across the basolateral membrane of 50 to 70 mV (the magnitude depending on the vertebrate class and species) (Fig. 6.8) is sufficient to sustain an intracellular organic cation activity of 10–15 times that in the peritubular fluid, a calculation compatible with the experimental measurements in the studies noted above.

In mammals, the transporters responsible for this basolateral entry step are members of the organic cation transporter (OCT) family of transporters (Pelis and Wright 2011; Wright and Dantzler 2004). OCT1, OCT2, and OCT3 have been identified in the basolateral membranes of renal proximal tubules, the specific ones and the exact tubule site varying with species (Pelis and Wright 2011; Wright and Dantzler 2004). Although these transporters can mediate electroneutral organic cation/organic cation exchange, they generally function as electrogenic uniporters and, thus, are not obligatory exchangers (Koepsell et al. 2007; Pelis and Wright 2011). Many studies are now focused on the molecular and structural characteristics of these transporters that enable them to bind and transport so many substrates of widely diverse structure.

Given the physiological similarity of the basolateral transport step in mammalian and nonmammalian renal proximal tubules, it seems highly likely that homologues of the mammalian OCTs are the relevant basolateral transporters in nonmammalian species. However, no information is available on the molecular identity of the basolateral transporters in nonmammalian vertebrates.

Luminal Organic Cation Exit Studies with brush border membrane vesicles and isolated, perfused renal tubules from mammals, birds, and reptiles have demonstrated that the transport step for organic cations out of the cells at the luminal membrane during transepithelial secretion involves obligatory 1:1 electroneutral exchange for hydrogen ions (Fig. 6.8) (Dantzler and Brokl 1988; Dantzler et al. 1989, 1991a; Holohan and Ross 1980, 1981; Kinsella et al. 1979; Pritchard and Miller 1993; Villalobos and Braun 1995). This organic cation-hydrogen ion exchanger is a secondary active transport system, maintaining hydrogen ions away from electrochemical equilibrium (Pelis and Wright 2011), and is rate-limiting in the transepithelial organic cation secretory process (Schali et al. 1983). However, the overall organic cation exit step at the luminal membrane, like the overall entry step for organic anions at the basolateral membrane (vide supra; Fig. 6.5), can be considered a tertiary active transport process (Fig. 6.8). In this process, the secondary active sodium-proton exchanger in the luminal membrane maintains the inwardly directed hydrogen ion gradient for the action of the organic cation-hydrogen ion exchanger, and the primary active Na-K-ATPase in the basolateral membrane, in turn, maintains the inwardly directed sodium gradient to drive the sodium-proton exchanger (Fig. 6.8). In the late portions of the mammalian proximal tubules, the V-type H^+-ATPase may also play a role at the luminal membrane in maintaining the inwardly directed hydrogen ion gradient (Pelis and Wright 2011), but there is no evidence for this in nonmammalian vertebrates.

In mammals, the organic cation-hydrogen ion exchanger in the luminal membrane has been identified as involving two members of the multidrug and toxin extruder (MATE) family, MATE1 and MATE2-K (Pelis and Wright 2011), and substantial work on understanding the structure and multispecificity of these transporters is ongoing. Because of the physiological similarity of this transport step in the proximal tubules of mammalian and nonmammalian vertebrates, it seems likely that homologues of mammalian MATEs play this same role in nonmammalian vertebrates, but no information is yet available on the molecular identity of this transport step in nonmammalian vertebrates.

Type II Organic Cations

As noted above, relatively few type II organic cations are secreted by the renal tubules, and the mechanism of their secretion is less well studied in any species than the mechanism of type I secretion. The basolateral entry step is not clear, but these large, bulky compounds are much more lipophilic than type I organic cations, and studies of the secretion of a number of fluorescent type II organic cations in isolated teleost (killifish, *F. heteroclitus*) renal tubules indicate that they enter the cells across the basolateral membrane by simple passive diffusion driven by the electrical potential difference (Fig. 6.8) (Miller 1995; Miller et al. 1997; Schramm et al. 1995). These same studies, involving the use of numerous transport inhibitors, indicate that these compounds are probably transported out of the cells across the luminal membrane by the primary, ATP-requiring multidrug resistant transporter (MDR1), which is expressed in the luminal membrane (Fig. 6.8). This overall process may well hold for other type II organic cations and other species (Pelis and Wright 2011).

6.10.2.2   Net Reabsorption

Net tubular reabsorption occurs, under some circumstances, for a number of usually endogenous type I organic cations. As noted above, this is particularly notable for choline, which clearance studies on mammals (dogs and rabbits) and birds (chickens, *G. domesticus*) have shown to be reabsorbed at physiological plasma levels (10–20 μM) and secreted when plasma levels are raised above 100 μM (Acara and Rennick 1973; Besseghir et al. 1981). In vivo renal microperfusion studies in rats (Ullrich and Rumrich 1996) and in vitro studies with rabbit renal brush border membrane vesicles (Wright et al. 1992) showed that there is an electrogenic uniporter with a high affinity for choline (and compounds of similar structure) in the luminal membrane. This transporter tends to move choline down its electrochemical gradient from the lumen into the cells (Fig. 6.8). In contrast, the organic cation-hydrogen ion transporter in the luminal membrane, which moves organic cations out of the cells into the lumen (discussed above; Fig. 6.8), has a low affinity (but large capacity) for choline. Therefore, at low (physiological) plasma levels and, consequently, low levels in the filtrate and proximal tubule lumen, choline will be reabsorbed (as observed in the clearance studies above). However, when the plasma levels are high, choline will be readily secreted in the manner described above.

As in mammalian kidneys, studies with brush border membrane vesicles from avian kidneys (chickens, *G. domesticus*) indicate that the secretory organic cation-hydrogen exchanger in the luminal membrane has a low affinity for choline (Villalobos and Braun 1998). It seems likely, given the similarity of the renal choline clearances in birds and mammals, that bird renal proximal tubules also have a high-affinity electrogenic choline uniporter in the luminal membrane, but this has yet to be examined. The renal tubular transport of choline has yet to be examined in detail in species from other vertebrate classes.

However, studies of transport of the endogenous organic cation N^1-methylnicotinamide (NMN) by isolated, perfused snake renal proximal tubules showed that this compound undergoes net reabsorption (Dantzler and Brokl 1986, 1987). The rate of net reabsorption is relatively low and reflects a small difference between a large transepithelial lumen-to-bath flux and an almost equally large bath-to-lumen flux. In these fluxes, NMN enters the cells across the luminal membrane and across the peritubular membrane down an electrochemical gradient by a carrier-mediated, sodium-dependent process that is not inhibited by TEA. It is then transported out of the cells against an electrochemical gradient at the opposite membrane. Inhibitor studies with analogues of NMN and with mepiperphenidol suggest that a ring configuration containing quaternary ammonium is essential and that the transporter into the cells down an electrochemical gradient at the luminal membrane, which appears to be the dominant one, has a greater specificity for the NMN structure than the transporter into the cells at the basolateral membrane (Dantzler and Brokl 1986, 1987). Current data suggest that transport of NMN out of the cells at the basolateral membrane, which apparently is greater than transport at the luminal membrane, involves countertransport for some other cation, but the physiological nature of this other cation is unclear (Dantzler and Brokl 1986, 1987). Moreover, although the bath-to-lumen flux for NMN could involve the same transporters as describe above for secretion of other type I organic cations, this has not been directly determined in these reptilian tubules.

References

Acara M, Rennick B (1973) Regulation of plasma choline by the renal tubule: bidirectional transport of choline. Am J Physiol 225:1123–1128

Althoff T, Hentschel H, Luig J, Schütz H, Kasch M, Kinne RKH (2006) Na$^+$-D-glucose cotransporter in the kidney of *Squalus acanthias*: molecular identification and intrarenal distribution. Am J Physiol Regul Integr Comp Physiol 290:R1094–R1104

Althoff T, Hentschel H, Luig J, Schütz H, Kasch M, Kinne RKH (2007) Na$^+$-D-glucose cotransporter in the kidney of *Leucoraja erinacea*: molecular identification and intrarenal distribution. Am J Physiol Regul Integr Comp Physiol 292:R2391–R2399

Austic RE, Cole RK (1972) Impaired renal clearance of uric acid in chickens having hyperuricemia and articular gout. Am J Physiol 223:525–530

Balinski JB, Baldwin E (1961) Comparative studies of nitrogen metabolism in amphibia. Biochem J 82:187–191

Barac-Nieto M, Murer H, Kinne R (1980) Lactate-sodium cotransport in rat renal brush border membranes. Am J Physiol Renal Physiol 239:F496–F506

Barac-Nieto M, Murer H, Kinne R (1982) Asymmetry in the transport of lactate by basolateral and brush border membranes of rat kidney cortex. Pflugers Arch 392:366–371

Barfuss DW, Dantzler WH (1976) Glucose transport in isolated perfused proximal tubules of snake kidney. Am J Physiol 231:1716–1728

Baze WB, Horne FR (1970) Ureogenesis in chelonia. Comp Biochem Physiol 34:91–100

Benyajati S, Dantzler WH (1986a) Plasma levels and renal handling of endogenous amino acids in snakes: a comparative study. J Exp Zool 238:17–28

Benyajati S, Dantzler WH (1986b) Renal secretion of amino acids in ophidian reptiles. Am J Physiol Regul Integr Comp Physiol 250:R712–R720

Benyajati S, Dantzler WH (1988) Enzymatic and transport characteristics of isolated snake renal brush-border membranes. Am J Physiol Regul Integr Comp Physiol 255:R52–R60

Berndt WO (1967) Probenecid binding by renal cortical slices and homogenates. Proc Soc Exp Biol Med 126:123–126

Berndt WO (1981) Organic base transport: a comparative study. Pharmacology 22:251–262

Berner W, Kinne R (1976) Transport of *p*-aminohippuric acid by plasma membrane vesicles isolated from rat kidney cortex. Pflugers Arch 361:269–277

Besseghir K, Pearce LB, Rennick B (1981) Renal tubular transport and metabolism of organic cations by the rabbit. Am J Physiol Renal Physiol 241:F308–F314

Beyer KH Jr, Gelarden RT (1988) Active transport of urea by mammalian kidney. Proc Natl Acad Sci U S A 85:4030–4031

Beyer KH Jr, Gelarden RT, Vary JE, Brown LE, Vesell ES (1990) Novel multivalent effects of pyrazinonylguanidine in patients with azotemia. Clin Pharmacol Ther 47:626–638

Beyer KH Jr, Gelarden RT, Vesell ES (1992) Inhibition of urea transport across renal tubules by pyrazinoylguanidine and analogs. Pharmacology 44:124–138

Blomstedt JW, Aronson PS (1980) pH gradient-stimulated transport of urate and *P*-aminohippurate in dog renal microvillus membrane vesicles. J Clin Invest 65:931–934

Boorman KN, Falconer IR (1972) The renal reabsorption of histidine, leucine and the cationic amino acids in the young cockeral (*Gallus domesticus*). Comp Biochem Physiol 42A:311–320

Boumendil-Podevin EF, Podevin R-A, Priol C (1979) Uric acid transport in brush border membrane vesicles isolated from rabbit kidney. Am J Physiol Renal Physiol 236:F519–F525

Boylan JW (1972) A model for passive urea reabsorption in the elasmobranch kidney. Comp Biochem Physiol A 42:27–30

Brand PH, Stansbury RS (1980a) Lactate absorption in Thamnophis proximal tubule: transport versus metabolism. Am J Physiol Renal Physiol 238:F218–F228

Brand PH, Stansbury R (1980b) Peritubular uptake of lactate by Thamnophis proximal tubule. Am J Physiol Renal Physiol 238:F296–F304

Brand PH, Stansbury RS (1981) Lactate transport by Thamnophis proximal tubule: sodium dependence. Am J Physiol Renal Physiol 240:F388–F394

Braun EJ (2009) Osmotic and ionic regulation in birds. In: Evans DH (ed) Osmotic and ionic regulation: cells and animals. CRC Press, Boca Raton, FL, pp 505–524

Braun EJ, Sweazea KL (2008) Glucose regulation in birds. Comp Biochem Physiol B 151:1–9

Brokl OH, Braun EJ, Dantzler WH (1994) Transport of PAH, urate, TEA, and fluid by isolated perfused and nonperfused avian renal proximal tubules. Am J Physiol Regul Integr Comp Physiol 266:R1085–R1094

Burg MB, Weller PF (1969) Iodopyracet transport by isolated perfused flounder proximal renal tubules. Am J Physiol 217:1053–1056

Burnham CE, Amlal H, Wang ZH, Shull GE, Soleimani M (1997) Cloning and functional expression of a human kidney Na$^+$: HCO$_3^-$ cotransporter. J Biol Chem 272:19111–19114

Busch AE, Quester S, Ulzheimer JC, Waldegger S, Gorboulev V, Arndt P, Lang F, Koepsell H (1996) Electrogenic properties and substrate specificity of the polyspecific rat cation transporter rOCT1. J Biol Chem 271:32599–32604

Cacini W, Quebbemann AJ (1978) The metabolism and active excretion of hypoxanthine by the renal tubules in the chicken. J Pharmacol Exp Ther 207:574–583

Cameron JN, Kormanik GA (1982) Acid-base responses of gills and kidneys to infused acid and base loads in channel catfish, *Ictalurus punctatus*. J Exp Biol 99:143–160

Casotti G, Braun EJ (1996) Functional morphology of the glomerular filtration barrier of *Gallus gallus*. J Morphol 228:327–334

Casotti G, Braun EJ (2004) Protein location and elemental composition of urine spheres in different avian species. J Exp Zool 301:579–587

Chambers R, Beck LV, Belkin M (1935) Secretion in tissue cultures. I. Inhibition of phenol red accumulation in the chick kidney. J Cell Comp Physiol 6:425–439

Chatsudthipong V, Dantzler WH (1991) PAH-a-KG countertransport stimulates PAH uptake and net secretion in isolated snake renal tubules. Am J Physiol Renal Physiol 261:F858–F867

Chatsudthipong V, Dantzler WH (1992) PAH/a-KG countertransport stimulates PAH uptake and net secretion in isolated rabbit renal tubules. Am J Physiol Renal Physiol 263:F384–F391

Chatsudthipong V, Jutabha P, Evans KK, Dantzler WH (1999) Effects of inhibitors and substitutes for chloride in lumen on *p*-aminohippurate transport by isolated perfused rabbit renal proximal tubules. J Pharmacol Exp Ther 288:993–1001

Chin TY, Quebbemann AJ (1978) Quantitation of renal uric acid synthesis in the chicken. Am J Physiol Renal Physiol 234:F446–F451

Chou C-L, Knepper MA (1993) In vitro perfusion of chinchilla thin limb segments: Urea and NaCl permeabilities. Am J Physiol Renal Physiol 264:F337–F343

Cohen JJ (1964) Specificity of substrate utilization by the dog kidney *in vivo*. In: Metcalf J (ed) Renal metabolism and epidemiology of some renal diseases. Nat Kidney Found, New York, pp 126–146

Cooper CA, Wilson JM, Wright PA (2013) Marine, freshwater and aerially acclimated mangrove rivulus (*Kryptolebias marmoratus*) use different strategies for cutaneous ammonia excretion. Am J Physiol Regul Integr Comp Physiol 304:R599–R612

Coulson RA, Hernandez T (1964) Biochemistry of the alligator. Louisiana State University Press, Baton Rouge, LA

Coulson RA, Hernandez T (1970) Nitrogen metabolism and excretion in the living reptile. In: Campbell JW (ed) Comparative biochemistry of nitrogen metabolism, vol 2, The vertebrates. Academic, New York, pp 640–710

Coulson RA, Hernandez T (1983) Alligator metabolism: studies on chemical reactions *in vivo*. Comp Biochem Physiol 74:1–182

Couriaud C, Leroy C, Simon M, Silberstein C, Bailly P, Ripoche P, Rousselet G (1999) Molecular and functional characterization of an amphibian transporter. Biochim Biophys Acta 1421:347–352

Craan AG, Lemieux G, Vinay P, Gougoux A (1982) The kidney of chicken adapts to chronic metabolic acidosis: in vivo and in vitro studies. Kidney Int 22:103–111

Craan AG, Vinay P, Lemieux G, Gougoux A (1983) Metabolism and transport of L-glutamine and L-alanine by renal tubules of chickens. Am J Physiol Renal Physiol 245:F142–F150

Cross RJ, Taggart JV (1950) Renal tubular transport: accumulation of p-aminohippurate by rabbit kidney slices. Am J Physiol 161:181–190

Danielson RA, Schmidt-Nielsen B (1972) Recirculation of urea analogs from renal collecting ducts of high and low-protein-fed rats. Am J Physiol 233:130–137

Dantzler WH (1968) Effect of metabolic alkalosis and acidosis on tubular urate secretion in water snakes. Am J Physiol 215:747–751

Dantzler WH (1970) Kidney function in desert vertebrates. In: Benson GK, Phillips JG (eds) Memoirs of the society of endocrinology, vol 18, Hormones and the environment. Cambridge University Press, London, pp 157–190

Dantzler WH (1973) Characteristics of urate transport by isolated perfused snake proximal renal tubules. Am J Physiol 224(2):445–453

Dantzler WH (1974a) PAH transport by snake proximal renal tubules: differences from urate transport. Am J Physiol 226:634–641

Dantzler WH (1974b) K^+ effects on PAH transport and membrane permeabilities in isolated snake renal tubules. Am J Physiol 227:1361–1370

Dantzler WH (1976a) Comparison of uric acid and PAH transport by isolated, perfused snake renal tubules. In: Silbernagl S, Lang F, Greger R (eds) Amino acid transport and uric acid transport. Georg Thieme, Stuttgart, pp 169–180

Dantzler WH (1976b) Renal function (with special emphasis on nitrogen excretion). In: Gans CG, Dawson WR (eds) Biology of Reptilia, vol 5, Physiology A. Academic, London, pp 447–503

Dantzler WH (1978) Urate excretion in nonmammalian vertebrates. In: Kelley WN, Weiner IM (eds) Uric acid, vol 51, Handbook of experimental pharmacology. Springer, Berlin, pp 185–210

Dantzler WH (1980) Renal mechanisms for osmoregulation in reptiles and birds. In: Gilles R (ed) Animals and environmental fitness. Pergamon Press, Oxford, pp 91–110

Dantzler WH (1981) Comparative physiology of the renal transport of organic solutes. In: Greger R, Lang F, Silbernagl S (eds) Renal transport of organic substances. Springer, Berlin, pp 290–308

Dantzler WH (1982) Studies on nonmammalian nephrons. Kidney Int 22:560–570

Dantzler WH (1992) Comparative aspects of renal function. In: Seldin DW, Giebisch G (eds) The kidney: physiology and pathophysiology, 2nd edn. Raven, New York, pp 885–942

Dantzler WH, Bentley SK (1976) Low Na^+ effects on PAH transport and permeabilities in isolated snake renal tubules. Am J Physiol 230:256–262

Dantzler WH, Bentley SK (1979) Effects of inhibitors in lumen on PAH and urate transport by isolated renal tubules. Am J Physiol Renal Physiol 236:F379–F386

Dantzler WH, Bentley SK (1980) Bath and lumen effects of SITS on PAH transport by isolated perfused renal tubules. Am J Physiol Renal Physiol 238:F16–F25

Dantzler WH, Bentley SK (1981) Effects of chloride substitutes on PAH transport by isolated perfused renal tubules. Am J Physiol Renal Physiol 241:F632–F644

Dantzler WH, Bradshaw SD (2009) Osmotic and ionic regulation in reptiles. In: Evans DH (ed) Osmotic and ionic regulation: cells and animals. CRC Press, Boca Raton, FL, pp 443–503

Dantzler WH, Brokl OH (1984a) Effects of low $[Ca^{2+}]$ and La^{3+} on PAH transport by isolated perfused renal tubules. Am J Physiol Renal Physiol 246:F175–F187

Dantzler WH, Brokl OH (1984b) Verapamil and quinidine effects on PAH transport by isolated perfused renal tubules. Am J Physiol Renal Physiol 246:F188–F200

Dantzler WH, Brokl OH (1986) N^1-methylnicotinamide transport by isolated perfused snake proximal renal tubules. Am J Physiol Renal Physiol 250:F407–F418

Dantzler WH, Brokl OH (1987) NMN transport by snake renal tubules: choline effects, countertransport, H^+-NMN exchange. Am J Physiol Renal Physiol 253:F656–F663

Dantzler WH, Brokl OH (1988) TEA transport by snake renal tubules: choline effects, countertransport, H^+-TEA exchange. Am J Physiol Renal Physiol 255:F167–F176

Dantzler WH, Brokl OH, Wright SH (1989) Brush-border TEA transport in intact proximal tubules and isolated membrane vesicles. Am J Physiol Renal Physiol 256:F290–F297

Dantzler WH, Evans KK (1996) Effect of α-KG in lumen on PAH transport by isolated perfused rabbit renal proximal tubules. Am J Physiol Renal Physiol 271:F521–F526

Dantzler WH, Evans KK, Wright SH (1995) Kinetics of interactions of p-aminohippurate, probenecid, cysteine conjugates and N-acetyl cysteine conjugates with basolateral organic anion transporter in isolated rabbit proximal renal tubules. J Pharmacol Exp Ther 272:663–672

Dantzler WH, Pannabecker TL, Layton AT, Layton HE (2011) Urine concentrating mechanism in the inner medulla of the mammalian kidney: role of three-dimensional architecture. Acta Physiol 202:361–378

Dantzler WH, Schmidt-Nielsen B (1966) Excretion in fresh-water turtle (*Pseudemys scripta*) and desert tortoise (*Gopherus agassizii*). Am J Physiol 210:198–210

Dantzler WH, Wright SH (1997) Renal tubular transport of organic anions and cations. In: Sipes IG, McQueen CA, Gandolfi AJ (eds) Comprehensive toxicology, vol 7, Renal toxicology. Elsevier (Pergamon), Oxford, pp 61–75

Dantzler WH, Wright SH, Brokl OH (1991a) Tetraethylammonium transport by snake renal brush-border membrane vesicles. Pflugers Arch 418:325–332

Dantzler WH, Wright SH, Chatsudthipong V, Brokl OH (1991b) Basolateral tetraethylammonium transport in intact tubules: specificity and trans-stimulation. Am J Physiol Renal Physiol 261 (30):F386–F392

Deetjen P, Maren T (1974) The dissociation between renal HCO_3^- reabsorption and H^+ secretion in the skate, *Rajo erinacea*. Pflugers Arch 346:25–30

Dessauer HC (1952) Biochemical studies on the lizard, *Anolis carolinensis*. Proc Soc Exp Biol Med 80:742–744

Dudas PL, Pelis RM, Braun EJ, Renfro JL (2005) Transepithelial urate transport by avian renal proximal tubule epithelium in primary culture. J Exp Biol 208:4305–4315

Eveloff J (1987) p-Aminohippurate transport in basal-lateral membrane vesicles from rabbit renal cortex: stimulation by pH and sodium gradients. Biochim Biophys Acta 897:474–480

Eveloff J, Field M, Kinne R, Murer H (1980) Sodium-cotransport systems in intestine and kidney of the winter flounder. J Comp Physiol 135:175–182

Eveloff J, Kinne R, Kinter WB (1979) p-Aminohippuric acid transport into brush-border vesicles isolated from flounder kidney. Am J Physiol Renal Physiol 237:F291–F298

Eveloff J, Morishige WK, Hong SK (1976) The binding of phenol red to rabbit renal cortex. Biochim Biophys Acta 448:167–180

Fange R, Krog J (1963) Inability of kidney of the hagfish to secrete phenol red. Nature 199:713

Flöge J, Stolte H, Kinne R (1984) Presence of a sodium-dependent D-glucose transport system in the kidney of the atlantic hagfish (*Myxine glutinosa*). J Comp Physiol B 154:355–364

Forster RP (1970) Active tubular transport of urea and its role in environmental physiology. In: Schmidt-Nielsen B, Kerr DWS (eds) Urea and the kidney. Excerpta Medica Found, Amsterdam, pp 229–237

Frazier LW, Vanatta JC (1972) Mechanism of acidification of the mucosal fluid by the toad urinary bladder. Biochim Biophys Acta 290:168–177

Freire CA, Kinne-Saffran E, Beyenbach KW, Kinne RKH (1995) Na-D-glucose cotransport in renal brush-border membrane vesicles of an early teleost (*Oncorhynchus mykiss*). Am J Physiol Regul Integr Comp Physiol 269:R592–R602

Friedman PA, Hebert SC (1985) Urea and water transport in the isolated perfused tubules of the dogfish, *Squalus acanthias*. Bull Mt Desert Isl Biol Lab 25:24–26

Friedman PA, Hebert SC (1990) Diluting segment in kidney of dogfish shark. I. Localization and characterization of chloride absorption. Am J Physiol Renal Physiol 258(2):R398–R408

Garvin JL, Burg MB, Knepper MA (1987) NH_3 and NH_4^+ transport by rabbit renal proximal straight tubules. Am J Physiol Renal Physiol 252:F232–F239

Giebisch G (1956) Measurements of pH, chloride, and insulin concentrations in proximal tubule fluid of *Necturus*. Am J Physiol 185:171–174

Goldstein L (1976) Ammonia production and excretion in the mammalian kidney. Int Rev Physiol 11:284–315

Goldstein L, Forster RP (1971) Osmoregulation and urea metabolism in the little skate *Raja erinacea*. Am J Physiol 220:742–746

Good DW, Knepper MA (1985) Ammonia transport in the mammalian kidney. Am J Physiol Renal Physiol 248:F459–F471

Grantham JJ (1982) Studies of organic anion and cation transport in isolated segments of proximal tubules. Kidney Int 22:519–525

Grassl SM (2002) Urate/a-ketoglutarate exchange in avian basolateral membrane vesicles. Am J Physiol Cell Physiol 283:C1144–C1154

Gudzent F (1908) Physikalisch-chemusche untersuchungen uber das verhakten der harnsauren salze losungen. Hoppe Seylers Zeitschrift fur Physiol Chemie 56:150–179

Halperin ML, Goldstein MB, Sinebaugh BJ, Jungas RL (1985) Biochemistry and physiology of ammonium excretion. In: Seldin DW, Giebisch G (eds) The kidney: physiology and pathophysiology. Raven, New York, pp 1471–1490

Hawk CT, Dantzler WH (1984) Tetraethylammonium transport by isolated perfused snake renal tubules. Am J Physiol Renal Physiol 246:F476–F487

Hernandez T, Coulson RA (1967) Amino acid excretion in the alligator. Comp Biochem Physiol 23:775–784

Hill L, Dawbin WH (1969) Nitrogen excretion in the tuatara, Sphenodon punctatus. Comp Biochem Physiol 31:453–468

Hodler J, Heinemann HO, Fishman AP, Smith HW (1955) Urine pH and carbonic anhydrase activity in marine dogfish. Am J Physiol 93:155–162

Holohan PD, Ross CR (1980) Mechanisms of organic cation transport in kidney plasma membrane vesicles. 1. Counter-transport studies. J Pharmacol Exp Ther 215:191–197

Holohan PD, Ross CR (1981) Mechanisms of organic cation transport in kidney plasma membrane vesicles: 2. pH studies. J Pharmacol Exp Ther 216:294–298

Hoshi T, Sudo K, Suzuki Y (1976) Characteristics of changes in the intracellular potential associated with transport of neutral, dibasic and acidic amino acids in Triturus proximal tubule. Biochim Biophys Acta 448:492–504

Husted RF, Cohen LH, Steinmetz PR (1979) Pathways for bicarbonate transfer across the serosal membrane of turtle urinary bladder: studies with a disulfonic stilbene. J Membr Biol 47:27–37

Hyodo S, Kakumura K, Takagi W, Hasegawa K, Yamaguchi Y (2014) Morphological and functional characteristics of the kidney of cartilaginous fishes: with special reference to urea reabsorption. Am J Physiol Regul Integr Comp Physiol 307:R1381–R1395

Hyodo S, Kato F, Kaneko T, Takai Y (2004) A facilitative urea transporter is localized in the renal collecting tubule of the dogfish Triakis scyllia. J Exp Biol 207:347–356

Irish JM III (1975) Selected characteristics of transport in isolated perfused renal proximal tubules of the bullfrog (Rana catesbeiana). Dissertation, University of Arizona, Tucson, AZ

Irish JM III, Dantzler WH (1976) PAH transport and fluid absorption by isolated perfused frog proximal renal tubules. Am J Physiol 230:1509–1516

Isozaki T, Lea JP, Tumlin JA, Sands JM (1994) Sodium-dependent net urea transport in rat initial inner medullary collecting ducts. J Clin Invest 94:1513–1517

Janech MG, Fitzgibbon WR, Chen R, Nowak MW, Miller DH, Paul RV, Ploth DW (2003) Molecular and functional characterization of a urea transporter from the kidney of the Atlantic stingray. Am J Physiol Renal Physiol 284:F996–F1005

Janech MG, Fitzgibbon WR, Nowak MW, Miller DH, Paul RV, Ploth DW (2006) Cloning and functional characteristics of a second urea transporter (strUT-2) from the kidney of the Atlantic stingray, Dasyatis sabina. Am J Physiol Regul Integr Comp Physiol 291:R844–R853

Janech MG, Gefroh H, Cwengros EE, Sulikowski JA, Ploth DW, Fitzgibbon WR (2008) Cloning of urea transporters from the kidneys of two batoid elasmobranchs: evidence for a common elasmobranch urea transporter isoform. Mar Biol 153:1173–1179

Kahn AM, Aronson PS (1983) Urate transport via anion exchange in dog renal microvillus membrane vesicles. Am J Physiol Renal Physiol 244:F56–F63

Kahn AM, Branham S, Weinman EJ (1983) Mechanism of urate and p-aminohippurate transport in rat renal microvillus membrane vesicles. Am J Physiol Renal Physiol 245:F151–F158

Kahn AM, Weinman EJ (1985) Urate transport in the proximal tubule: in vivo and vesicle studies. Am J Physiol Renal Physiol 249:F789–F798

Kasher JS, Holohan PD, Ross CR (1983) Na$^+$ gradient-dependent p-aminohippurate (PAH) transport in rat basolaterial membrane vesicles. J Pharmacol Exp Ther 227:122–129

Kato K, Sands JM (1998) Evidence for sodium-dependent active urea secretion in the deepest subsegment of the rat inner medullary collecting duct. J Clin Invest 101:423–428

Khalil F (1948a) Excretion in reptiles. II. Nitrogen constituents of urinary concretions of the oviparous snake Zamensis diadema. J Biol Chem 172:101–103

Khalil F (1948b) Excretion in reptiles. III. Nitrogen constituents of urinary concretions of the viviparous snake Eryx thebaicus. J Biol Chem 172:105–106

Khalil F (1951) Excretion in reptiles. IV. Nitrogenous constituents of the excreta of lizards. J Biol Chem 189:443–445

Khalil F, Haggag G (1958) Nitrogenous excretion in crocodiles. J Exp Biol 35:552–555

Khuri RN, Flanigan WJ, Oken DE, Solomon AK (1966) Influence of electrolytes on glucose absorption in *Necturus* kidney proximal tubules. Fed Proc 25:899–902

Kim YK, Brokl OH, Dantzler WH (1997) Regulation of intracellular pH in avian renal proximal tubules. Am J Physiol Regul Integr Comp Physiol 272:R341–R349

King PA, Beyenbach KW, Goldstein L (1982) Taurine transport by isolated flounder renal tubules. J Exp Zool 223:103–114

King PA, Goldstein L (1983a) Renal ammonia excretion and production in goldfish, *Carassium auratus*, at low environmental pH. Am J Physiol Regul Integr Comp Physiol 245:R590–R599

King PA, Goldstein L (1983b) Renal ammoniagenesis and acid excretion in the dogfish, *Squalus acanthias*. Am J Physiol Regul Integr Comp Physiol 245:R581–R598

King PA, Goldstein L (1985) Renal excretion of nitrogenous products in vertebrates. Renal Physiol 8:261–278

King PA, Kinne R, Goldstein L (1985) Taurine transport by brush border membrane vesicles isolated from the flounder kidney. J Comp Physiol B 155:185–193

Kinne R, Kinne-Saffran E (1985) Renal metabolism: coupling of luminal and antiluminal transport processes. In: Seldin DW, Giebisch G (eds) The kidney: physiology and pathophysiology. Raven, New York, pp 719–737

Kinsella JL, Holohan PD, Pessah NI, Ross CR (1979) Transport of organic ions in renal cortical luminal and antiluminal membrane vesicles. J Pharmacol Exp Ther 209:443–450

Kipp H, Kinne-Saffran E, Bevan C, Kinne RKH (1997) Characteristics of renal Na^+-D-glucose cotransport in the skate (*Rajo erinacea*) and shark (*Squalus acanthias*). Am J Physiol Regul Integr Comp Physiol 273:R134–R142

Kippen I, Hirayama B, Klinenberg JR, Wright EM (1979) Transport of p-aminohippuric acid, uric acid and glucose in highly purified rabbit renal brush border membranes. Biochim Biophys Acta 556:161–174

Knepper MA, Mindell JA (2009) Molecular coin slots for urea. Nature 462:733–734

Koepsell H, Lips K, Volk C (2007) Polyspecific organic cation transporters: structure, function, physiological roles, and biopharmaceutical inplications. Pharm Res 24:1227–1251

Kono T, Nishida M, Nishiki Y, Seki Y, Sato K, Akiba Y (2005) Characterisation of glucose transporter (GLUT) gene expression in broiler chickens. Br Poultry Sci 46:510–515

Larsen EH, Deaton LE, Onken H, O'Donnell M, Grosell M, Dantzler WH, Weihrauch D (2014) Osmoregulation and excretion. Compr Physiol 4:405–573

Laverty G, Alberici M (1987) Micropuncture study of proximal tubule pH in avian kidney. Am J Physiol Renal Physiol 253:R587–R591

Laverty G, Dantzler WH (1982) Micropuncture of superficial nephrons in avian (*Sturnus vulgaris*) kidney. Am J Physiol Renal Physiol 243:F561–F569

Laverty G, Dantzler WH (1983) Micropuncture study of urate transport by superficial nephrons in avian (*Sturnus vulgaris*) kidney. Pflugers Arch 397:232–236

Lee S-H, Pritchard JB (1983) Proton-coupled L-lysine uptake by renal brush border membrane vesicles from mullet (*Mugil cephalus*). J Membr Biol 75:171–178

Lemieux G, Berkofsky J, Quenneville A, Lemieux C (1985) Net tubular secretion of bicarbonate by the alligator kidney. Antimammalian response to acetazolamide. Kidney Int 28:760–766

Lemieux G, Craan AG, Quenneville A, Lemieux C, Berkofsky J, Lewis VS (1984) Metabolic machinery of the alligator kidney. Am J Physiol Renal Physiol 247:F686–F693

Leslie BR, Schwartz JH, Steinmetz PR (1973) Coupling between Cl^- absorption and HCO_3^- secretion in turtle urinary bladder. Am J Physiol 225:610–617

Levin EJ, Cao Y, Enkavi G, Quick M, Pan Y, Tajkhorshid E, Zhou M (2012) Structure and permeation mechanism of a mammalian urea transporter. Proc Natl Acad Sci U S A 109:11194–11199

Levin EJ, Quick M, Zhou M (2009) Crystal structure of a bacterial homologue of the kidney urea transporter. Nature 462:757–762

Lim S-W, Han K-H, Jung J-Y, Kim W-Y, Yang C-W, Sands JM, Knepper MA, Madsen KM, Kim J (2006) Ultrastructural localization of UT-A and UT-B in rat kidneys with different hydration status. Am J Physiol Regul Integr Comp Physiol 290:R479–R492

Long S, Skadhauge E (1983) Renal acid excretion in the domestic fowl. J Exp Biol 104:51–58

Long WS (1973) Renal handling of urea in *Rana catesbeiana*. Am J Physiol 224:482–490

Love JK, Lifson N (1958) Transtubular movements of urea in the doubly perfused bullfrog kidney. Am J Physiol 193:662–668

Loveridge JP (1970) Observations on nitrogenous excretion and water relations of *Chiromantis xerampelina* (Amphibia, Anura). Arnoldia (Rhodesia) 5:1–6

Malvin RL, Cafruny EJ, Kutchai H (1965) Renal transport of glucose by the aglomerular fish *Lophius americanus*. J Cell Comp Physiol 65:381–384

Mandal AK, Mount DB (2015) The molecular physiology of uric acid homeostasis. Annu Rev Physiol 77:323–345

Marini AM, Matassi G, Raynal V, Andre B, Cartron JP, Cherif-Zahar B (2000) The human Rhesus-associated RhAG protein and a kidney homologue promote ammonium transport in yeast. Nat Genet 26:341–344

Marshall EK Jr, Grafflin AL (1928) The structure and function of the kidney of *Lophius piscatorius*. Bull Johns Hopkins Hosp 43:205–230

Martinez-Guerrero LJ, Evans KK, Dantzler WH, Wright SH (2016) The multidrug transporter, MATE1, sequesters OCs within an intracellular compartment that has no influence on OC secretion in renal proximal tubules. Am J Physiol Renal Physiol 310(1):F57–67

Maruyama T, Hoshi T (1972) The effect of D-glucose on the electrical potential profile across the proximal tubule of newt kidney. Biochim Biophys Acta 282:214–225

Maxild J (1978) Effect of externally added ATP and related compounds on active transport of *p*-aminohippurate and metabolism in cortical slices of the rabbit kidney. Arch Int Physiol Biochim 86:509–530

McBean RL, Goldstein L (1967) Ornithine-urea cycle activity in *Xenopus laevis*: adaptation in saline. Science 157:931–932

McDonald DG, Wood CM (1981) Branchial and renal acid and ion fluxes in the rainbow trout, *Salmo gairdneri*, at low environmental pH. J Exp Biol 93:101–118

McNabb FMA, McNabb RA (1975) Proportions of ammonia, urea, urate and total nitrogen in avian urine and quantitative methods for their analysis on a single urine sample. Poultry Sci 54:1498–1505

McNabb FMA, McNabb RA, Prather ID, Conner RN, Adkisson CS (1980) Nitrogen excretion by turkey vultures. The Condor 82:219–223

Miller DS (1995) Daunomycin secretion by killifish renal proximal tubules. Am J Physiol Regul Integr Comp Physiol 269:R370–R379

Miller DS, Fricker G, Drewe J (1997) *p*-Glycoprotein-mediated transport of a fluorescent rapamycin derivative in renal proximal tubule. J Pharmacol Exp Ther 282:440–444

Miller DS, Letcher S, Barnes DM (1996) Fluorescence imaging study of organic anion transport from renal proximal tubule cell to lumen. Am J Physiol Renal Physiol 271:F508–F520

Miller DS, Pritchard JB (1991) Indirect coupling of organic anion secretion to sodium in teleost (*Paralichthys lethostigma*) renal tubules. Am J Physiol Regul Integr Comp Physiol 261: R1470–R1477

Miller DS, Pritchard JB (1994) Nocodazole inhibition of organic anion secretion in teleost renal proximal tubules. Am J Physiol Regul Integr Comp Physiol 267:R695–R704

Miller DS, Stewart DE, Pritchard JB (1993) Intracellular compartmentation of organic anions within renal cells. Am J Physiol Regul Integr Comp Physiol 264:R882–R890

Minnich JE (1972) Excretion of urate salts by reptiles. Comp Biochem Physiol 41A:535–549

Minnich JE (1976) Adaptations in the reptilian excretory system for excreting insoluble urates. Isr J Med Sci 12:854–861

Minnich JE, Piehl PA (1972) Spherical precipitates in the urine of reptiles. Comp Biochem Physiol 41A:551–554

Mistry AC, Chen G, Kato A, Nag K, Sands JM, Hirose S (2005) A novel type of urea transporter, UT-C, is highly expressed in proximal tubule of seawater eel kidney. Am J Physiol Regul Integr Comp Physiol 288:F455–F465

Mistry AC, Honda S, Hirata T, Kato A, Hirose S (2001) Eel urea transporter is localized to chloride cells and is salinity dependent. Am J Physiol Regul Integr Comp Physiol 281:R1594–R1604

Moe OW, Wright SH, Palacin M (2012) Renal handling of organic solutes. In: Taal MW, Chertow GM, Marsden PA, Skorecki K, Yu ASL, Brenner BM (eds) Brenner and Rector's the kidney, 9th edn. Elsevier (Saunders), Philadelphia, pp 252–292

Montgomery H, Pierce JA (1937) The site of acidification of the urine within the renal tubule in amphibia. Am J Physiol 118:144–152

Morgan C, Braun EJ (2001) Glucose handling by the kidney of the domestic fowl. FASEB J 15: A854 (Aabstract)

Morgan RL, Ballantyne JS, Wright PA (2003) Regulation of a renal urea transporter with reduced salinity in a marine elasmobranch, *Raja erinacea*. J Exp Biol 206:3285–3292

Moyle V (1949) Nitrogenous excretion in chelonian reptiles. Biochem J 44:581–584

Mudge GH, Berndt WO, Valtin H (1973) Tubular transport of urea, glucose, phosphate, uric acid, sulfate, and thiosulfate. In: Orloff J, Berliner RW (eds) Handbook of physiology: renal physiology. American Physiological Society, Washington, DC, pp 558–652

Mukherjee SK, Dantzler WH (1985) Effects of SITS on urate transport by isolated, perfused snake renal tubules. Pflugers Arch 403:35–40

Munro AF (1953) The ammonia and urea excretion of different species of amphibia during their development and metamorphosis. Biochem J 54:29–36

Nakada T, Hoshijima K, Esaki M, Nagayoshi S, Kawakami K, Hirose S (2007) Localization of ammonia transporter Rhcg1 in mitochondrion-rich cells of yolk sac, gill, and kidney of zebrafish and its ionic strength-dependent expression. Am J Physiol Regul Integr Comp Physiol 263:R1743–R1753

Nawata CM, Evans KK, Dantzler WH, Pannabecker TL (2014) Transepithelial water and urea permeabilities of isolated perfused Munich-Wistar rat inner medullary thin limbs of Henle's loop. Am J Physiol Renal Physiol 306:F123–F129

O'Dell RM, Schmidt-Nielsen B (1961) Retention of urea by frog and mammalian kidney slices in vitro. J Cell Comp Physiol 57:211–219

O'Regan MG, Malnic G, Giebisch G (1982) Cell pH and luminal acidification in *Necturus* proximal tubule. J Membr Biol 69:99–106

Oken DE, Weise M (1978) Micropuncture studies of the transport of individual amino acids by the *Necturus* proximal tubule. Kidney Int 13:445–451

Pajor AM (2000) Molecular properties of sodium/dicarboxylate cotransporters. J Membr Biol 175:1–8

Pegg DG, Hook JB (1977) Glutathione S-transferases: an evaluation of their role in renal organic anion transport. J Pharmacol Exp Ther 200:65–74

Pelis RM, Wright SH (2011) Renal transport of organic anions and cations. Compr Physiol 1:1795–1835

Perschmann C (1956) Über die Bedeutum der Nierenpfortader insbesondere für die Ausschiedung von Harnstoff und Harnsäure bei *Testudo hermanni* Gml. und *Lacerta viridis* Laur. sowie über die Funktion der Harnblase bei *Lacerta viridis* Laur. Zool Beitr 2:447–480

Porter P (1963) Physico-chemical factors involved in urate calculus formation. II. Colloidal flocculation. Res Vet Sci 4:592–602

Preest MR, Beuchat CA (1997) Ammonia excretion by hummingbirds. Nature 386:561–562

Pritchard JB (1987) Luminal and peritubular steps in the renal transport of *p*-aminohippurate. Biochim Biophys Acta 906:295–308

Pritchard JB (1988) Coupled transport of *p*-aminohippurate by rat kidney basolateral membrane vesicles. Am J Physiol Renal Physiol 255:F597–F604

Pritchard JB (1990) Rat renal cortical slices demonstrate *p*-aminohippurate/glutarate exchange and sodium/glutarate coupled *p*-aminohippurate transport. J Pharmacol Exp Ther 255:969–975

Pritchard JB, Miller DS (1993) Mechanisms mediating renal secretion of organic anions and cations. Physiol Rev 73:765–796

Pritchard JB, Sykes DB, Walden R, Miller DS (1994) ATP-dependent transport of tetraethylammonium by endosomes isolated from rat renal cortex. Am J Physiol Renal Physiol 266:F966–F976

Rall DP, Burger JW (1967) Some aspects of hepatic and renal excretion in *Myxine*. Am J Physiol 212:354–356

Randle HW, Dantzler WH (1973) Effects of K^+ and Na^+ on urate transport by isolated perfused snake renal tubules. Am J Physiol 255(5):1206–1214

Rennick BR (1981a) Renal tubular transport of organic cations. In: Greger R, Lang F, Silbernagl S (eds) Renal transport of organic substances. Springer, Berlin, pp 178–188

Rennick BR (1981b) Renal tubule transport of organic cations. Am J Physiol Renal Physiol 240: F83–F89

Roch-Ramel F, Peters G (1981) Renal transport of urea. In: Greger R, Lang F, Silbernagl S (eds) Renal transport of organic substances. Springer, Berlin, pp 134–153

Roch-Ramel F, Werner D, Guisan B (1994) Urate transport in brush-border membrane of human kidney. Am J Physiol Renal Physiol 266:F797–F805

Roch-Ramel F, White F, Vowles L, Simmonds HA, Cameron JS (1980) Micropuncture study of tubular transport of urate and PAH in the pig kidney. Am J Physiol Renal Physiol 239:F107–F112

Romero MF, Fong PY, Berger UV, Hediger MA, Boron WF (1998) Cloning and functional expression of rNBC, an electrogenic Na^+-HCO_3^- cotransporter from rat kidney. Am J Physiol Renal Physiol 274:F425–F432

Romero MF, Hediger MA, Boulpaep EL, Boron WF (1997) Expression cloning and characterization of a renal electrogenic Na^+/HCO_3^- cotransporter. Nature 387:409–413

Ross CR, Weiner IM (1972) Adenine nucleotides and PAH transport in slices of renal cortex: effects of DNP and CN^-. Am J Physiol 222:356–359

Sala-Rabanal M, Hirayama BA, Loo DDF, Chaptal V, Abramson J, Wright EM (2012) Bridging the gap between structure and kinetics of human SGLT1. Am J Physiol Cell Physiol 302: C1293–C1305

Sands JM, Layton HE, Fenton RA (2012) Urine concentration and dilution. In: Taal MW, Chertow GM, Marsden PA, Skorecki K, Yu ASL, Brenner BM (eds) Brenner and Rector's the kidney, 9th edn. Elsevier (Saunders), Philadelphia, pp 326–352

Schali C, Roch-Ramel F (1981) Uptake of [^{3}H]PAH and [^{14}C]urate into isolated proximal tubular segments of the pig kidney. Am J Physiol Renal Physiol 241:F591–F596

Schali C, Schild L, Overney J, Roch-Ramel F (1983) Secretion of tetraethylammonium by proximal tubules of rabbit kidneys. Am J Physiol Renal Physiol 245:F238–F246

Schmidt-Nielsen B (1955) Urea excretion in white rats and kangaroo rats as influenced by excitement and by diet. Am J Physiol 181:131–139

Schmidt-Nielsen B (1972) Renal transport of urea in elasmobranchs. In: Ussing HH, Thorn NA (eds) Transport mechanisms in epithelia. Alfred Benzon Symposium V. Munksgaard, Copenhagen, pp 608–621

Schmidt-Nielsen B, Forster RP (1954) The effect of dehydration and low temperature on renal function in the bullfrog. J Cell Comp Physiol 44:233–246

Schmidt-Nielsen B, Shrauger CR (1963) Handling of urea and related compounds by the renal tubules of the frog. Am J Physiol 205:483–488

Schmidt-Nielsen B, Truniger B, Rabinowitz L (1972) Sodium-linked urea transport by the renal tubule of the spiny dogfish, *Squalus acanthias*. Comp Biochem Physiol 42A:13–25

Schmitt BM, Biemesderfer D, Romero MF, Boulpaep EL, Boron WF (1999) Immunolocalization of the electrogenic Na^+-HCO_3^- cotransporter in mammalian and amphibian kidney. Am J Physiol Renal Physiol 276:F27–F36

Schramm U, Fricker G, Wenger R, Miller DS (1995) P-glycoprotein-mediated secretion of a fluorescent cyclosporin analogue by teleost renal proximal tubules. Am J Physiol Renal Physiol 268:F46–F52

Schrock H, Forster RP, Goldstein L (1982) Renal handling of taurine in marine fish. Am J Physiol Renal Physiol 242:F64–R69

Sheikh MI, Moller JV (1983) Nature of Na^+-independent stimulation of renal transport of p-aminohippurate by exogenous metabolites. Biochem Pharmacol 32:2745–2749

Shideman JR, Zmuda MJ, Quebbemann AJ (1981) The acute effects of furosemide, ethycrinic acid and chlorothiazide on the renal tubular handling of uric acid in chickens. J Pharmacol Exp Ther 216:441–446

Shimada H, Moewes B, Burckhardt G (1987) Indirect coupling to Na^+ of p-aminohippuric acid uptake into rat renal basolateral membrane vesicles. Am J Physiol Renal Physiol 253:F795–F801

Shimomura A, Chonko AM, Grantham JJ (1981) Basis for heterogeneity of para-aminohippurate secretion in rabbit proximal tubules. Am J Physiol Renal Physiol 240:F430–F436

Shoemaker VH, McClanahan LL Jr (1975) Evaporative water loss, nitrogen excretion and osmoregulation in phyllomedusine frogs. J Comp Physiol 100:331–345

Shuprisha A, Lynch RM, Wright SH, Dantzler WH (1999) Real-time assessment of a-ketoglutarate effect on organic anion secretion in perfused rabbit proximal tubules. Am J Physiol Renal Physiol 277:F513–F523

Shuprisha A, Wright SH, Dantzler WH (2000) Method for measuring luminal efflux of fluorescent organic compounds in isolated, perfused renal tubules. Am J Physiol Renal Physiol 279:F960–F964

Silbernagl S (1985) Amino acids and oligopeptides. In: Seldin DW, Giebisch G (eds) The kidney: physiology and pathophysiology. Raven, New York, pp 1677–1701

Silbernagl S (1988) The renal handling of amino acids and oligopeptides. Physiol Rev 68:911–1007

Silbernagl S, Scheller D (1986) Formation and excretion of NH_3:NH_4^+. New aspects of an old problem. Klin Wochenschr 64:862–870

Silbernagl S, Swenson ER, Maren TH (1986) Proximal tubule acidification in the isolated perfused kidney of the skate, *Raja erinacea*. Bull Mt Desert Isl Biol Lab 26:156–158

Smith CP, Wright PA (1999) Molecular characterization of an elasmobranch urea transporter. Am J Physiol Regul Physiol 276:R622–R626

Smith PM, Pritchard JB, Miller DS (1988) Membrane potential drives organic cation transport into teleost renal proximal tubules. Am J Physiol Regul Integr Comp Physiol 255(3):R492–R499

Sokol PP, McKinney TD (1990) Mechanism of organic cation transport in rabbit renal basolateral membrane vesicles. Am J Physiol Renal Physiol 258(6):F1599–F1607

Sperber I (1960) Excretion. In: Marshall AJ (ed) Biology and physiology of birds. Academic, New York, pp 469–492

Stanton B, Omerovic A, Koeppen B, Giebisch G (1987) Electroneutral H^+ secretion in distal tubule of *Amphiuma*. Am J Physiol Renal Physiol 252:F691–F699

Stetson DL (1978) Renal alterations induced by osmotic stress and metabolic acidosis in *Xenopus laevis*. J Exp Zool 206:157–166

Stewart DJ, Holmes WN, Fletcher G (1969) The renal excretion of nitrogenous compounds by the duck (*Anas platyrhynchos*) maintained on freshwater and on hypertonic saline. J Exp Biol 50:527–539

Sweazea KL, Braun EJ (2006) Glucose transporter (GLUT) expression in English sparrows (*Passer domesticus*). Comp Biochem Physiol B 144:263–270

Sykes AH (1971) Formation and composition of urine. In: Bell DJ, Freeman BM (eds) Physiology and biochemistry of the domestic fowl. Academic, New York, pp 233–278

Tannen RL (1978) Ammonia metabolism. Am J Physiol Renal Physiol 235:F265–F277

Tanner GA (1967) Micropuncture study of PAH and diodrast transport in *Necturus* kidney. Am J Physiol 212:1341–1346

Tanner GA, Carmines PK, Kinter WB (1979) Excretion of phenol red by the *Necturus* kidney. Am J Physiol Renal Physiol 236:F442–F447

Tanner GA, Kinter WB (1966) Reabsorption and secretion of *p*-aminohippurate and Diodrast in *Necturus* kidney. Am J Physiol 210:221–231

Thorson TB (1970) Fresh water stingrays, *Potamotrygon* spp.: failure to concentrate urea when exposed to saline medium. Life Sci 9:893–900

Thorson TB, Cowan CM, Watson DE (1967) *Potamotrygon* spp.: elasmobranchs with low urea content. Science 158:375–377

Truniger B, Schmidt-Nielsen B (1964) Intrarenal distribution of urea and related compounds: effects of nitrogen intake. Am J Physiol 207:971–978

Tune BM, Burg MB (1971) Glucose transport by proximal renal tubules. Am J Physiol 221:580–585

Tune BM, Burg MB, Patlak CS (1969) Characteristics of p-aminohippurate transport in proximal renal tubules. Am J Physiol 217:1057–1063

Ullrich KJ, Rumrich G (1996) Luminal transport system for choline$^+$ in relation to the other organic cation transport systems in the rat proximal tubule. Kinetics, specificity: alkyl/arylamines, alkylamines with OH, O, SH, NH$_2$, ROCO, RSCO and H$_2$PO$_4$-groups, methylaminostyryl, rhodamine, acridine, phenanthrene and cyanine compounds. Pflugers Arch 432:471–485

Ullrich KJ, Rumrich G, Fritzsch G, Kloss S (1987a) Contraluminal para-aminohippurate (PAH) transport in the proximal tubule of the rat kidney. I. Kinetics, influence of cations, anions, and capillary preperfusion. Pflugers Arch 409:229–235

Ullrich KJ, Rumrich G, Fritzsch G, Kloss S (1987b) Contraluminal para-aminohippurate (PAH) transport in the proximal tubule of the rat kidney. II. Specificity: aliphatic dicarboxylic acids. Pflugers Arch 408:38–45

Ullrich KJ, Rumrich G, Kloss S (1987c) Contraluminal *para*-aminohippurate (PAH) transport in the proximal tubule of the rat kidney. III. Specificity: monocarboxylic acids. Pflugers Arch 409:547–554

Ullrich KJ, Rumrich G, Schmidt-Nielsen B (1967) Urea transport in the collecting duct of rats on normal and low protein diet. Pflugers Arch 295:147–156

Villalobos AR, Braun EJ (1995) Characterization of organic cation transport by avian renal brush-border membrane vesicles. Am J Physiol Regul Integr Comp Physiol 269:R1050–R1059

Villalobos AR, Braun EJ (1998) Substrate specificity of organic cation/H$^+$ Exchange in avian renal brush-border membranes. J Pharmacol Exp Ther 287:944–951

Vogel G, Kurten M (1967) Untersuchungen zur Na$^+$-Abhangigkeit der renal-tubularen Harnstoff-Sekretion bei *Rana ridibunda*. Pflugers Arch 295:42–55

Vogel G, Lauterbach F, Kroger W (1965) Die Bedeutung des Natriums fur die renalen Transporte von Glucose und para-Aminohippursaure. Pflugers Arch 283:151–159

Vogel G, Stoeckert I (1966) Die Bedeutung des Anions fur den renal tubularen Transport von Na$^+$ und die Transporte von Glucose und PAH. Pflugers Arch 292:309–315

Von Baeyer H, Deetjen P (1985) Renal glucose transport. In: Seldin DW, Giebisch G (eds) The kidney: physiology and pathophysiology. Raven, New York, pp 1663–1675

Walker AM, Bott PA, Oliver J, MacDowell MC (1941) The collection and analysis of fluid from single nephrons of the mammalian kidney. Am J Physiol 134:580–595

Walker AM, Hudson CL (1937a) The reabsorption of glucose from the renal tubule in amphibia and the action of phlorhizin upon it. Am J Physiol 118:130–141

Walker AM, Hudson CL (1937b) The role of the tubule in the excretion of urea by the amphibian kidney. with an improved technique for the ultramicro determination of urea nitrogen. Am J Physiol 118:153–166

Weiner ID, Verlander JW (2012) Renal acidification mechanisms. In: Taal MW, Chertow GM, Marsden PA, Skorecki K, Yu ASL, Brenner BM (eds) Brenner and Rector's the kidney, 9th edn. Elsevier (Saunders), Philadelphia, pp 293–325

Weiner IM (1973) Transport of weak acids and bases. In: Orloff J, Berliner RW (eds) Handbook of physiology. Sect. 8. Renal physiology. American Physioliological Society, Washington, DC, pp 521–554

Weiner IM (1985) Organic acids and bases and uric acid. In: Seldin DW, Giebisch G (eds) The kidney: physiology and pathophysiology. Raven, New York, pp 1703–1724

Welborn JR, Shpun S, Dantzler WH, Wright SH (1998) Effect of a-ketoglutarate on organic anion transport in single rabbit renal proximal tubules. Am J Physiol Renal Physiol 274:F165–F174

Wheatly MG, Hobe H, Wood CM (1984) The mechanisms of acid-base and ionoregulation in the freshwater rainbow trout during environmental hyperoxia and subsequent normoxia. II. The role of the kidney. Respir Physiol 55:155–173

Wolff NA, Kinne R, Elger B, Goldstein L (1987) Renal handling of taurine, L-alanine, L-glutamate and D-glucose in *Opsanus tau*: studies on isolated brush border membrane vesicles. J Comp Physiol B 157:573–581

Wolff NA, Perlman DF, Goldstein L (1986) Ionic requirements of peritubular taurine transport in *Fundulus* kidney. Am J Physiol Regul Integr Comp Physiol 250:R984–R990

Wolff NA, Werner A, Burkhardt S, Burckhardt G (1997) Expression cloning and characterization of a renal organic anion transporterfrom winter flounder. FEBS Lett 417:287–291

Wood CM, Milligan CL, Walsh PJ (1999) Renal responses of trout to chronic respiratory and metabolic acidoses and metabolic alkalosis. Am J Physiol Regul Integr Comp Physiol 277: R482–R492

Woodhall PB, Tisher CC, Simonton CA, Robinson RR (1978) Relationship between para-aminohippurate secretion and cellular morphology in rabbit proximal tubules. J Clin Invest 51:1320–1329

Wright PA, Wood CM, Wilson JM (2014) Rh versus pH: the tole of Rhesus glycoproteins in renal ammonia excretion during metabolic acidosis in a freshwater teleost fish. J Exp Biol 217:2855–2865

Wright SH, Dantzler WH (2004) Molecular and cellular physiology of renal organic cation and anion transport. Physiol Rev 84:987–1049

Wright SH, Wunz TM, Wunz TP (1992) A choline transporter in renal brush-border membrane vesicles: energetics and structural specificity. J Membr Biol 126:51–65

Yokota SD, Benyajati S, Dantzler WH (1985) Comparative aspects of glomerular filtration in vertebrates. Renal Physiol 8:193–221

Yoshimura H, Yata M, Ŷuasa M, Wolbach RA (1961) Renal regulation of acid-base balance in the bullfrog. Am J Physiol 201:980–986

Yucha CB, Stoner LC (1986) Bicarbonate transport by amphibian nephron. Am J Physiol 251: F865–F872

Zmuda MJ, Quebbemann AJ (1975) Localization of renal tubular uric acid transport defect in gouty chickens. Am J Physiol 229:820–825

Chapter 7
Diluting and Concentrating Mechanism

Abstract Although only mammals and, to a lesser extent, birds can produce a urine significantly hyperosmotic to the plasma, many nonmammalian vertebrates can produce a ureteral urine that varies from significantly hypoosmotic to the plasma to isosmotic with the plasma. This chapter initially considers the maximum range of ureteral urine osmolalities produced by representatives of all vertebrate classes, the ability of these species to dilute their urine, and the possible physiological significance of differences in this ability. It then discusses the mechanism by which urine is diluted and the renal tubule sites at which dilution occurs. Next the chapter considers the hyperosmotic urine concentrations found in a broad range of birds and mammals. It then discusses the most recent models of the concentrating mechanism in these two vertebrate classes and indicates major weaknesses in these models and important areas for future research. Finally, the chapter considers the role of the tubular and glomerular actions of antidiuretic hormone in the regulation of urine osmolality and, as far as possible, the cellular and molecular mechanisms involved in these actions.

Keywords Urine diluting mechanism • Avian urine concentrating mechanism • Mammalian urine concentrating mechanism • Species differences in urine dilution and concentration • Antidiuretic hormone • Arginine vasotocin • Arginine vasopressin • Lysine vasopressin

7.1 Introduction

The production of a dilute urine with an osmolality below that of the plasma enables an animal to excrete excess water. The greater the ability to dilute the urine, the greater is the ability to eliminate excess water rapidly. The ability to change the urine from one with an osmolality below that of the plasma to one isosmotic with the plasma enables an animal to conserve water. Of course, the ability to produce a urine significantly hyperosmotic to the plasma—an ability found only in mammals and birds—enables an animal to conserve even more water while still eliminating excess inorganic and organic ions and nitrogenous waste. The ability to conserve water by these means is also related to the excretory end products of nitrogen metabolism. This chapter covers the differences among vertebrate classes and

© The American Physiological Society 2016   237
W.H. Dantzler, *Comparative Physiology of the Vertebrate Kidney*,
DOI 10.1007/978-1-4939-3734-9_7

species in ability to dilute and concentrate the urine, the process involved in diluting the urine, the process involved in concentrating the urine, and the regulation of these processes.

7.2 Range of Urine Osmolality

The kidneys of many species of fishes, amphibians, and reptiles, although incapable of producing a urine substantially hyperosmotic to the plasma, are capable of producing a urine with an osmolality ranging from about one-tenth that of the plasma to isosmotic with the plasma (7.1). Those kidneys capable of producing very dilute urine are generally found in animals with a major need to excrete excess water, e.g., stenohaline freshwater fishes, some euryhaline fishes adapted to freshwater, and freshwater amphibians and reptiles (Table 7.1). However, the kidneys of some species appear only to be capable of producing urine hypoosmotic to the plasma regardless of the specific need to conserve or to excrete water (Table 7.1). For a number of these species, this production of hypoosmotic urine under all circumstances appears to be a true functional limitation, e.g., the desert tortoise (*Gopherus agassizii*), the blue spiny lizard (*S. cyanogenys*), the South African clawed toad (*X. laevis*), the stenohaline marine fishes, and, probably, the stenohaline freshwater fishes (Dantzler and Schmidt-Nielsen 1966; Forster 1953; McBean and Goldstein 1970; Stolte et al. 1977b). For other species, e.g., the bullfrog (*R. catesbeiana*), the maximum urine osmolality may simply not yet have been determined.

The kidneys of some other species appear to produce a urine nearly isosmotic with the plasma or with very little variation around isosmolality (Table 7.1). This appears reasonable for stenohaline marine fishes, xerophilic lizards (e.g., horned lizard, *Phrynosoma cornutum*), and marine snakes (e.g., olive sea snake, *A. laevis*) that rarely or never obtain excess water to excrete, but it is also the case for the euryhaline aglomerular toadfish (*Opsanus tau*), which is found in freshwater (Table 7.1) (Lahlou et al. 1969).

Although it is certainly true that urates can be excreted with very little water (vide infra) whereas the excretion of urea or ammonia in significant amounts requires a large urine volume or a significantly concentrated urine, the above-discussed differences in the ability to produce hypoosmotic urine or isosmotic urine or to regulate the urine osmolality cannot be related simply to differences in the primary excretory end products of nitrogen metabolism for there are species that excrete primarily urates, primarily urea, or primarily ammonia in each group (Table 6.2). However, in many species of amphibians and reptiles and even in some species of fishes, additional regulation of the urine osmolality almost certainly takes place distal to the kidney—in the bladder, cloaca, or colon (Dantzler 1970; Lahlou et al. 1969).

Even though the kidneys of fishes, amphibians, and reptiles are incapable of producing a urine substantially hyperosmotic to the plasma, as suggested by the

Table 7.1 Examples of range of osmolal urine-to-plasma ratios (U/P)

Species	Osmolal U/P (approximate maximum Range)	Environment and mode of existence	References
Fishes			
Myxinoidea			
Atlantic hagfish,	1.0	Marine	Stolte and Schmidt-Nielsen
Myxine glutinosa			(1978)
Petromyzonta			
River lamprey,	0.1	Freshwater	Logan et al. (1980)
Lampetra fluviatilis			
Elasmobranchii			
Little skate,		Marine	Stolte et al. (1977a)
Raja erinacea	0.96	(placed in 100 % seawater)	
	0.80	(placed in 75 % seawater)	
Teleostei			
Goldfish,	0.14	Freshwater	Hickman and Trump
Carassius auratus			(1969)
American eel,		Euryhaline	
Anguilla rostrata	0.15	(adapted to freshwater)	Schmidt-Nielsen and
	0.60	(adapted to freshwater)	Renfro (1975)
Southern flounder,		Euryhaline	
Paralicthys lethostigma	0.96	(adapted to sea water)	Hickman and Trump (1969)
Toadfish (aglomerular),		Euryhaline	
Opsanus tau	0.85	(adapted to freshwater)	Lahlou et al. (1969)
	0.90	(adapted to seawater)	
Goosefish (aglomerular),	0.84	Marine	Forster (1953)
Lophius americanus			
Longhorn sculpin,	0.84–0.94	Marine	Forster (1953)
Myoxocephelus octodecimspinosus			
Amphibia			
Anura			
Bullfrog,	0.1–0.3	Semi-aquatic (freshwater)	Long (1973), Schmidt-
Rana catesbeiana			Nielsen and Forster (1954)
South African clawed toad,		Aquatic (freshwater)	
Xenopus laevis	0.14	(placed in freshwater)	
	0.65	(placed in hyperosmotic saline)	
Urodela			
Mudpuppy,	0.2–0.9	Aquatic (freshwater)	Garland et al. (1975)
Necturus maculosus		(TF/P osmolal ratio in distal tubule, before and after administration of arginine vasotocin)	
Reptilia			
Testudinea			
Desert tortoise,	0.3–0.7	Terrestrial, xeric	Dantzler and Schmidt-
Gopherus agassizii			Nielsen (1966)
Freshwater turtle	0.3–1.0	Semi-aquatic, freshwater	Dantzler and Schmidt-
Pseudemys scripta			Nielsen (1966)
Crocodilia			
Crocodile,	0.35–0.95	Semi-aquatic, freshwater	Schmidt-Nielsen and
Crocodylus acutus		and marine	Skadhauge (1967)

Table 7.1 (continued)

Species	Osmolal U/P (approximate maximum Range)	Environment and mode of existence	References
Squamata			
Ophidia			
Bull snake,	0.5–1.0	Terrestrial, xeric	Komadina and Solomon
Pituophis melanoleucus			(1970)
Freshwater snake,	0.1–1.0	Semi-aquatic, freshwater	Dantzler (1967)
Nerodia sipedon			
Olive sea snake,	0.8–1.2	Aquatic, marine	Yokota et al. (1985)
Aipysurus laevis			
Sauria			
Horned lizard,	0.8–1.0	Terrestrial, xeric	Roberts and Schmidt-
Phrynosoma cornutum			Nielsen (1966)
Blue spiny lizard,	0.3–0.7	Terrestrial, xeric	Stolte et al. (1977b)
Sceloporus cyanogenys			
Sand goanna,	0.4–1.0	Terrestrial, xeric	Bradshaw and Rice (1981)
Varanus gouldii			
Aves			
Galliformes			
Domestic fowl,	0.1–2.0	Terrestrial, moist	Ames et al. (1971),
Gallus gallus domesticus			Dantzler (1966),
			Skadhauge and Schmidt-
			Nielsen (1967a)
Gambel's quail,	0.5–2.5	Terrestrial, xeric	Braun and Dantzler (1972,
Callipepla gambelii			1975)
Mammalia			
White rat,	0.2–8.9	Terrestrial, moist	Dantzler (1970)
Rattus norvegicus alb.			

arrangement and structure of their nephrons (see Chap. 2), some still appear capable of producing a slightly hyperosmotic urine. The production of a urine slightly hyperosmotic to the plasma, i.e., urine/plasma osmolal ratio of about 1.2–1.3, has been reported for a species of euryhaline teleost (*F. kansae*) (Fleming and Stanley 1965; Stanley and Fleming 1964), marine turtle (*Chelonia mydas*) (Prange and Greenwald 1979), xerophilic lizard (*Amphibolurus maculosus*) (Braysher 1976), and marine snake (*A. laevis*) (Yokota et al. 1985). In some cases, this may involve modification of ureteral urine in the bladder or cloaca because ureteral urine was not collected directly. However, such is not the case for the sea snakes in which ureteral urine was collected directly and found to be slightly hyperosmotic to the plasma at low urine flows (Table 7.1) (Yokota et al. 1985). It appears most likely, although by no means proven, that this hyperosmolality results from secretion of solutes, i.e., sodium, potassium, magnesium, or ammonia, into a small volume of tubule fluid (Yokota et al. 1985). Similarly, the sodium chloride secretion that has been demonstrated in killifish proximal tubules (vide supra; Chap. 4) may account for the hyperosmotic urine observed when these animals are abruptly transferred from freshwater to seawater (Beyenbach 1986; Stanley and Fleming 1964). Although the secretion of ions by the renal tubules may be important for any of these animals in terms of regulating the plasma levels of these substances (vide supra; Chaps. 4

and 6), the production of a urine slightly hyperosmotic to the plasma can be of little adaptive significance in the conservation of water for marine reptiles or teleosts because the osmolality of their plasma is so far below that of seawater.

One possible exception to the forgoing discussion is the marine catfish (*Cnidoglanis microcephalus*). A few measurements suggest that the bladder urine in these animals can be not only hyperosmotic to the plasma but also hyperosmotic to the surrounding seawater (Kowarsky 1973). If the production of urine hyperosmotic to the seawater by these animals is ever confirmed and if it is shown to emanate from the renal tubules, it may result from hyperosmotic sodium chloride secretion by the apparent chloride cells in the collecting tubules identified by ultrastructural analysis (vide supra; Chap. 4) (Hentschel and Elger 1987).

Both avian and mammalian kidneys are capable of producing a urine hypoosmotic to the blood (Table 7.1). Among the birds, it appears that this diluting ability may be slightly better developed in those species that have most ready access to water (Table 7.1), but this possibility has not been examined systematically.

As noted in Chap. 2, the structural arrangement of the nephrons and blood supply in the medullary cones of the avian kidney suggests, by analogy with the mammalian kidney, that the avian kidney would be capable of producing a urine hyperosmotic to the plasma (see below for discussion of the details of this process). Indeed, the concentrating ability of the avian kidney (Tables 7.1 and 7.2) was recognized long before anyone understood the meaning of these structural relationships (Dantzler 1987). However, as shown in Tables 7.1, 7.2, and 7.3, the concentrating ability of birds is quite limited compared with that of mammals. At best,

Table 7.2 Maximum osmolal urine-to-plasma ratio (U/P) and relative medullary thickness (RMT)

Species	Osmolal U/P	RMT[f]
Domestic fowl (*Gallus gallus domesticus*)	2.1[a]	
Bobwhite quail (*Colinus virginianus*)	1.6[a]	
California quail (*Callipepla californica*)	1.7[a]	
Gambel's quail (*Callipepla gambelii*)[b]	2.5[a]	
House finch (*Carpodacus mexicanus*)	2.3[a]	3.59
Zebra finch (*Poephila guttata*)[b]	2.8[c]	4.71
Senegal dove (*Streptopelia senegalensis*)	1.7[c]	
Kookaburra (*Dacelo gigas*)	2.7[c]	
Emu (*Dromaius novaehollandiae*)[b]	1.4[c]	
Crested pigeon (*Ocyphaps lophotes*)[b]	1.8[c]	
Galah (*Cacatua roseicapilla*)[b]	2.6[c]	
Budgerigar (*Melopsittacus undulatus*)[b]	3.0[d]	5.06
Savannah sparrow (*Passerculus sandwichensis beldingi*)	1.6[e]	3.24

[a]Reviewed in Dantzler (1970)
[b]Birds from xeric environment
[c]Skadhauge (1974)
[d]Krag and Skadhauge (1972)
[e]Goldstein et al. (1990)
[f]Johnson (1974)
Relative medullary thickness equals ten times the mean length of the medullary cones divided by the cube root of the kidney volume

Table 7.3 Maximum osmolality and osmolal urine-to-plasma ratio (U/P) and relative medullary thickness (RMT) for a number of mammals

Species	Urine osmolality[a] (mosmol/kg H_2O)	Osmolal U/P[a]	RMT[a]
Human	1430	4.2	3.0
Chinchilla (*Chinchilla laniger*)[b]	2000	6.7	9.4
Pack rat (*Neotoma albigula*)[b]	2700	7.0	6.6
Camel (*Camelus dromedarius*)[b]	2800	8.0	
White rat (*Rattus norvegicus alb.*)[b]	2900	8.9	5.8
Cat (*Felix domestica*)	3250	9.9	4.8
Ground squirrel (*Citellus leucurus*)[b]	3900	9.5	
Hopping mouse (*Notomys cervinus*)[b]	4920	14.2	12.0
Kangaroo rat (*Dipodomys merriami*)[b]	5500	14.0	8.5
Gerbil (*Gerbillus gerbillus*)[b]	5500	14.0	
Sand rat (*Psammomys obesus*)[b]	6340	17.0	10.7
Jerboa (*Jaculus jaculus*)[b]	6500	16.0	9.3
Pocket mouse (*Perognathus penicillatus*)[b]	7600	23.5	16.8
Hopping mouse (*Notomys alexus*)[b]	9370	24.6	12.2
Hopping mouse (*Leggadina hermansburgensis*)[b]	8970	26.8	

[a]Reviewed in Dantzler (1970) except for *C. laniger* and *P. penicillatus*. Values for *C. laniger* are from Gutman and Beyth (1970). Values for *P. penicillatus* are from E. J. Braun, personal communication
[b]Mammals from xeric habitat
Relative medullary thickness equals ten times the longest axis of the medulla divided by the cube root of the product of the renal dimensions of length, width, and breadth (Sperber 1944)

birds can produce a maximally concentrated urine only two to three times the osmolality of the plasma (Table 7.2). But it should be noted that a direct comparison of the maximum U/P osmolal ratios of birds and mammals can be somewhat misleading because the plasma osmolality of birds tends to increase more than that of mammals with dehydration (Braun 1985). The reason for this greater lability of the plasma osmolality in birds than in mammals is unknown.

A slight gradation in the apparent maximum concentrating ability does exist among three closely related species of quail—bobwhite (*Colinus virginianus*), California quail (*Callipepla californica*), and Gambel's quail (*Callipepla gambelii*) (Table 7.2)—the maximum urine concentration increasing with the aridity of the habitat (Dantzler 1970). However, even the xerophilic Gambel's quail is incapable of producing a urine much more concentrated that that of the mesophilic domestic chicken (Table 7.2). And other desert-dwelling birds are even less able to concentrate their urine (Table 7.2).

The details of the concentrating process and structural and other differences between birds and mammals are discussed below. However, the fact that birds excrete uric acid as the primary excretory end product of nitrogen metabolism, rather than urea, makes it possible for them to conserve water without producing as concentrated a urine as mammals. In addition, in birds, significant modification of the ionic composition of the ureteral urine can take place by the transport of ions in structures distal to the kidney, primarily the coprodeum and colon (Larsen et al. 2014; Laverty and Skadhauge 1999; Skadhauge 1981). In this case, the production of a ureteral urine only modestly more concentrated than the plasma would reduce the tendency for water to move from the plasma into the colon and permit additional solute-linked water reabsorption (Larsen et al. 2014). Moreover, uric acid as the major end product of nitrogen metabolism in the ureteral urine will not diffuse back into the plasma across the intestinal wall. It is not readily converted to ammonia and can be stored in the intestine during the postrenal modification period.

The observation that the low solubility of uric acid permits the excretion of nitrogenous waste with very little water is an old one, but it deserves further discussion in relation to the osmolality of the urine. Homer W. Smith (1953) noted that if the nitrogen metabolism of 1 g of protein is excreted as urea in a solution isosmotic with the plasma, it requires 20 ml of water. However, as Skadhauge (1981) noted, reptiles and birds can excrete this much nitrogen as uric acid in 1 ml of urine without significantly concentrating the urine. Urea must remain in solution whereas uric acid almost certainly does not. Much of the uric acid found in the ureteral urine of uricotelic vertebrates is in the form of precipitates. This can be as much as 90 % of that excreted at the highest concentrations in domestic fowl (McNabb and Poulson 1970). These precipitates do not, of course, contribute to the osmotic pressure of the urine. In addition, as noted in Chap. 6, in many uricotelic vertebrates, significant amounts of inorganic cations and ammonium may be included with these precipitates (Dantzler 1978). The actual amounts may be as high as 80–90 % of the excreted cations and may vary with the diet, including ionic intake, and with hydration (Dantzler 1978; Skadhauge 1981). These cations also do not contribute to the osmotic pressure of the urine. However, the degree to which such ion trapping occurs and the extent to which the molar ratios of these cations to urate exceed those expected for simple salts of urate are still controversial, primarily because of the technical difficulty of determining the extent to which ions actually accompany urate precipitates in ureteral urine (Dantzler 1978; Skadhauge 1981). Clearly, this problem is worthy of further study.

However, as noted in Chap. 6, even the amount of urate in the liquid phase of the ureteral urine in many uricotelic vertebrates exceeds the solubilities of uric acid (0.384 mM), sodium urate (6.76 mM), and potassium urate (12.06 mM) (Dantzler 1978). Therefore, much of the urate in the liquid phase must be in a colloidal state. This can amount to three-fourths of the urate in the liquid phase of the ureteral urine in birds (McNabb and Poulson 1970). The property of urates to form lyophobic colloids has been known for many years (Bechhold and Ziegler 1914; Porter 1963; Schade and Boden 1913; Young and Musgrave 1932). However, the concentration of urates in the liquid phase of the urine of uricotelic vertebrates often exceeds the stability limits of lyophobic colloids in aqueous solutions (Dantzler 1978). These findings suggest that the lyophobic urate colloids are converted to a lyophilic state by adsorption to lyophilic macromolecules (Porter 1963). Such lyophilic colloids can exist in the liquid phase of the urine at concentrations above the stability limits for lyophobic colloids. Mucoid materials that may serve as appropriate lyophilic macromolecules have been identified by histochemical techniques in the kidneys of birds (Longley et al. 1963; McNabb et al. 1973); similar materials may exist in the urine of reptiles (Dantzler 1978). The apparent formation of lyophilic colloids permits much more urate to remain in the liquid phase of the urine than would otherwise be the case. This may be important in protecting the urinary system by reducing the amount of urate precipitation. In any case, neither lyophobic nor lyophilic colloids contribute significantly to the osmotic pressure of the ureteral urine. Although, as noted above, the ability to excrete urate as the major end product of nitrogen metabolism may not correlate exactly with the ability of the kidneys to vary the urine osmolality, it does permit the conservation of water and certainly reduces the need of terrestrial species to produce a concentrated urine.

7.3 Process and Sites of Dilution

Formation of a urine hypoosmotic to the plasma requires reabsorption of solute (primarily sodium and chloride) in excess of water somewhere along the nephrons. As pointed out in Chap. 4, sodium and chloride reabsorption in excess of water appears to occur by a similar mechanism in the early distal tubules of teleost nephrons, amphibian nephrons, reptilian-type avian nephrons, the thick ascending limb of Henle's loop of mammalian-type avian nephrons and mammalian nephrons, and possibly the thin intermediate segment or early distal tubule of reptilian nephrons (see Chap. 4 for details on the transport mechanism in most of these regions). In vivo micropuncture studies document this early distal site for dilution in freshwater lampreys (*L. fluviatilis*) (Logan et al. 1980), aquatic freshwater urodeles (*N. maculosus*) (Garland et al. 1975), and xerophilic lizards (*S. cyanogenys*) (Stolte et al. 1977b) and indicate that dilution continues throughout the length of the distal tubules and collecting ducts. Also, as noted in Chap. 4, extremely low water permeability and significant solute reabsorption have been demonstrated by in vivo or in vitro microperfusion of early distal segments of amphibian nephrons (Oberleithner et al. 1983; Stoner 1977, 1985), freshwater teleost nephrons (Nishimura et al. 1983a), and reptilian-type avian nephrons (Miwa and Nishimura 1986; Nishimura et al. 1983b) and of the thick ascending limbs of mammalian-type avian nephrons (Nishimura et al. 1983b) and mammalian nephrons (Burg and Green 1973; Rocha and Kokko 1973). Preliminary in vitro microperfusion studies suggest that this also may be the case for the thin intermediate segment of reptilian nephrons (garter snakes, *Thamnophis* spp.) (S. D. Yokota and W. H Dantzler, unpublished observations), but this is far from certain. Also, as discussed in Chap. 4, the late portion of the distal tubules in these reptiles may be specialized to permit additional dilution of tubular fluid in which the sodium concentration is already low (Beyenbach 1984; Beyenbach and Dantzler 1978; Beyenbach et al. 1980).

   As noted above, dilution of the tubular fluid occurs in some nonmammalian vertebrates despite an apparent need to conserve water (Table 7.1). In marine elasmobranchs, which remain isosmotic with their environment by retention of urea and trimethylamine oxide (TMAO), some dilution occurs by reabsorption of filtered sodium and chloride in excess of water, which becomes more marked when the animals are maintained on 75% seawater (Table 7.1) (Stolte et al. 1977a). Micropuncture studies indicate that in elasmobranchs, in contrast to those nonmammalian vertebrates in which formation of a hypoosmotic urine is physiologically significant, the primary site of dilution is the collecting duct, although some dilution does occur in the distal tubule (Stolte et al. 1977a). This dilution, which is not compensated by fluid reabsorption at more distal nephron sites, appears to serve no adaptive function. In the blue spiny lizard (*S. cyanogenys*), dilution of the tubular fluid also occurs throughout the distal tubule and collecting ducts regardless of the state of hydration (Stolte et al. 1977b). However, in these animals, further modification of the ureteral urine may occur in the cloaca.

The slight dilution of the urine observed in stenohaline marine teleosts, e.g., in the longhorn sculpin, *Myxocephalis octodecimspinosus* (Table 7.1), presents an even more curious situation than the above cases. The dilution apparently results from net reabsorption of filtered sodium and chloride in excess of water because secreted magnesium and sulfate and some filtered chloride account for most of the osmotic activity in the urine (Hickman and Trump 1969). Not only is this dilution of no obvious adaptive advantage, but, because these animals lack distal tubules, it must occur in the proximal tubules. However, no direct studies of the site or mechanism have been made.

The process involved in the changes in dilution by the renal tubules of euryhaline teleosts adapted to different salinities has not been studied in detail. When these animals are adapted to seawater, secreted magnesium and sulfate along with filtered sodium and chloride [and possibly some secreted sodium and chloride (see Chap. 4)] are the principal osmotically active substances in the urine (Hickman and Trump 1969; Schmidt-Nielsen and Renfro 1975). However, substantial net tubular reabsorption of filtered sodium and chloride apparently still occurs during adaptation to seawater (Schmidt-Nielsen and Renfro 1975). With adaptation to freshwater, the tubular secretion of magnesium and sulfate ceases and the epithelium of the distal tubules apparently becomes impermeable to water (Hickman and Trump 1969). However, the distal reabsorption of sodium and chloride continues, producing a tubular fluid of much lower osmolality than that formed during adaptation to seawater (Table 7.1) (Hickman and Trump 1969; Schmidt-Nielsen and Renfro 1975).

7.4 Process of Concentration

The ability of mammalian and avian kidneys to produce urine hyperosmotic to the plasma is related to the presence of nephrons with loops of Henle arranged parallel to collecting ducts and loops of capillaries (see Chap. 2) and the resulting establishment of an increasing osmotic gradient along the length of the inner and outer medulla in mammalian kidneys (Dantzler et al. 2011) and along the length of the medullary cones in avian kidneys (Emery et al. 1972; Skadhauge and Schmidt-

Nielsen 1967b). However, in mammalian kidneys, this gradient consists primarily of sodium chloride and urea (Dantzler et al. 2011) whereas, in avian kidneys, this gradient consists only of sodium chloride and to a very slight extent of potassium chloride (Skadhauge and Schmidt-Nielsen 1967b). Urea is essentially absent in birds. In the presence of antidiuretic hormone—arginine vasopressin (AVP) in the case of mammals and arginine vasotocin (AVT) in the case of birds—the permeability of the collecting ducts to water increases and the fluid in the collecting ducts equilibrates with the hyperosmotic medullary interstitium to produce urine hyperosmotic to the plasma (see below for more detail on hormonal regulation). In mammalian kidneys, arginine vasopressin also activates UT-A1 and UT-A3 urea channels in the inner medullary collecting ducts (vide supra; Chap. 6), allowing urea to move into the interstitium. Thus, during a maximum antidiuresis, urea contributes about 40 % of the inner medullary interstitial osmotic gradient; sodium chloride contributes the other 60 %. During a maximum water diuresis, in the absence of AVP, urea contributes less than 10 % of the inner medullary interstitial osmolality.

The process by which the medullary osmotic gradient in the mammalian kidney is generated is far from completely understood. In the period between 1940 and 1960, a number of theoretical and experimental studies (Gottschalk and Mylle 1959; Hargitay and Kuhn 1951; Kuhn and Ramel 1959; Kuhn and Ryffel 1942) appeared to provide a mechanism by which this gradient might be generated. These studies indicated that a small osmotic pressure difference between the descending and ascending limbs of Henle's loops, generated by the net transport of solute (presumably sodium chloride), unaccompanied by water, out of the ascending limbs, could be multiplied by the countercurrent flow in the loops to generate the increasing interstitial osmolality from cortex to papilla tip. Vascular washout of this gradient would be delayed by countercurrent exchange in the loops of the vasa recta.

This classic concept of countercurrent multiplication, with minor modifications, has been widely accepted as generating the gradient in the outer medulla (OM). In most mammalian species studied, this gradient leads only to about a doubling of the interstitial osmolality of the cortex at the border between the outer and inner medulla (IM). In the OM, active transport of sodium chloride out of the water-impermeable thick ascending limb of Henle's loop has been well established (see Chap. 4) (Burg and Green 1973; Rocha and Kokko 1973) and shown by modeling studies to be theoretically sufficient to generate the observed gradient (Layton and Layton 2005).

The classic concept of countercurrent multiplication, however, cannot explain the osmotic gradient established along the length of the mammalian IM. Although the ascending thin limbs of Henle's loops in the IM, like the ascending thick limbs in the OM, have essentially no transepithelial osmotic water permeability (Chou and Knepper 1992; Imai and Kokko 1974; Yool et al. 2002), they also lack any active transepithelial transport of sodium chloride or any other known solute (Imai and Kokko 1974; Imai and Kusano 1982; Marsh and Azen 1975; Marsh and Solomon 1965). Nevertheless, the IM is the region in which the steepest osmotic gradient is generated, a gradient that can lead to a maximum osmolality at the papilla tip of about 2–15 times the osmolality at the OM–IM border, depending on the species (Table 7.3). If this steep osmotic gradient in the IM cannot be generated by the same classic countercurrent multiplication system as the gradient in the OM, how then is it generated? Most scientists studying this process believe that it involves some form of countercurrent multiplication by the thin limbs of Henle's loops. However, there is at present no accepted mechanism for how such countercurrent multiplication could work.

Currently, the most influential and frequently cited explanation of the concentrating process in the mammalian IM is the "passive mechanism" proposed independently in 1972 by Stephenson (1972) and by Kokko and Rector (1972). These authors hypothesized that the separation of sodium chloride from urea by the active transport of sodium chloride (without water or urea) out of the thick ascending limbs in the OM provides a source of potential energy for generating an osmotic gradient in the IM. They suggested that urea, which has been concentrated in the cortical and outer medullary collecting ducts by the reabsorption of sodium chloride and water, diffuses out of the inner medullary collecting ducts (especially in the presence of antidiuretic hormone) into the surrounding interstitium. This increased IM interstitial urea concentration draws water from the descending thin limbs and inner medullary collecting ducts, thereby tending to reduce the IM interstitial concentration of sodium chloride. These processes result in a urea concentration in the IM interstitium that is higher than the urea concentration in the ascending thin limbs and a sodium chloride concentration in the ascending thin limbs that is higher than the sodium chloride concentration in the surrounding IM interstitium. Sodium chloride will then tend to diffuse out of the ascending thin limbs into the IM interstitium, and urea will tend to diffuse from the IM interstitium into the ascending thin limbs. If the permeability of the ascending thin limbs to sodium chloride is sufficiently high and to urea sufficiently low, then much sodium chloride will diffuse out of the ascending thin limbs while little urea diffuses in. This will mean that the fluid in the IM interstitium will be concentrated while the fluid in the ascending thin limbs will be diluted. The increased interstitial osmolality will, in turn, draw water from the collecting ducts, thereby concentrating the urine. This mechanism is only "passive" with regard to diffusion of sodium chloride out of the ascending thin limbs and urea diffusion out of the collecting ducts in the IM because it depends critically on the active transport of sodium chloride out of the thick ascending limbs and cortical and outer medullary collecting ducts.

This model, although elegant and widely disseminated, depends critically on the permeabilities of the IM ascending thin limbs to sodium chloride and urea. Unfortunately, using measured urea permeabilities, renal researchers have been unable to develop a mathematical model that generates a significant axial osmotic gradient in the IM using this initial version of the passive model (Sands and Layton 2007). This failure has led to a number of additional suggestions of hypothetical mechanisms and associated mathematical models to explain the development of the IM osmotic gradient. These have included (1) attempts to consider increasing structural complexity in mathematical models (Wang and Wexler 1996; Wang et al. 1998; Wexler et al. 1991), which, however, have failed to produce a physiological gradient; (2) muscular contractions of the renal pelvic wall, possibly acting through hyaluronan as a transducer of mechanical energy into a concentrating effect (Knepper et al. 2003; Schmidt-Nielsen 1995), which has yet to be experimentally verified or even shown to be thermodynamically feasible; and (3) accumulation of a newly produced or external osmotically active substance in the IM (Hervy and Thomas 2003; Jen and Stephenson 1994; Thomas 2000), which has yet to be demonstrated experimentally or to be shown theoretically able to account for the entire gradient.

Over the past decade, a series of studies has provided substantial new information on the three-dimensional arrangement of all the tubular and vascular structures in the inner medulla and on the permeabilities of the inner medullary thin limbs of Henle's loops that has been used in new mathematical models of the concentrating mechanism that have produced results closer to those observed physiologically. The details of these studies, on Munich-Wistar rats, are beyond the scope of this discussion but they have been extensively reviewed by Dantzler et al. (2014, 2011) and Pannabecker et al. (2008).

However, a few of the principal findings on the inner medulla in rats are briefly considered as follows for purposes of further comparative analyses among mammalian species: (1) The upper 40 % of each descending thin limb expresses the water channel AQP1 and is highly permeable to water whereas the lower 60 % of each descending thin limb lacks AQP1 and is impermeable to water. Therefore, this lower portion does not readily equilibrate osmotically with the surrounding interstitium by the loss of water. (2) The chloride channel ClC-K1 begins in a prebend segment of the descending thin limb and continues throughout the ascending thin limb, accounting for high chloride permeability throughout this entire region. (3) Both descending thin limbs and ascending thin limbs are highly permeable to urea. (4) Clusters of coalescing collecting ducts form the organizing motif for the three-dimensional arrangement of the thin limbs and vasa recta, with the descending thin limbs and descending vasa recta being outside the collecting duct clusters and the ascending thin limbs and ascending vasa recta being both inside and outside the clusters. (5) The arrangement of ascending vasa recta, ascending thin limbs, and collecting ducts within the clusters form interstitial nodal spaces bordered above and below by lateral interstitial cells, thereby creating stacks of microdomains along the length of the IM for solute and water mixing (Fig. 7.1). (6) Finally, the longest loops that turn at the tip of the papilla have wide lateral bends providing a large surface area for the delivery of solute to this region.

The most recent mathematical model takes advantage of these new structural and physiological findings (Layton et al. 2012). Briefly, in this model, urea and water still diffuse out of the inner medullary collecting ducts, as in the original passive model, thereby reducing the interstitial sodium chloride concentration and establishing a gradient for sodium chloride diffusion out of the loops of Henle. But this gradient is not enhanced by movement of water out of the descending thin limbs as in the original model. In addition, the diffusion of sodium chloride out of the loops occurs primarily out of the prebend portion of the descending thin limbs and

Fig. 7.1 Cross section through the upper inner medulla of rat kidney where tubules and vessels are organized around a collecting duct cluster. *Inset:* Schematic configuration of a collecting duct, ascending vasa recta (AVR), an ascending thin limb, and an interstitial nodal space (INS). Modified from Layton et al. (2009) with permission

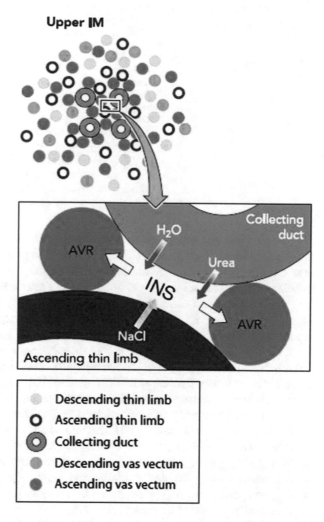

to a similar distance up the ascending thin limbs. It also occurs from the loops with wide bends over a very short axial distance at the tip of the papilla. Sodium chloride from the loops and urea and water from the collecting ducts mix in the interstitial nodal spaces surrounding the collecting ducts, and the resulting hyperosmotic solution is moved to higher regions by the ascending vasa recta (Fig. 7.1). Because both the descending thin limbs and the ascending thin limbs have high urea permeabilities, they act as countercurrent exchangers for urea. The mathematical analysis of this model (Layton et al. 2012) results in a urine concentration and tubule fluid concentration in loop bends at the tip of the papilla in good agreement with those actually measured in moderately antidiuretic rats (Table 7.4) (Pennell et al. 1974). Therefore, it is superior to previous models and appears to be closer to the actual inner medullary concentrating process than those models. However, the

Table 7.4 Comparison of values from model of urine concentrating mechanism with measurements on rat

Variable	Model[a]	Rat[b]
Urine		
Osmolality (mOsmol/kg H_2O)	1155	1216
Na^+ concentration (mM)	254	100
Urea concentration (mM)	554	345
Flow rate for whole kidney (µl/min)	3.58	2.27
Bend of Henle's loop		
Osmolality (mOsmol/kg H_2O)	1235	1264
Na^+ concentration (mM)	384	475
Urea concentration (mM)	544	287

Model loop values are for the longest loop of Henle; rat loop values are for very long loop visible on surface of papilla tip

[a]Data are from Layton et al. (2012)

[b]Mean values or estimated mean values from Pennell et al. (1974)

model does not yet reflect the true physiologic operation of the concentrating mechanism in the renal inner medulla, for it fails to produce a maximally concentrated urine or the appropriate interstitial axial sodium gradient. Consequently, this whole process is the subject of continuing active research.

At the time when countercurrent multiplication by the loops of Henle was becoming accepted as underlying the urine concentrating process, it was well understood that, other things being equal, the multiplier effect would increase with the length of the loop. Thus, it was thought that a relationship was likely to exist between that the length of the renal papilla relative to kidney size (calculated as relative medullary thickness, RMT) (Sperber 1944) and the maximum concentrating ability in mammalian species, and it was one of the first aspects of this process to be examined (Schmidt-Nielsen and O'Dell 1961). Indeed, a general correlation is observed (Table 7.3). However, it is obvious from the data in Table 7.3 that this relationship does not hold for all species. For example, the chinchilla has a relatively long papilla, but cannot concentrate its urine as well as a laboratory rat with a much shorter papilla. Similarly, the gundi (*Ctenodactylus vali*), another xerophilic species, can only concentrate its urine to about 1400 mosmol/kg H_2O, a value similar to that of humans (Table 7.3), despite a papilla about the length of the gerbil (Bankir and De Rouffignac 1985; De Rouffignac et al. 1981). And two other xerophilic species, an Australian hopping mouse (*Notomys cervinus*) and the sand rat (*Psammomys obesus*), do not concentrate their urine nearly as much as would be expected from the length of their papillae (Table 7.3). These observations all indicate that a long papilla with a comparable length of some loops of Henle is insufficient to assure production of a highly concentrated urine. In view of current knowledge that indicates that the concentrating process in the IM, where the long thin loops are found, does not involve a classical countercurrent multiplier system, these findings are no longer surprising.

The number of long-looped and short-looped nephrons is also highly variable and does not necessarily correlate with the concentrating ability of the mammalian kidney (Bankir and De Rouffignac 1985). Many xerophilic rodents capable of producing highly concentrated urine have fewer long-looped nephrons than mesophilic species of comparable body and kidney mass that are not capable of producing highly concentrated urine. Fortunately, for one of these xerophilic rodents that produces a highly concentrated urine (Table 7.3), the kangaroo rat (*Dipodomys merriami*), studies on the three-dimensional structure and permeabilities of the tubular and vascular elements of the inner medulla, similar to those being carried out with Munich-Wistar rats, have recently begun (Issaian et al. 2012; Urity et al. 2012). So far these studies indicate that the tubulovascular architecture of the inner medulla of the kidney of the kangaroo rat is remarkably similar to that of the Munich-Wistar rat. However, there are a few differences that may contribute to the ability of the kangaroo rat kidney to establish a steeper inner medullary osmotic gradient and thus a higher urine concentration than the Munich-Wistar rat kidney. First, the upper 60 % of each descending thin limb in the inner medulla of the kangaroo rat kidney expresses AQP1 and thus has a high permeability to water compared to only the upper 40 % in the Munich-Wistar rat kidney. Water reabsorption over this longer length of descending thin limb in the kangaroo rat may lead to a higher concentration of sodium chloride in the descending limb fluid at the loop bend and a greater reduction of the sodium chloride concentration in the surrounding interstitium than in the Munich-Wistar rat. Second, the ClC-K1 positive prebend region of each descending thin limb in the kangaroo rat kidney is only half as long as in the Munich-Wistar rat kidney. This may promote a greater driving force for sodium chloride reabsorption at the tip of each loop in the kangaroo rat kidney than in the Munich-Wistar rat kidney. Third, stacks of interstitial nodal spaces like those described above for the Munich-Wistar rat kidney (Fig. 7.1) are found in the kangaroo rat kidney. However, because the papilla of the kangaroo rat kidney is relatively much longer than that of the Munich-Wister rat kidney, the number in each stack is greater. If, indeed, as suggested in the model above, these act as mixing chambers for urea and sodium chloride, the greater axial number may enhance the axial osmotic gradient. In this regard, studies on the human kidney, which has a relatively short papilla and a limited concentrating ability, reveal relatively few interstitial nodal spaces (Wei et al. 2015). Clearly, a great deal more work on concentrating process in inner medulla of the mammalian kidney is required.

In further regard to the urine concentrating mechanism, at least one xerophilic species mentioned above, the sand rat (*P. obesus*), has a large kidney mass and a very large number of nephrons for its body size (Bankir and De Rouffignac 1985). Because this animal eats succulent plants with a high salt and water content, in contrast to many desert species that eat dry seeds, it must excrete a large volume of concentrated urine with a high salt content. Indeed, the kidneys of *P. obesus* are more effective in concentrating sodium and chloride and less effective in concentrating urea than those of other desert species. This may also be reflected in the high permeability of the thin descending limb of Henle to sodium chloride in this species (Jamison et al. 1979). This species may have unique requirements because of its diet and is certainly not a typical desert rodent. It would be extremely informative to explore the three-dimensional relationships and permeabilities of the tubular and vascular components in the inner medulla of these animals as is now being done for ordinary laboratory rats and kangaroo rats.

As noted above, generation of the osmotic gradient in the outer medulla of mammalian kidneys is reasonably well understood. However, the size and complexity of the vascular bundles of the inner stripe of the outer medulla, the degree of incorporation of the thin descending limbs of Henle's loops in the vascular bundles, the arrangement of the collecting ducts and thick ascending limbs outside the vascular bundles, and the thickness of the inner stripe all may be correlated with the concentrating ability in some species but not others (Bankir and De Rouffignac 1985). Bankir and de Rouffignac (1985), from their anatomical and functional analysis, also suggest that the complex vascular–tubular relations of the inner stripe of the outer medulla may be particularly important in generating or, especially, in maintaining the osmotic gradient and in enhancing the concentrating ability. These relationships may be particularly important in the recycling of urea and, thus, in influencing its accumulation in the inner, as well as the outer, medulla and, thereby, enhancing the concentrating ability of the mammalian kidney (Layton et al. 2009; Lei et al. 2011). This may vary in different species and certainly merits continued study.

In the avian kidney, as already noted above, the concentrating process requires establishment of an increasing osmotic gradient consisting almost entirely of sodium chloride along the length of the medullary cones (Emery et al. 1972; Skadhauge and Schmidt-Nielsen 1967b). There is no urea involved. This gradient appears to be established by classical countercurrent multiplication in the loops of Henle of the mammalian-type nephrons. Based primarily on studies of tubular transport in Japanese quail, *Coturnix coturnix*, Nishimura and colleagues (1989) proposed that this countercurrent multiplication occurs by single-solute recycling. They suggested that active transport of sodium chloride without water (vide supra, Chap. 4) out of the entire thick ascending limb (which begins prior to the bend of the loop, Chap. 2) is the single effect driving the process. This sodium chloride is then recycled by entering the thin descending limb, which has very high sodium and chloride permeabilities and very low water permeability (Nishimura et al. 1989). Furthermore, Nishimura et al. (1989) suggested that this recycling process combined with a cascade of loops of increasing lengths could establish the observed sodium-chloride based osmotic gradient in the medullary cones. As in the outer medulla of the mammalian kidney, the fluid diluted by sodium chloride reabsorption without water in the thick ascending limbs of these avian mammalian-type nephrons would be carried from the medullary cones into the collecting ducts of the cortical regions. In the absence of antidiuretic hormone, this fluid would remain dilute and be excreted as dilute urine. However, as noted above, in the presence of antidiuretic hormone (arginine vasotocin), the osmotic gradient in the medullary cones would cause water to move out of the parallel collecting ducts, thereby concentrating the final urine up to the level established by the osmotic gradient (see below for regulation by antidiuretic hormone). In support of this model of the avian concentrating mechanism, a mathematical simulation (using the anatomical and transport characteristics of quail nephrons) generated a maximum U/P osmolality ratio of about 2.26, a value in good agreement with the ratios actually measured in birds (Table 7.2) (Layton et al. 2000). This mathematical simulation also indicated that active transport of sodium chloride out of the prebend region of the thick ascending limbs is of great importance in generating the osmotic gradient along the medullary cones. Moreover, additional mathematical modeling (Layton 2005) indicated that the effectiveness of the avian concentrating mechanism depends in part on the cross-sectional arrangement of the tubules and vessels within the medullary cones (a central core of capillaries and descending thin limbs surrounded by a ring of collecting ducts with thick ascending limbs distributed both inside and outside the ring of collecting ducts) described by Braun (1985) and Nishimura et al. (1989). However, there appears to be little or no relationship between the length of the medullary cones (or relative medullary thickness) and the concentrating ability of the avian kidney (Johnson 1974).

7.5 Regulation of Urine Osmolality

As discussed above, the kidneys of many species of fishes, amphibians, and reptiles are capable of producing urine that varies from distinctly hypoosmotic to the plasma during hydration or adaptation to a moist environment or freshwater to isosmotic with the plasma during dehydration or adaptation to a dry environment or seawater (Table 7.1). The production of an isosmotic urine by the kidneys of amphibian and reptilian species apparently results from the equilibration of tubular fluid that has been diluted in one portion of the nephrons (apparently, early distal tubules or, possibly, the thin intermediate segment) with the interstitium surrounding a more distal portion of those same nephrons or the collecting ducts. It is generally assumed that this equilibration is facilitated by an increase in permeability of the distal portions of the nephrons or collecting ducts to water when the production of a urine of higher osmolality is required. As described above, the increased osmolality of the urine of euryhaline teleosts adapted to seawater apparently results largely from the tubular secretion of magnesium and sulfate (and possible sodium and chloride), but some increase in the permeability of the nephrons to water also may occur.

If an increase in the permeability of the distal tubules or collecting ducts to water occurs, it is generally considered to be produced by the action of arginine vasotocin (AVT), which appears to be the physiological antidiuretic hormone in all nonmammalian tetrapods and apparently some fishes (vide supra; Chap. 3). That a change in permeability to water does occur and that it is produced by AVT is supported indirectly in reptiles and amphibians by a number of in vivo renal clearance studies. The administration of apparently physiological quantities of AVT to some reptiles and amphibians during a water diuresis produces a prolonged increase in urine osmolality and decrease in relative free-water clearance (Bradshaw and Rice 1981; Butler 1972; Dantzler 1967; Jard and Morel 1963). As discussed above (Chap. 3), these doses of AVT also produce a reduction in the whole-kidney GFR, apparently via a reduction in the number of filtering nephrons. However, the decrease in relative free-water clearance is generally prolonged beyond the decrease in GFR, suggesting that there is a true change in the permeability of the tubules to water (Butler 1972; Dantzler 1967; Jard and Morel 1963). In addition, in one reptile species (*V. gouldii*), a decrease in relative free-water clearance during dehydration is clearly correlated with an increase in plasma osmolality and AVT level (Bradshaw and Rice 1981).

The most direct observation indicating an effect of AVT on water permeability of the tubular epithelium of amphibians has been made on one urodele species (*N. maculosus*). In these animals, micropuncture collections at the same distal tubule site before and after the administration of apparently physiological doses of AVT indicate that the tubular fluid-to-plasma osmolality ratio can increase from 0.2 to 0.9 (Table 7.1) (Garland et al. 1975). This striking change, which only occurs in this distal tubule segment, appears to indicate a direct effect of the hormone on the epithelium. However, it should be noted that the volume flow rate along the nephrons is markedly reduced following the administration of the hormone, and its effect on the equilibration of the lumen and plasma osmolalities has not been evaluated. Moreover, in another urodele species (*A. tigrinum*), AVT has no effect on the water permeability of segments of distal tubules or collecting ducts isolated and perfused in vitro (Stoner 1977).

Increasing evidence now suggests that in those amphibian species in which AVT increases the permeability of tubule epithelium to water the mechanism of hormone action is similar to that by which arginine vasopressin (AVP) increases the permeability of mammalian collecting ducts to water. This involves the binding of AVP to V2 receptors on the basolateral membrane and, via an adenylate cyclase pathway, the insertion of aquaporin 2 (AQP2) water channels in the luminal membrane (see below for more detail on mammals). Homologs of the mammalian V2 receptors have now been identified in the kidneys of an anuran species (Japanese tree frog, *Hyla japonica*) (Kohno et al. 2003) and a urodele species (Japanese red-bellied newt, *Cynops pyrrhogaster*) (Hasunuma et al. 2007). Moreover, a homolog of the mammalian AQP2, AQP-h2K, has also been identified in the kidney of *H. japonica* (Ogushi et al. 2007). In addition, AVT appears to stimulate the production of cyclic AMP (cAMP) in segments of connecting tubules (between distal tubule and collecting duct) isolated from the kidneys of an anuran (bullfrog, *R. catesbeiana*) (Uchiyama 1994).

Despite the clearance data suggesting that AVT increases the permeability of reptilian nephrons to water, more direct studies do not support this effect. For example, isolated, perfused snake (*Thamnophis* spp.) (Beyenbach 1984) distal tubules do not respond to AVT despite clearance data suggesting that snake renal tubules do respond (Dantzler 1967). Of course, AVT may increase the permeability to water of a more distal region of the nephron than that examined or of the cloaca. In addition, the effects of volume flow rate through the distal nephrons or collecting ducts on the equilibration of tubule fluid with the surrounding interstitium have yet to be evaluated. In those lizards that always produce a dilute urine, such as the blue spiny lizard, *S. cyanogenys* (Table 7.1), in vivo micropuncture studies actually show that dilution continues throughout the distal tubules and collecting ducts even following the administration of very large, possibly pharmacological, amounts of AVT (Stolte et al. 1977b).

With regard to fishes, AVT has no effect on distal tubules of freshwater-adapted rainbow trout (*O. mykiss*) isolated and perfused in vitro (Nishimura et al. 1983a). However, it seems likely that it might have an effect on water permeability of the nephrons of lungfish to help conserve water during their estivation in subterranean mud cocoons. Early in vivo studies of the effects of exogenous AVT on South American lungfish (*Lepidosiren paradoxa*) suggested that it causes primarily vaso-constriction and has no effect on tubular permeability to water (Sawyer et al. 1982). However, more recent studies have identified a homolog of the mammalian V2 receptor in the kidneys of African lungfish (*Protopterus annectens*) (Konno et al. 2009). Moreover, exposure of these cloned V2-type receptors, expressed in a heterologous cell system, to AVT leads to cellular production and accumulation of cyclic AMP (Konno et al. 2009). All these data suggest that AVT may have a physiological role in altering tubular permeability to water in these primitive fish.

As indicated, despite new data, an effect of AVT on the permeability to water of distal tubules and collecting ducts has yet to be directly documented in many fish, amphibian, and reptilian species and clearly does not occur in some species. However, as discussed in Chap. 3, AVT plays an important role in regulating the number of filtering nephrons in all species studied during dehydration or adaptation to a marine environment. A reduction in glomeruli filtering is practical for fishes, amphibians, and reptiles, in which the nephrons do not function in concert to produce a urine hyperosmotic to the plasma. It conserves water at the expense of excreting a reduced quantity of ions and nitrogenous waste. Of additional impor-tance, it reduces the volume flow through the collecting ducts into which the nephrons drain. In euryhaline teleost fishes, amphibians, and reptiles, this could promote the production of a urine isosmotic to the plasma if a smaller volume of fluid in the collecting ducts has more time to equilibrate with the surrounding interstitium. As noted above, in many species in these vertebrate classes, the number of filtering nephrons appears to be altered more readily than the perme-ability of the distal nephrons to water in response to AVT. In other species, however, the prolonged depression of the relative free-water clearance suggests that that there may be a change in the permeability of the distal nephrons to water and that this may be influenced more readily by AVT than the number of filtering nephrons. More detailed information on more species is required on the effect of AVT on the tubular permeability to water and on the relative importance of an increase in tubular permeability to water and a decrease in the number of filtering nephrons in the total antidiuretic effect of AVT in fishes, amphibians, and reptiles.

The kidneys of birds and mammals, as discussed above, are capable of producing a urine that varies from markedly hypoosmotic to the blood to hyperosmotic to the blood, depending on the degree of hydration (Tables 7.1, 7.2, and 7.3). In mammals, as noted above, the effect of arginine vasopressin (AVP) on the permeability of the collecting ducts to water is well documented. The cellular mechanism is now largely understood, although the details of the pathway are still the subject of intensive study. The details of this mechanism are extensively reviewed elsewhere (Brown and Fenton 2012) and are beyond the scope of the present discussion. Briefly, AVP binds to G-protein-coupled V2 receptors on the basolateral membrane of collecting duct principal cells, thereby activating adenylate cyclase. This, in turn, leads to increased cytoplasmic cAMP levels. This activates protein kinase A (PKA), which leads to phosphorylation of a number of proteins including the water channel AQP2. Through some other steps, these AQP2 water channels are inserted into the luminal cell membrane, thereby increasing its permeability to water and permitting osmotically driven transepithelial water reabsorption.

In birds, the increase in urine osmolality, decrease in urine volume flow rate, and decrease in free-water clearance observed with dehydration can be mimicked by the administration of single injections of AVT (Ames et al. 1971) and by an infusion of AVT that reproduces the plasma levels of the hormone measured during dehydration (Stallone and Braun 1985). This antidiuretic effect is generally assumed, as in mammals, to result from an increase in the permeability of the collecting ducts to water, permitting the fluid in them to equilibrate with the surrounding hyperosmotic interstitium of the medullary cones (vide supra) (Ames et al. 1971; Skadhauge 1964). Indeed, a homolog of the mammalian AVP-responsive AQP2 water channel has been identified in the luminal and subluminal regions of the cells of the cortical and medullary collecting ducts of Japanese quail (*C. coturnix*), and its presence in the luminal membrane appears to increase with dehydration or AVT administration (Yang et al. 2004). However, the osmotic water permeability of quail medullary collecting ducts isolated and perfused in vitro is low and only increases a modest amount in the presence of AVT (Nishimura et al. 1996). The response may be greater in vivo, but direct in vivo measurements on osmotic water permeability of avian collecting ducts have yet to be made and would be technically difficult. In addition, because all the reptilian-type nephrons contribute to the fluid flowing through the collecting ducts (Fig. 2.6), a decrease in the number of these nephrons filtering in response to AVT (vide supra; Chap. 3) will reduce volume flow through the collecting ducts, should enhance equilibration with the surrounding interstitium, and may contribute significantly to the production of a urine hyperosmotic to the plasma (Braun and Dantzler 1974). However, this possible effect of volume flow through the collecting ducts on urine concentrating ability has also yet to be examined directly.

Although neither the effects of AVT on the permeability of avian collecting ducts to water nor the effects of volume flow rate through avian collecting ducts on equilibration with the interstitium have been examined directly in vivo, indirect in vivo evidence suggests that there is a significant change in permeability to water and that the permeability to water is more sensitive to AVT than the number of filtering nephrons. Stallone and Braun (1985) infused AVT into unanesthetized domestic fowl (*G. gallus domesticus*) receiving a constant hypoosmotic saline infusion to mimic the plasma levels observed during dehydration and measured whole-kidney GFR and free-water clearance. They found that both whole-kidney GFR and free-water clearance decreased with increasing plasma levels of AVT. However, the initial decrease in free-water clearance occurred at a lower plasma level of AVT than the initial decrease in GFR. In fact, about 90 % of the maximum decrease in free-water clearance occurred before there was any significant decrease in GFR. At plasma concentrations of AVT equivalent to those found with 24 h or more of dehydration, there was a 30 % decrease in GFR but little additional change in free-water clearance.

Gerstberger et al. (1985) examined the effects of perfusions of the third ventricle of unanesthetized, salt-water-adapted ducks (*Anas platyrhynchos*) on plasma immunoreactive AVT levels, whole-kidney GFR, and free-water clearance. These animals, like the domestic fowl examined by Stallone and Braun (1985), were receiving a hypoosmotic saline infusion during the experiments. Although both the plasma AVT levels and the renal function in these salt-water-adapted animals with functioning salt glands are certainly related to the function of these glands and may not be quantitatively applicable to birds without salt glands, some of the observations are relevant to the sensitivity of the glomeruli and tubules to AVT. Perfusion of the third ventricle with a solution hyperosmotic to native cerebrospinal fluid resulted in an increase in the plasma level of AVT and a decrease in whole-kidney GFR and free-water clearance. However, the whole-kidney GFR returned to the control level while the perfusion continued and the plasma AVT level remained elevated, but the free-water clearance remained depressed. Perfusion of the third ventricle with a solution isosmotic to the native cerebrospinal fluid had no effect on plasma AVT or renal function. However, perfusion of the third ventricle with a solution hypoosmotic to the native cerebrospinal fluid resulted in a decrease in the plasma level of AVT and an increase in the whole-kidney GFR and free-water clearance. In this case, GFR tended to return to the control level while perfusion continued and the plasma AVT level was depressed, but free-water clearance remained elevated.

Single intravenous injections of small, apparently physiological doses of AVT in these same animals depressed both whole-kidney GFR and free-water clearance (Gerstberger et al. 1985). However, as in some of the studies with reptiles and amphibians (vide supra), the whole-kidney GFR returned to the control level before the free-water clearance.

The data on domestic fowl and ducks all suggest that, in vivo, AVT produces a significant increase in the permeability of the collecting ducts to water and that this permeability is more readily altered by AVT than the whole-kidney GFR, and, thus, the number of filtering nephrons and the volume flow through the collecting ducts. It must be remembered, however, that changes in the collecting duct permeability to water have not yet been measured directly in vivo in birds. Moreover, this apparent pattern of greater tubular than glomerular sensitivity to AVT may not hold for all avian species, not even for all gallinaceous species. Finally, it should be noted that in some avian species, e.g., the passerine European starling (*S. vulgaris*), an increase in the number of filtering mammalian-type nephrons during the administration of a salt load (vide supra; Chap. 3), when the plasma AVT levels are presumably increased, may also enhance the concentrating ability by permitting more long-looped nephrons to produce a greater osmotic gradient along the medullary cones.

References

Ames E, Steven K, Skadhauge E (1971) Effects of arginine vasotocin on renal excretion of Na^+, K^+, Cl^-, and urea in the hydrated chicken. Am J Physiol 221:1223–1228

Bankir L, De Rouffignac C (1985) Urinary concentrating ability: insights from comparative anatomy. Am J Physiol Regul Integr Comp Physiol 249:R643–R666

Bechhold H, Ziegler J (1914) Vorstudien über Gicht. III. Biochem Z 64:471–489

Beyenbach KW (1984) Water-permeable and -impermeabile barriers of snake distal tubules. Am J Physiol Renal Physiol 246:F290–F299

Beyenbach KW (1986) Secretory NaCl and volume flow in renal tubules. Am J Physiol Regul Integr Comp Physiol 250:R753–R763

Beyenbach KW, Dantzler WH (1978) Generation of transepithelial potentials by isolated perfused reptilian distal tubules. Am J Physiol Renal Physiol 234(3):F238–F246

Beyenbach KW, Koeppen BM, Dantzler WH, Helman SI (1980) Luminal Na concentration and the electrical properties of the snake distal tubule. Am J Physiol Renal Physiol 239(8):F412–F419

Bradshaw SD, Rice GE (1981) The effects of pituitary and adrenal hormones on renal and postrenal reabsorption of water and electrolytes in the lizard, *Varanus gouldii* (gray). Gen Comp Endocrinol 44:82–93

Braun EJ (1985) Comparative aspects of the urinary concentrating process. Renal Physiol 8:249–260

Braun EJ, Dantzler WH (1972) Function of mammalian-type and reptilian-type nephrons in kidney of desert quail. Am J Physiol 222(3):617–629

Braun EJ, Dantzler WH (1974) Effects of ADH on single-nephron glomerular filtration rates in the avian kidney. Am J Physiol 226:1–8

Braun EJ, Dantzler WH (1975) Effects of water load on renal glomerular and tubular function in desert quail. Am J Physiol 229:222–228

Braysher MI (1976) The excretion of hyperosmotic urine and other aspects of the electrolyte balance in the lizard *Amphibolurus maculosus*. Comp Biochem Physiol A 54:341–345

Brown D, Fenton RA (2012) The cell biology of vasopressin action. In: Taal MW, Chertow GM, Marsden PA, Skorecki K, Yu ASL, Brenner BM (eds) Brenner and Rector's the kidney, 9th edn. Elsevier (Saunders), Philadelphia, pp 353–383

Burg MB, Green N (1973) Function of thick ascending limb of Henle's loop. Am J Physiol 224:659–668

Butler DG (1972) Antidiuretic effect of arginine vasotocin in the Western painted turtle (*Chrysemys picta belli*). Gen Comp Endocrinol 18:121–125

Chou C-L, Knepper MA (1992) In vitro perfusion of chinchilla thin limb segments: segmentation and osmotic water permeability. Am J Physiol Renal Physiol 263(3):F417–F426

Dantzler WH (1966) Renal response of chickens to infusion of hyperosmotic sodium chloride solution. Am J Physiol 210:640–646

Dantzler WH (1967) Glomerular and tubular effects of arginine vasotocin in water snakes (*Natrix sipedon*). Am J Physiol 212:83–91

Dantzler WH (1970) Kidney function in desert vertebrates. In: Benson GK, Phillips JG (eds) Memoirs of the society of endocrinology. 18. Hormones and the environment. Cambridge University Press, London, pp 157–190

Dantzler WH (1978) Urate excretion in nonmammalian vertebrates. In: Kelley WN, Weiner IM (eds) Uric acid, vol 51, Handbook of experimental pharmacology. Springer, Berlin, pp 185–210

Dantzler WH (1987) Comparative renal physiology. In: Gottschalk CW, Berliner RW, Giebisch G (eds) Renal physiology: people and ideas. American Physiological Society, Washington, DC, pp 437–481

Dantzler WH, Layton AT, Layton HE, Pannabecker TL (2014) Urine-concentrating mechanism in the inner medulla: function of the thin limbs of the loops of Henle. Clin J Am Soc Nephrol 9:1781–1789

Dantzler WH, Pannabecker TL, Layton AT, Layton HE (2011) Urine concentrating mechanism in the inner medulla of the mammalian kidney: role of three-dimensional architecture. Acta Physiol 202:361–378

Dantzler WH, Schmidt-Nielsen B (1966) Excretion in fresh-water turtle (Pseudemys scripta) and desert tortoise (Gopherus agassizii). Am J Physiol 210:198–210

De Rouffignac C, Bankir L, Roinel N (1981) Renal function and concentrating ability in a desert rodent: the gundi (Ctenodactylus vali). Pflugers Arch 390:138–144

Emery N, Poulson TL, Kinter WB (1972) Production of concentrated urine by avian kidneys. Am J Physiol 223:180–187

Fleming WR, Stanley JG (1965) Effects of rapid changes in salinity on the renal function of a euryhaline teleost. Am J Physiol 209:1025–1030

Forster RP (1953) A comparative study of renal function in marine teleosts. J Cell Comp Physiol 42:487–510

Garland HO, Henderson IW, Brown JA (1975) Micropuncture study of the renal responses of the urodele amphibian Necturus maculosus to injections of arginine vasotocin and an anti-aldosterone compound. J Exp Biol 63:249–264

Gerstberger R, Kaul R, Gray DA, Simon E (1985) Arginine vasotocin and glomerular filtration rate in saltwater-acclimated ducks. Am J Physiol Renal Physiol 248:F663–F667

Goldstein DL, Williams JB, Braun EJ (1990) Osmoregulation in the field by salt-marsh savannah sparrows Passerculus sandwichensis beldingi. Physiol Zool 63:669–682

Gottschalk CW, Mylle M (1959) Micropuncture study of the mammalian urinary concentrating mechanism: evidence for the countercurrent hypothesis. Am J Physiol 196:927–936

Gutman Y, Beyth Y (1970) Chinchilla laniger: discrepancy between concentrating ability and kidney structure. Life Sci 9:37–42

Hargitay B, Kuhn W (1951) Das Multiplikationsprinzip als Grundlage der Harnkonzentrierung in der Niere. Z Electrochem u ang physikal Chem 55:539–558

Hasunuma I, Sakai T, Nakada T, Toyoda F, Namiki H, Kikuyama S (2007) Molecular cloning of three types of arginine vasotocin receptor in the newt, Cynops pyrrhogaster. Gen Comp Endocrinol 151:252–258

Hentschel H, Elger M (1987) The distal nephron in the kidney of fishes. Adv Anat Embryol Cell Biol 108:1–151

Hervy S, Thomas SR (2003) Inner medullary lactate production and urine-concentrating mechanism: a flat medullary model. Am J Physiol Renal Physiol 284:F65–F81

Hickman CP, Trump BF (1969) The kidney. In: Hoar WS, Randall DJ (eds) Fish physiology, vol I, Excretion, ion regulation, and metabolism. Academic, New York, pp 91–239

Imai M, Kokko JP (1974) Sodium chloride, urea, and water transport in the thin ascending limb of Henle. Generation of osmotic gradients by passive diffusion of solutes. J Clin Invest 53:393–402

Imai M, Kusano E (1982) Effects of arginine vasopressin on the thin ascending limb of Henle's loop of hamsters. Am J Physiol Renal Physiol 243:F167–F172

Issaian T, Urity V, Dantzler WH, Pannabecker TL (2012) Architecture of vasa recta in the renal inner medulla of the desert rodent Dipodomys merriami: potential impact on the urine concentrating mechanism. Am J Physiol Regul Integr Comp Physiol 303:R748–R756

Jamison RL, Roinel N, De Rouffignac C (1979) Urinary concentrating mechanism in the desert rodent Psammomys obesus. Am J Physiol Renal Physiol 236:F448–F453

Jard S, Morel F (1963) Actions of vasotocin and some of its analogues on salt and water excretion by the frog. Am J Physiol 204:222–226

Jen JF, Stephenson JL (1994) Externally driven countercurrent multiplication in a mathematical model of the urinary concentrating mechanism of the renal inner medulla. Bull Math Biol 56:491–514

Johnson OW (1974) Relative thickness of the renal medulla in birds. J Morphol 142:277–284

Knepper MA, Saidel GM, Hascall VC, Dwyer T (2003) Concentration of solutes in the renal inner medulla: interstitial hyaluronan as a mechano-osmotic transducer. Am J Physiol Renal Physiol 284:F433–F446

Kohno S, Kamishima Y, Iguchi T (2003) Molecular cloning of an anuran V_2 type [Arg^8] vasotocin receptor and mesotocin receptor: functional characterization and tissue expression in the Japanese tree frog (Hyla japonica). Gen Comp Endocrinol 132:485–498

Kokko JP, Rector FC (1972) Countercurrent multiplication system without active transport in inner medulla. Kidney Int 2:214–223

Komadina S, Solomon S (1970) Comparison of renal function of bull and water snakes (Pituophis melanoleucus and Natrix sipedon). Comp Biochem Physiol 32:333–343

Konno N, Hyodo S, Yamaguchi Y, Kaiya H, Miyazato M, Matsuda K, Uchiyama M (2009) African lungfish, Protopterus annectens, possess an arginine vasotocin receptor homologous to the tetrapod V2-type receptor. J Exp Biol 212:2183–2193

Kowarsky J (1973) Extra-branchial pathways of salt exchange in a teleost fish. Comp Biochem Physiol 46A:477–486

Krag B, Skadhauge E (1972) Renal salt and water excretion in the budgerygah (Melopsittacus undulatus). Comp Biochem Physiol 41A:667–683

Kuhn W, Ramel A (1959) Activer Salztransport als möglicher (und wahrscheinlicher) Einzeleffekt bei der Harnkonzentrierung in der Niere. Helv Chim Acta 42:628–660

Kuhn W, Ryffel K (1942) Herstellung konzentrierter Losüngen aus verdünten durch blosse Membranwirkung: ein Modellversuch zur Funktion der Niere. Hoppe-Seylers Z Physiol Chem 276:145–178

Lahlou B, Henderson IW, Sawyer WH (1969) Renal adaptations by Opsanus tau, a euryhaline aglomerular teleost, to dilute media. Am J Physiol 216:1266–1272

Larsen EH, Deaton LE, Onken H, O'Donnell M, Grosell M, Dantzler WH, Weihrauch D (2014) Osmoregulation and excretion. Compr Physiol 4:405–573

Laverty G, Skadhauge E (1999) Physiological roles and regulation of transport activities in the avian lower intestine. J Exp Zool 283:480–494

Layton AT (2005) Role of structural organization in the urine concentrating mechanism of an avian kidney. Math Biosci 197:211–230

Layton AT, Dantzler WH, Pannabecker TL (2012) Urine concentrating mechanism: impact of vascular and tubular architecture and a proposed descending limb urea-Na^+ cotransporter. Am J Physiol Renal Physiol 302:F591–F605

Layton AT, Layton HE (2005) A region-based mathematical model of the urine concentrating mechanism in the rat outer medulla. I. Formulation and base-case results. Am J Physiol Renal Physiol 289:F1346–F1366

Layton AT, Layton HE, Dantzler WH, Pannabecker TL (2009) The mammalian urine concentrating mechanism: hypotheses and uncertainties. Physiology 24:250–256

Layton HE, Davies JM, Casotti G, Braun EJ (2000) Mathematical model of an avian urine concentrating mechanism. Am J Physiol Renal Physiol 279:F1139–F1160

Lei T, Zhou L, Layton AT, Zhou H, Zhao X, Bankir L, Yang B (2011) Role of thin descending limb urea transport in renal urea handling and the urine concentrating mechanism. Am J Physiol Renal Physiol 301:F1251–F1259

Logan AG, Moriarty RJ, Rankin JC (1980) A micropuncture study of kidney function in the river lamprey, Lampetra fluviatilis, adapted to fresh water. J Exp Biol 85:137–147

Long WS (1973) Renal handling of urea in Rana catesbeiana. Am J Physiol 224:482–490

Longley JB, Burtner HJ, Monis B (1963) Mucous substances in excretory organs: a comparative study. Ann NY Acad Sci 106:493–501

Marsh DJ, Azen SP (1975) Mechanism of NaCl reabsorption by hamster thin ascending limbs of Henle's loop. Am J Physiol 228:71–79

Marsh DJ, Solomon S (1965) Analysis of electrolyte movement in thin Henle's loops of hamster papilla. Am J Physiol 208:1119–1128

McBean RL, Goldstein L (1970) Renal function during osmotic stress in the aquatic toad *Xenopus laevis*. Am J Physiol 219:1115–1123

McNabb FMA, McNabb RA, Steeves HR (1973) Renal mucoid materials in pigeons fed high and low protein diets. The Auk 90:14–18

McNabb FMA, Poulson TL (1970) Uric acid excretion in pigeons, (*Columba livia*). Comp Biochem Physiol 33:933–939

Miwa T, Nishimura H (1986) Diluting segment in avian kidney. II. Water and chloride transport. Am J Physiol Regul Integr Comp Physiol 250:R341–R347

Nishimura H, Imai M, Ogawa M (1983a) Sodium chloride and water transport in the renal distal tubule of the rainbow trout. Am J Physiol Renal Physiol 244:F247–F254

Nishimura H, Imai M, Ogawa M (1983b) Transepithelial voltage in the reptilian- and mammalian-type nephrons from Japanese quail. Fed Proc 42:304 (Abstract)

Nishimura H, Koseki C, Imai M, Braun EJ (1989) Sodium chloride and water transport in the thin descending limb of Henle of the quail. Am J Physiol Renal Physiol 257:F994–F1002

Nishimura H, Koseki C, Patel TB (1996) Water transport in collecting ducts of Japanese quail. Am J Physiol Regul Integr Comp Physiol 271:R1535–R1543

Oberleithner H, Guggino W, Giebisch G (1983) The effect of furosemide on luminal sodium, chloride and potassium transport in the early distal of *Amphiuma* kidney. Pflugers Arch 396:27–33

Ogushi Y, Mochida H, Nakakura T, Suzuki M, Tanaka S (2007) Immunocytochemical and phylogenetic analyses of an arginine vasotocin-dependent aquaporin, AQP-H2K, specifically expressed in the kidney of the tree frog, *Hyla japonica*. Endocrinology 148:5891–5901

Pannabecker TL, Dantzler WH, Layton HE, Layton AT (2008) Role of three-dimensional architecture in the urine concentrating mechanism of the rat renal inner medulla. Am J Physiol Renal Physiol 295:F1271–F1285

Pennell JP, Lacy FB, Jamison RL (1974) An *in vivo* study of the concentrating process in the descending limb of Henle's loop. Kidney Int 5:337–347

Porter P (1963) Physico-chemical factors involved in urate calculus formation. II. Colloidal flocculation. Res Vet Sci 4:592–602

Prange HD, Greenwald L (1979) Concentrations of urine and salt gland secretions in dehydrated and normally hydrated sea turtles. Fed Proc 38:970 (abstract)

Roberts JS, Schmidt-Nielsen B (1966) Renal ultrastructure and excretion of salt and water by three terrestrial lizards. Am J Physiol 211:476–486

Rocha AS, Kokko JP (1973) Sodium chloride and water transport in the medullary thick ascending limb of Henle. J Clin Invest 52:612–623

Sands JM, Layton HE (2007) The urine concentrating mechanism and urea transporters. In: The kidney: physiology and pathophysiology. Elsevier, Philadelphia, pp 1143–1177

Sawyer WH, Uchiyama M, Pang PKT (1982) Control of renal function in lungfishes. Fed Proc 41:2361–2364

Schade H, Boden E (1913) Ober die anomalie der harnsaureloslichkeit (kolloide harnsaure). Hoppe Seylers Zeitschrift fur Physiol Chemie 83:347–380

Schmidt-Nielsen B (1995) The renal concentrating mechanism in insects and mammals: a new hypothesis involving hydrostatic pressures. Am J Physiol Regul Integr Comp Physiol 268: R1087–R1100

Schmidt-Nielsen B, Forster RP (1954) The effect of dehydration and low temperature on renal function in the bullfrog. J Cell Comp Physiol 44:233–246

Schmidt-Nielsen B, O'Dell R (1961) Structure and concentrating mechanism in the mammalian kidney. Am J Physiol 200:1119–1124

Schmidt-Nielsen B, Renfro JL (1975) Kidney function of the american eel *Anguilla rostrata*. Am J Physiol 228:420–431

Schmidt-Nielsen B, Skadhauge E (1967) Function of the excretory system of the crocodile (*Crocodylus acutus*). Am J Physiol 212:973–980

Skadhauge E (1964) Effects of unilateral infusion of arginine vasotocin into the portal circulation of the avian kidney. Acta Endocrinol 47:321–330

Skadhauge E (1974) Renal concentrating ability in selected West Australian birds. J Exp Biol 61:269–276

Skadhauge E (1981) Osmoregulation in birds. Springer, Berlin

Skadhauge E, Schmidt-Nielsen B (1967a) Renal function in domestic fowl. Am J Physiol 212:793–798

Skadhauge E, Schmidt-Nielsen B (1967b) Renal medullary electrolyte and urea gradient in chickens and turkeys. Am J Physiol 212:1313–1318

Smith HW (1953) From fish to philosopher. Little, Brown, Boston

Sperber I (1944) Studies on the mammalian kidney. Zoologiska Bidrag Fran Uppsala 22:249–435

Stallone JN, Braun EJ (1985) Contributions of glomerular and tubular mechanisms to antidiuresis in conscious domestic fowl. Am J Physiol Renal Physiol 249:F842–F850

Stanley JG, Fleming WR (1964) Excretion of hypertonic urine by a teleost. Science 144:63–64

Stephenson JL (1972) Concentration of urine in a central core model of the renal counterflow system. Kidney Int 2:85–94

Stolte H, Galaske RG, Eisenbach GM, Lechene C, Schmidt-Nielsen B, Boylan JW (1977a) Renal tubule ion transport and collecting duct function in the elasmobranch little skate, *Raja erinacea*. J Exp Zool 199:403–410

Stolte H, Schmidt-Nielsen B (1978) Comparative aspects of fluid and electrolyte regulation by cyclostome, elasmobranch, and lizard kidney. In: Barker Jorgenson C, Skadhhauge E (eds) Osmotic and volume regulation. Alfred Benzon symposium XI. Munksgaard, Copenhagen, pp 209–220

Stolte H, Schmidt-Nielsen B, Davis L (1977b) Single nephron function in the kidney of the lizard, *Sceloporus cyanogenys*. Zool J Physiol 81:219–244

Stoner LC (1977) Isolated, perfused amphibian renal tubules: the diluting segment. Am J Physiol Renal Physiol 233:F438–F444

Stoner LC (1985) The movement of solutes and water across the vertebrate distal nephron. Renal Physiol 8:237–248

Thomas SR (2000) Inner medullary lactate production and accumulation: a vasa recta model. Am J Physiol Renal Physiol 279:F468–F481

Uchiyama M (1994) Sites of action of arginine vasotocin in the nephron of the bullfrog kidney. Gen Comp Endocrinol 94:366–373

Urity V, Issaian T, Braun EJ, Dantzler WH, Pannabecker TL (2012) Architecture of kangaroo rat inner medulla: segmentation of descending thin limb of Henle's loop. Am J Physiol Regul Integr Comp Physiol 302:R720–R726

Wang X, Thomas SR, Wexler AS (1998) Outer medullary anatomy and the urine concentrating mechanism. Am J Physiol Renal Physiol 274:F413–F424

Wang XQ, Wexler AS (1996) The effects of collecting duct active NaCl reabsorption and inner medulla anatomy on renal concentrating mechanism. Am J Physiol Renal Physiol 270:F900–F911

Wei G, Rosen S, Dantzler WH, Pannabecker TL (2015) Architecture of the human renal inner medulla and functional implications. Am J Physiol Renal Physiol 309(7):F627–F637

Wexler AS, Kalaba RE, Marsh DJ (1991) Three-dimensional anatomy and renal concentrating mechanism. I. Modeling results. Am J Physiol Renal Physiol 260:F368–F383

Yang Y, Cui Y, Wang W, Zhang L, Bufford L, Sasaki S, Fan Z, Nishimura H (2004) Molecular and functional characterization of a vasotocin-sensitive aquaporin water channel in quail kidney. Am J Physiol Regul Integr Comp Physiol 287:R915–R924

Yokota SD, Benyajati S, Dantzler WH (1985) Renal function in sea snakes. I. Glomerular filtration rate and water handling. Am J Physiol Regul Integr Comp Physiol 249:R228–R236

Yool AJ, Brokl OH, Pannabecker TL, Dantzler WH, Stamer WD (2002) Tetraethylammonium block of water flux in Aquaporin-1 channels expressed in kidney thin limbs of Henle's loop and a kidney-derived cell line. BMC Physiol 2:4

Young EG, Musgrave FF (1932) The formation and decomposition of urate gels. Biochem J 26:941–953

Chapter 8
Integrative Summary of Renal Function

Abstract This chapter provides integrative generalizations of kidney function in fishes, amphibians, reptiles, birds, and mammals based on the information provided in the previous chapters. These are broad, somewhat simplified generalizations. Following these integrative sections, the chapter provides generalized comparisons and contrasts between vertebrate classes for several renal functions: the glomerular filtration rate in regard to short-term stability versus lability; diluting and concentrating processes; and sites and mechanisms of tubular transport. These comparisons involve only a few major points for which differences or similarities between and among vertebrate classes are clear and need emphasis.

Keywords Integrative renal function • Species differences • Species adaptations • GFR stability • GFR lability • Integration of urine dilution and concentration • Species variation in tubular transport

8.1 Introduction

Understanding of renal glomerular and tubular functions discussed in the preceding chapters is far from complete, even for mammals. It is much less complete for nonmammalian vertebrates, which are the primary subjects of this volume. Nevertheless, it is possible to make some integrative generalizations about renal function in each major group of vertebrates and some general comparisons and contrasts of renal function among groups of vertebrates. However, these are just broad, somewhat simplified generalizations. They do not include all the functional details discussed earlier. Moreover, they may differ somewhat when considered for individual species within a given vertebrate group.

8.2   Integrative Summary Within Each of the Major Vertebrate Groups

8.2.1   Fishes

8.2.1.1   Cyclostomes

Myxini (hagfishes) These primitive marine animals are in osmotic equilibrium with their seawater environment, the osmolality of their body fluids being determined primarily by the concentrations of sodium and chloride. Their kidneys have relatively few large glomeruli which empty via a short neck segment into a primitive archinephric duct (ureter). Both the whole-kidney and single-nephron filtration rates are high, possibly because of the large area available for filtration. As the filtrate passes along the archinephric duct, glucose is reabsorbed and urea, potassium, phosphate, magnesium, and sulfate are secreted, but essentially no filtered sodium or water is reabsorbed. The final urine is isosmotic with the body fluids and seawater. Thus, the kidneys function to help regulate the body content of some organic compounds, divalent ions, and potassium and, possibly, the total body fluid volume. They do not, however, play any role in regulating the osmolality or sodium concentrations of the body fluids.

Petromyzones (lampreys) These primitive euryhaline animals, unlike hagfishes, do not conform to their environment. Instead, they maintain their body fluids hyperosmotic to a freshwater and hypoosmotic to a seawater environment. Their kidneys have nephrons with a gross structure similar to those of more advanced vertebrates. Each nephron has a relatively large glomerulus, short neck segment, proximal tubule, intermediate segment, distal tubule, and collecting duct. Animals adapted to freshwater have moderately high whole-kidney and single-nephron glomerular filtration rates. Filtration equilibrium is not reached along the glomerular capillaries; therefore, the single nephron filtration rate is particularly sensitive to changes in hydrostatic pressure but not to changes in plasma flow rate. Only about 40 % of the filtered water is reabsorbed, most of it in the collecting ducts. However, over 90 % of the filtered sodium, chloride, and even potassium is reabsorbed, again primarily along the distal tubules and collecting ducts. This reabsorption of ions, far in excess of the reabsorption of water, results in a final urine only one-tenth the osmolality of the plasma, thereby permitting the animals to eliminate excess water and helping them to maintain the osmolality of their body fluids well above that of the surrounding freshwater.

Much less information is available about lampreys adapted to seawater than about those adapted to freshwater. However, seawater-adapted lampreys have whole-kidney glomerular filtration rates far below those of freshwater-adapted animals, apparently resulting from a decrease in the filtration rate of each nephron, not from a decrease in the number of nephrons filtering. Under these circumstances, filtration equilibrium is achieved along the glomerular capillaries, and the single nephron glomerular filtration rate is particularly sensitive to changes in the glomerular capillary plasma flow rate. In contrast to the freshwater-adapted lampreys, 90 % of the filtered water is reabsorbed by the renal tubules. As in the freshwater-adapted animals, however, this reabsorption of filtered water in seawater-adapted animals occurs primarily in the distal tubules and collecting ducts. With the decrease in filtration rate and the reabsorption of most of the filtered water, the kidneys of the seawater-adapted lampreys help to retain water and maintain the osmolality of the body fluids below that of the surrounding seawater. However, the factors regulating these changes in renal function between freshwater and seawater are unknown.

8.2.1.2   Elasmobranchs

Marine elasmobranchs, like the hagfishes, conform to the osmolality of their environment, but, unlike hagfishes, much of the osmolality of the extracellular fluid is determined not by the concentrations of sodium and chloride but by the concentrations of urea and TMAO. Therefore, these animals that ingest seawater must excrete absorbed divalent ions, e.g., magnesium and sulfate, and monovalent ions, e.g., sodium and chloride, and retain urea and TMAO. Stenohaline freshwater elasmobranchs, in contrast to marine elasmobranchs, do not conform to their environment. They have a much lower plasma osmolality than marine elasmobranchs (reflecting very low plasma concentrations of urea and TMAO) but still well above the osmolality of the surrounding freshwater. The low plasma urea levels in these animals are a result of both a very low rate of urea synthesis and a failure of the renal tubules to reabsorb filtered urea.

Although the marine elasmobranch nephrons contain all the standard components found in advanced vertebrates, the glomeruli are quite large and the proximal and distal tubules are long with a highly complex arrangement. This arrangement may play a role in urea reabsorption and retention, possibly by permitting some form of passive but apparently concentrative transport that is coupled indirectly to the transport of sodium. This process seems possible, and there are some data to support it, including the observation that the renal tubules of stenohaline freshwater elasmobranchs, which lack some segments of this complex structure, do not reabsorb filtered urea. However, urea reabsorption by renal tubules in marine elasmobranchs may also involve sodium-coupled secondary active transport or perhaps even primary active transport.

Whole-kidney glomerular filtration rates in marine elasmobranchs are somewhat higher than those of marine teleosts, and a large fraction of the filtrate can be reabsorbed by the renal tubules. However, the apparent fraction of filtrate reabsorbed can vary over an extremely wide range. Moreover, substantial fluid secretion by the renal tubules can occur, perhaps at a rate equaling the rate of glomerular filtration. And, although fluid secretion may be the most important when glomerular filtration is low, providing a medium for the excretion of ions such as magnesium and sulfate, the exact integration and control of the processes of fluid filtration and fluid secretion are not understood.

Because the renal tissues of marine elasmobranchs have proved useful for physiological studies of transport in vitro, much is known about a number of specific transport processes but little about the overall integration of these processes. Sodium chloride is secreted in the second segment of the proximal tubule by a secondary active transport process which also provides the driving force for the secretion of water. However, sodium chloride is apparently reabsorbed in this segment as well as in other tubule segments. For both the reabsorptive and secretory processes, the primary energy is provided by the Na-K-ATPase on the basolateral membrane. Although the secretory process may be controlled in part by regulation of chloride channels in the luminal membrane by cAMP, the actual way in which the secretory and reabsorptive processes are integrated is not understood. In general, what appears to be two-thirds of the filtered sodium chloride is reabsorbed by the renal tubules, although this may not be a true value because of the sodium chloride secretion. This reabsorption produces some variable dilution of the tubular fluid, apparently along the collecting ducts, but the control and physiological significance of this dilution in these marine animals is unclear. Much of the sodium chloride reabsorbed by the renal tubules may be secreted by the rectal gland, a structure specialized to secrete these ions and to rid the animals of excess ingested sodium chloride. However, whether renal reabsorption of filtered and secreted sodium chloride is regulated in any way relative to the secretion by the rectal gland is unknown.

Magnesium, sulfate, calcium, and phosphate, although filtered, are also secreted in the second segment of the proximal tubule. Such net tubular secretion (and ultimate excretion) is important for the removal of these divalent ions ingested in seawater. In fact, this excretion of divalent ions is generally considered to be quantitatively more important physiologically than any renal excretion of other ions and water.

Finally, the urine of marine elasmobranchs is always acid with an unchanging pH of about 5.8. The total acidification takes place in the early proximal tubule by a process involving direct bicarbonate reabsorption as well as hydrogen ion secretion. Since the pH of the final urine does not change, this acidification in the early proximal tubule may be less significant for the regulation of acid excretion than for the maintenance of the solubility of the divalent ions secreted in the second segment of the proximal tubule.

8.2.1.3   Marine Teleosts

The stenohaline marine teleosts, which maintain the osmolality of their body fluids well below that of the surrounding seawater, ingest seawater and must excrete reabsorbed divalent ions, e.g., magnesium and sulfate, and monovalent ions, e.g., sodium and chloride. Their kidneys have relatively few small or no glomeruli and the nephrons lack distal tubules. The filtration rate for glomerular species is low and the urine flow rate for all species is low. Most of the filtered water is reabsorbed, apparently secondarily to reabsorption of sodium chloride. Magnesium, sulfate, calcium, and phosphate, for all of which the kidney is the primary excretory route, undergo net secretion apparently along the second segment of the proximal tubule. Net secretion of fluid, primarily dependent on secretion of sodium chloride by the same mechanism as in the elasmobranchs, also occurs along the second segment of the proximal tubule. This fluid secretion determines the actual urine flow rate in aglomerular and, probably, glomerular species. The final urine is close to isosmotic with the plasma, but the concentrations of sodium and chloride are usually low, and the urine may be slightly hypoosmotic to the plasma. The excess sodium chloride reabsorbed by the nephrons and the gastrointestinal tract is excreted by the gills.

8.2.1.4 Freshwater Teleosts

The true stenohaline freshwater teleosts, which maintain the osmolality of their body fluids well above that of the surrounding freshwater, must excrete excess water and retain ions. The kidneys of most species have larger and more numerous glomeruli than those of stenohaline marine teleosts, and the nephrons have well-developed distal tubules. Whole-kidney glomerular filtration rates are much higher than those of marine teleosts, and the urine is copious and dilute. There is no net tubular secretion of divalent ions. Most of the filtered sodium chloride is reabsorbed; the portion of this reabsorption along the distal tubules, which are always essentially impermeable to water, can produce a urine with an osmolality one-tenth that of the plasma. The primary diluting process takes place in the early distal tubule where sodium chloride reabsorption involves a Na-K-2Cl cotransporter in the luminal membrane and Na-K-ATPase in the basolateral membrane. The few aglomerular freshwater teleosts apparently produce urine by net tubular secretion of ions (most likely sodium and chloride) and water in one nephron segment and reabsorption of the ions without the water in a more distal nephron segment or bladder.

8.2.1.5 Euryhaline Teleosts

The euryhaline teleosts, which can maintain the osmolality of their body fluids below that of seawater when adapted to seawater and above that of freshwater when adapted to freshwater, must excrete excess ions and retain water in the former case and must excrete excess water and retain ions in the latter. Like the stenohaline freshwater teleosts, these animals have kidneys with numerous moderately large glomeruli and nephrons with distal tubules. In seawater-adapted animals, the number of glomeruli filtering is low and, thus, the whole-kidney glomerular filtration rate is low. The low number of filtering glomeruli may result from enhanced α-adrenergic nerve activity, enhanced release of arginine vasotocin, enhanced release of angiotensin II, suppressed release of prolactin, or some combination of these factors. With adaptation to seawater, net secretion of magnesium, sulfate, calcium, and phosphate occurs primarily in the second segment of the proximal tubule. Net secretion of sodium chloride and water by the same mechanism as in elasmobranchs and stenohaline marine teleosts also occurs in the second segment of euryhaline teleosts adapted to seawater. As in marine teleosts, this net fluid secretion may determine the actual rate of urine flow. Much of the filtered and secreted sodium chloride is apparently reabsorbed, but the control of the balance between secretion and reabsorption, as in other marine fishes, is unknown. The reabsorption can result in some dilution of the tubular fluid. However, this dilution is slight because the nephrons are quite permeable to water. The mechanism controlling this permeability is unknown. The final urine in these seawater-adapted animals tends to be close to isosmotic with the plasma and contains significant concentrations of divalent ions as well as sodium and chloride.

In freshwater-adapted animals, the number of glomeruli filtering is high, and, consequently, whole-kidney glomerular filtration is high. The large number of filtering glomeruli may result from depressed α-adrenergic nerve activity, suppressed release of arginine vasotocin, suppressed release of angiotensin II, enhanced release of prolactin, or some combination of these factors. Net secretion of divalent ions ceases. Net secretion of sodium chloride and water in the second segment of the proximal tubule may continue, but it apparently has little influence on the final rate of urine flow as long as the glomerular filtration rate is high. The epithelium of the distal tubules becomes impermeable to water, but reabsorption of sodium chloride continues and a dilute urine is produced, thus enabling the animals to excrete excess water. Dilution in the early distal tubule apparently involves sodium chloride reabsorption by the same process, using a Na-K-2Cl cotransporter in the luminal membrane and Na-K-ATPase in the basolateral membrane, as in stenohaline freshwater teleosts. The mechanism regulating the permeability to water of the epithelium of the distal tubules is unknown.

As these renal changes occur with adaptation to different aqueous environments, simultaneous appropriate changes in ion and water movements across the gills and bladder occur. In fact, the adaptation of all the fishes to their particular environment involves the coordinated regulation of ion and water transport through the kidneys, gills, integument, bladder, gastrointestinal tract, and specialized ion-transporting structures, such as the elasmobranch rectal gland.

8.2.2   Amphibians

Amphibians live and indeed thrive in habitats ranging from completely aqueous—usually freshwater, but occasionally seawater—to arid terrestrial. Their kidneys contain nephrons with large glomeruli and all the standard segments. However, the nephrons do not contain loops of Henle and are not arranged in a manner to permit them to produce a urine hyperosmotic to the plasma.

For the wholly aquatic species, the number of glomeruli filtering and, consequently, the whole-kidney glomerular filtration rate decrease with adaptation to seawater and increase with adaptation to freshwater. For semiaquatic and terrestrial species, they decrease with dehydration and increase with hydration. The number of glomeruli filtering is controlled primarily by arginine vasotocin and secondarily by α-adrenergic nerves. The number filtering decreases with increased levels of arginine vasotocin and α-adrenergic activity and increases with decreased levels of the hormone and neural activity. The rates of filtration by individual glomeruli may also be controlled by a distal tubule-glomerular feedback mechanism in each nephron. The single nephron glomerular filtration rates of well-hydrated animals or animals in freshwater are, like those of the hagfishes, quite high for the observed net filtration pressure. This may reflect the large area available for filtration and may be important for excretion of excess water.

Although the structure and arrangement of amphibian nephrons do not permit them to produce a urine hyperosmotic to the plasma, they are capable of producing urine varying from one-tenth the osmolality of the plasma during hydration, when excretion of excess water is required, to nearly isosmotic with the plasma during dehydration, when retention of water is required. Luminal fluid hypoosmotic to the plasma is always generated in the early portion of the distal tubule where sodium chloride is reabsorbed without water. This reabsorptive process involves the same process, using a Na-K-2Cl cotransporter in the luminal membrane and Na-K-ATPase in the basolateral membrane, as in the early distal tubules of freshwater and euryhaline teleosts. During dehydration, arginine vasotocin apparently acts at a tubule site distal to this diluting area to increase the permeability of the epithelium to water and permit the fluid in the tubules to equilibrate with the surrounding interstitium, thus forming a final urine isosmotic with the plasma. The decrease in the number of glomeruli filtering during dehydration, in addition to conserving water by reducing the quantity of initial filtrate, also reduces the volume flow rate through the collecting ducts, and this may contribute to the equilibration process.

Although the bulk of the filtered sodium is reabsorbed by the renal tubules of amphibians, some 10–15 % is not, the remainder apparently being reabsorbed in the bladder or cloaca depending on the requirements of the animals. Of the sodium reabsorbed by the renal tubules, some one-half to two-thirds is reabsorbed in the distal tubules and collecting ducts. Sodium reabsorption, at least in the diluting segment and probably in other regions of the distal tubule, is stimulated by aldosterone, which may also play an important role in regulating sodium reabsorption by the bladder or cloaca. Sodium reabsorption, most likely in the distal portions of the nephrons, also may be stimulated by arginine vasotocin and angiotensin II. Exactly, how these endocrine controls are coordinated in maintenance of sodium balance in amphibians is unknown.

Much of the filtered potassium is reabsorbed, but net secretion can occur throughout the distal tubule, including the diluting segment, and possibly along the collecting duct. This net secretion is apparently under the control of aldosterone.

Amphibians, like birds and mammals, can significantly acidify the fluid flowing along their renal tubules, but all such acidification occurs in the distal tubules. This acidification process, to a large extent, may involve a sodium-hydrogen countertransport mechanism for hydrogen ion secretion by these tubules, controlled in part by aldosterone. In addition, acidification, in contrast to that in mammals, may involve reabsorption of a large fraction of the filtered bicarbonate in the distal portions of the nephron.

For most amphibians, urea is the major excretory end product of nitrogen metabolism and may even be secreted by a primary or secondary active transport process in the renal tubules of some anuran species. However, for some xerophilic anuran amphibians, urates form the major excretory end products and undergo net secretion by the renal tubules. Urates have low aqueous solubilities and can be excreted with very little water. Moreover, urate precipitates also may contain large quantities of inorganic cations that can therefore be excreted without contributing to the osmolality of the urine. Finally, as in the case of fishes, maintenance of fluid, electrolyte, and acid–base balance in amphibians adapted to various environments requires the coordinated regulation of renal function with ion and water movements across extrarenal structures—skin and colon, cloaca, or bladder, depending on the species.

8.2.3 Reptiles

Habitats of reptiles, like those of amphibians, range from completely aqueous, both freshwater and seawater, to arid terrestrial. Their nephrons contain glomeruli of modest size and the standard segments. As in amphibians, these nephrons lack loops of Henle, although the ciliated intermediate segment may vary greatly in length among species, and they are not arranged in a manner to permit them to produce a urine hyperosmotic to the plasma.

In general, as in amphibians, the number of glomeruli filtering and therefore the whole-kidney glomerular filtration rate tend to increase with hydration and decrease with dehydration. However, the changes do not always appear to be as rapid or as striking as in amphibians, possibly simply because of differences in the changes in hydration under which the measurements were made. The number of glomeruli filtering is apparently controlled primarily by arginine vasotocin but possibly also by prolactin; the number filtering decreases with stimulated release of arginine vasotocin and suppressed release of prolactin and increases with suppressed release of arginine vasotocin and stimulated release of prolactin.

Although the structure and arrangement of the nephrons in the reptilian kidney do not permit production of a urine hyperosmotic to the plasma, nephrons in many species are capable of producing a urine varying from one-tenth the osmolality of the plasma to isosmotic with it, depending on the degree of hydration of the animal. In some species, however, the nephrons produce a ureteral urine that is always isosmotic with the plasma. Apparently, when additional regulation of urine osmo-lality is necessary in these species, it takes place distal to the kidney—in the colon, cloaca, or bladder.

In those species capable of producing a urine of low osmolality, the initial dilution apparently occurs in the early distal tubule or, possibly, in the thin intermediate segment between the proximal and distal tubules. In some species, this diluting process may continue throughout the distal tubules and collecting ducts. The site of the initial dilution and the mechanism of sodium chloride transport involved in that dilution are not as clearly defined as in fishes, amphibians, birds, and mammals.

In those species capable of varying urine osmolality, the urine diluted in one portion of a nephron apparently can equilibrate to varying degrees with the interstitium surrounding a more distal portion of that nephron or the collecting ducts. The extent of the equilibration may be determined in part by the permeability of the distal tubule or the collecting ducts to water and in part by the volume flow through the collecting ducts. Permeability is determined by arginine vasotocin and volume flow and by the number of filtering nephrons. The latter is also controlled primarily by arginine vasotocin. Therefore, arginine vasotocin, either directly through its action on epithelial permeability or indirectly through its regulation of the number of filtering nephrons and thus the volume flow through the collecting ducts, regulates the urine osmolality.

In reptiles, as in amphibians, although most of the filtered sodium is reabsorbed by the renal tubules, some 5–10 % is not. The remainder apparently is reabsorbed by structures distal to the kidneys—colon, cloaca, or bladder—depending on the requirements of the animal. Again, as in the case of the amphibians, only about one-third to one-half the sodium reabsorbed by the nephrons is reabsorbed by the proximal tubule. The rest is reabsorbed by more distal nephron segments. Although there may be some regulation of distal reabsorption by aldosterone and possibly by arginine vasotocin, hormonal control of reabsorption in reptiles is not at all clear. Little is known about sodium transport processes in reptilian renal tubules, but there appears to be intrinsic cellular regulation of late distal sodium reabsorption in some species. This regulation appears poised to prevent additional reabsorption from relatively high luminal concentrations, thus permitting excretion of excess sodium when required, and poised to support continued reabsorption from very low concentrations, thereby permitting further dilution of already dilute urine and excretion of a water load.

Either net secretion or net reabsorption of potassium by the tubules of living reptiles can be observed. Net reabsorption occurs along the proximal tubules and overall net secretion, when it occurs, results from secretion along the distal tubules. However, the transport mechanisms and the manner in which they are controlled in reptiles are not at all understood. The nephrons of reptiles may not be able to acidify tubular fluid as much as those of mammals, birds, and amphibians, but further acidification apparently can occur in the bladder or cloaca.

In many species, net tubular secretion of phosphate, which is often quite high in the diet, can occur. This process is dependent on and under the control of parathyroid hormone.

Urates form the major excretory end-products of nitrogen metabolism in all reptiles except some testudines from aquatic or mesic terrestrial habitats. In these latter species, urea is the major excretory end-product of nitrogen metabolism, but urates are still important. Although urates are filtered, their excretion depends primarily on net secretion by the proximal tubules. As noted above for uricotelic amphibians, the low solubility of urates permits them to be excreted with very little water. Also inorganic cations contained in urate precipitates can be excreted without contributing to the osmolality of the urine. However, many of these cations may be reabsorbed in regions distal to the kidney—colon, cloaca, or bladder—where hydrogen ions may be secreted and urate complexes may be converted to uric acid. Indeed, maintenance of fluid, electrolyte, and acid–base balance in reptiles, as in fishes and amphibians, requires the coordinated regulation of renal function and the movements of ions and water across the integument, across the colon, cloaca, or bladder, and in some species, through structures specialized for the hyperosmotic excretion of ions, e.g., salt glands.

8.2.4 Birds

Habitats of birds, like those of amphibians and reptiles, range from aqueous—both freshwater and seawater—to arid terrestrial, although for many birds flight determines the intensity of their exposure to a specific habitat. Avian kidneys contain a majority of nephrons resembling those of reptiles and most other nonmammalian vertebrates, i.e., reptilian-type (or loopless) nephrons, and a minority resembling those of mammals, i.e., mammalian-type (or looped) nephrons. The parallel arrangement of the loops of Henle of the mammalian-type nephrons, the vasa recta arising from the efferent glomerular arterioles of these nephrons, and the collecting ducts draining all nephrons apparently permits birds, like mammals, to produce a urine hyperosmotic to the plasma. Birds also can produce a urine hypoosmotic to the plasma, apparently by reabsorbing sodium chloride without accompanying water along the thick ascending limbs of Henle's loops of the mammalian-type nephrons and the early distal tubules of the reptilian-type nephrons. This sodium chloride reabsorptive process involves a Na-K-2Cl cotransporter in the luminal membrane and Na-K-ATPase in the basolateral membrane, as in the early distal diluting segment of teleost and amphibian nephrons, possibly in the diluting region of reptilian nephrons, and the thick ascending limbs of mammalian nephrons. Moreover, in birds the ascending limb and a short prebend region of the mammalian-type nephrons consists entirely of a thick segment so that not only dilution but also development of an osmolar gradient along the medullary cones and thus the concentration of the urine depend on sodium chloride reabsorption by this segment. Urea plays no role in the production of a urine hyperosmotic to the plasma in birds.

The osmolality of the avian urine varies from about one-tenth that of the plasma during maximum hydration to about two to three times that of the plasma during maximum tolerable dehydration. This variation in urine osmolality with hydration is apparently controlled by arginine vasotocin, which, as in amphibians and reptiles and apparently in some teleosts, regulates both the permeability of the collecting ducts to water and the number of filtering nephrons. Both a change in the permeability of the collecting ducts to water and a change in the number of filtering nephrons and, hence, the volume flow through the collecting ducts, can influence the equilibration of fluid in the collecting ducts with the surrounding interstitium and thus the osmolality of the final urine. It appears, however, that the permeability of the collecting ducts to water, although not as great as in amphibians and mammals, is more sensitive to arginine vasotocin than the number of filtering nephrons.

In birds, as in mammals, the bulk of the filtered sodium and chloride is reabsorbed along the proximal tubule. The reabsorption of sodium by the distal tubules may be stimulated by arginine vasotocin, aldosterone, and corticosterone, and inhibited by angiotensin II, but the direct effects of these hormones on the transport process and the integration of their effects in the adaptation of birds to their environment have yet to be evaluated. Although net potassium secretion may occur along the proximal tubules of some superficial reptilian-type nephrons, major excretion of potassium appears to result primarily from net secretion along the distal nephrons. This secretion may be stimulated by aldosterone and corticosterone, but, again, this control process has not been studied in detail.

Birds, like both amphibians and mammals, can produce a highly acidified ureteral urine in response to an acid load. However, like amphibians but unlike mammals, little acidification appears to occur along the proximal tubules. Apparently, both significant bicarbonate reabsorption and significant hydrogen ion secretion can occur in the distal portions of the nephrons when acid excretion is required. Neither the exact tubule sites nor the mechanisms of bicarbonate and hydrogen ion transport have yet to been studied.

Either net tubular reabsorption or net tubular secretion of phosphate occurs in birds, apparently depending on existing requirements. Both processes occur along the proximal tubules of the superficial reptilian-type nephrons, and net secretion requires the presence of parathyroid hormone.

Urates are the major excretory end-products of nitrogen metabolism in birds. Although urates are filtered, their excretion, as in reptiles, depends primarily on net secretion by the renal tubules. This certainly occurs along the proximal tubules of the superficial reptilian-type nephrons, but whether other tubule sites are involved is unknown. As in the case of uricotelic amphibians and reptiles, the combination of inorganic cations with urate precipitates permits the excretion of a larger quantity of these cations than could be accommodated independently by the maximum urine osmolality. Finally, some birds of marine or arid terrestrial habitats have specialized extrarenal structures, i.e., salt glands, for the hyperosmotic excretion of inorganic ions. The maintenance of fluid, electrolyte, and acid–base balance in birds in the process of adaptation to their environment, as in other nonmammalian vertebrates, requires the coordinated regulation of renal function, the transport of ions by these salt glands, and the transport of ions and water across the lower gastrointestinal tract, primarily colon and coprodeum.

8.2.5 Mammals

Mammals, like amphibians, reptiles, and birds, have habitats ranging from completely aqueous to arid terrestrial. Nephrons of mammalian kidneys contain rather large glomeruli with highly complex multibranched capillary tufts, proximal convoluted and straight tubules, loops of Henle with thin and thick limbs, distal convoluted tubules, connecting tubules, and collecting ducts. The loops of Henle, collecting ducts, and vasa recta are arranged in parallel, permitting production of a urine hyperosmotic to the plasma. Mammals also can produce a urine hypoosmotic to the plasma by reabsorbing sodium chloride without accompanying water primarily along the thick ascending limbs of Henle's loops. This reabsorptive process involves a Na-K-2Cl cotransporter in the luminal membrane and Na-K-ATPase in the basolateral membrane, as along the early distal diluting segment of teleost and amphibian nephrons, possibly along the diluting region of reptilian nephrons, and along the thick ascending limbs of Henle's loops of mammalian-type avian nephrons and the early distal tubules of reptilian-type avian nephrons. Sodium chloride reabsorption by the thick ascending limbs of Henle's loops in mammals is important not only for diluting the urine but also, as in birds, for generating the osmotic gradient necessary for production of a concentrated urine. However, in contrast to the avian kidney, the concentrating process also depends on the presence of urea for the formation of the medullary osmotic gradient. Moreover, the sodium chloride reabsorption in the thick ascending limb works directly to generate the gradient only in the outer medulla of the mammalian kidney. It contributes indirectly to the process which generates the much greater gradient in the inner medulla of the mammalian kidney, a process, certainly involving urea, that is not yet understood.

The osmolality of the mammalian urine varies from about two-tenths that of the plasma during maximum hydration to as much as 25 times that of the plasma in some desert species that have no access to free water. This variation in urine osmolality with hydration is apparently controlled primarily by arginine vasopressin, which regulates the permeability of the collecting ducts to water. In contrast to nonmammalian species, the number of filtering glomeruli does not normally vary; therefore, alterations in volume flow rate along the collecting ducts as a result of changes in glomerular filtration rate do not normally play a role in the concentrating or diluting processes.

The mammalian nephrons, like those of birds but unlike those of other nonmammalian vertebrates, are capable of reabsorbing almost all the filtered sodium and chloride. Mammals, unlike most nonmammalian vertebrates, do not have any extrarenal mechanisms for significant regulation of sodium chloride excretion. In mammals, as in birds but not as in other vertebrates, most of the filtered sodium and chloride is reabsorbed along the proximal tubules. Although the reabsorptive process for sodium chloride in the proximal tubules of mammals is quantitatively different from that in the proximal tubules of most nonmammalian vertebrates, some aspects of sodium reabsorption, i.e., transport across the luminal membrane coupled to organic solutes or by sodium-hydrogen exchange and transport across the basolateral membrane via Na-K-ATPase, are the same in mammals and amphibians. Distal sodium reabsorption, which is of critical importance in mammals in determining the final sodium excretion, is clearly stimulated by aldosterone, primarily in the cortical collecting ducts. This is in marked contrast to the limited or unclear effect of adrenocorticosteroids in most nonmammalian vertebrates. Arginine vasopressin apparently stimulates sodium reabsorption in the thick ascending limb of Henle's loop in some mammalian species.

As in most nonmammalian vertebrates, either net tubular reabsorption or net tubular secretion of potassium can occur in mammals. Net reabsorption always occurs along the proximal tubules and overall net secretion, when it occurs, results from net secretion along the distal nephrons, primarily the late distal tubule or connecting tubule and the cortical collecting duct. This net secretion appears to be under the control of aldosterone.

Mammals, like amphibians and birds, can produce a highly acidified ureteral urine in response to an acid load. However, in contrast to amphibians and birds, significant acidification occurs along the proximal tubules where most of the filtered bicarbonate is reabsorbed. This process apparently involves primarily the sodium-hydrogen exchanger and secondarily vacuolar type H-ATPase, both in the luminal membrane. Additional acidification, apparently involving active hydrogen ion secretion by both vacuolar H-ATPase and H-K-ATPase in the luminal membranes of intercalated cells, can occur in the connecting tubules and collecting ducts. Although the final remainder of the filtered bicarbonate can be reabsorbed along the distal tubules and cortical collecting ducts, net bicarbonate secretion also can occur along the cortical collecting ducts.

Urea is the major excretory end product of nitrogen metabolism in all mammals. Its excretion involves filtration and variable degrees of overall net tubular reabsorption (although secretion in the inner medullary collecting ducts can occur). As noted above, urea contributes to the medullary osmotic gradient involved in the mammalian urine concentrating process. Such a concentrating process is required for the excretion of urea with little water. Finally, it must be stressed that there are no extrarenal or postrenal routes contributing to the regulation of ion and fluid excretion in mammals adapted to different environments.

8.3 Summary Comparisons and Contrasts of Renal Function Between Groups of Vertebrates

8.3.1 Glomerular Filtration Rate: Stability Versus Lability

The importance of the ultrafiltration of the arterial plasma as the primary step in urine formation in almost all vertebrates has been discussed in Chap. 3. The differences in normal filtration rates among species and classes are directly dependent physically on the differences in pressures, plasma flows, surface areas, and hydraulic conductivities, but they also may relate to the balance between the conflicting requirements to eliminate water and metabolic wastes and to conserve water, salts, metabolites, and energy placed on any animal group by its habitat and mode of existence. Of particular interest in comparing vertebrate groups, however, is the short-term lability or stability of the glomerular filtration rate with environmental changes, particularly changes in hydration.

In general, the whole-kidney glomerular filtration rate of mammals is remarkably stable during rapid moderate changes in hydration. The stability during moderate dehydration probably reflects the ability of the renal tubules of most mammals to reabsorb almost all filtered water and to produce a urine far more concentrated than the plasma. There is no need to reduce filtration to conserve water in response to moderate dehydration that might readily be experienced by these animals. However, a few species from arid habitats do show a decrease in whole-kidney GFR with what appears to be a degree of dehydration that they could routinely experience. Whether this degree of dehydration, although "physiological" in the sense that it may be regularly experienced, is simply too great to be compensated by tubular reabsorption of water is not completely clear, but a number of these species do not have an exceptional ability to concentrate their urine.

Of course, all mammals will show a reduction in whole-kidney GFR with severe dehydration, but this is hardly physiological. Moreover, when changes in whole-kidney GFR do occur, whether physiological or not, they generally reflect changes in the filtration rates of individual nephrons with all nephrons filtering rather than changes in the number of nephrons filtering. Only in extreme circumstances— probably only very severe dehydration—does the number of nephrons filtering decrease.

In contrast to mammals, most nonmammalian vertebrates have highly labile whole-kidney glomerular filtration rates. These tend to increase with hydration or adaptation to freshwater and to decrease with dehydration or adaptation to seawater. Modest changes in hydration usually can produce highly significant changes in whole-kidney GFR. These changes in whole-kidney GFR are most striking and occur most readily in amphibians and reptiles, but significant changes occur even in those fishes that adapt easily to seawater or freshwater and in birds. The changes, particularly in fishes, amphibians, and reptiles, appear to reflect an inability to alter to as large an extent as mammals the amount of filtered water reabsorbed by the renal tubules. Decreases in whole-kidney GFR with dehydration certainly reflect the inability to produce a urine hyperosmotic to the plasma. Such changes in whole-kidney GFR with hydration may occur at the expense of regulating the excretion of certain ions and metabolic wastes, but this depends on the degree to which tubular and extrarenal transport processes can compensate for changes in filtration rate. The degree of coordination, if any, between rates of tubular and extrarenal transport and whole-kidney glomerular filtration rate is currently unknown and merits further study.

Changes in the whole-kidney GFR of nonmammalian vertebrates, also in contrast to mammals, reflect primarily changes in the number of nephrons filtering, although changes in the single nephron glomerular filtration rates also can occur.

Such alterations in the number of nephrons filtering are practical for most nonmammalian vertebrates because the nephrons do not function in concert to produce a urine hyperosmotic to the plasma. Also, almost all nonmammalian vertebrates showing glomerular intermittence have renal venous portal systems that supply oxygen and nutrients to the tubules in the absence of a post-glomerular arterial blood supply. Those few fishes that apparently show glomerular intermittence without a renal venous portal system may have some form of collateral circulation from post-glomerular arterioles of filtering nephrons. The one nonmammalian species that definitely does not show glomerular intermittence, the primitive lamprey, does not have a renal venous portal system.

Changes in the number of filtering nephrons, often under the control of arginine vasotocin, alter the volume flow through the collecting ducts. Whether or not permeability of these collecting ducts to water is increased by arginine vasotocin, the changes in volume flow rate may help to determine the degree to which tubular fluid diluted in an earlier portion of the filtering nephrons, usually the early distal tubules, equilibrates with the interstitium surrounding the collecting ducts.

Birds, with their reptilian-type and mammalian-type nephrons, lie somewhere between the mammals and the other nonmammalian vertebrates. Although capable of producing a urine with an osmolality varying from about one-tenth to two to three times the osmolality of the plasma, birds still show changes in whole-kidney GFR with hydration. These changes, controlled mainly by arginine vasotocin, apparently involve primarily changes in the number of filtering reptilian-type nephrons, which do not function in concert and have a renal venous portal blood supply. As in other nonmammalian vertebrates, such changes alter the volume flow through the collecting ducts in the medullary cones and could influence equilibration with the surrounding interstitium and thereby the dilution and concentration of the urine. However, it appears that tubular permeability to water is more sensitive to arginine vasotocin and the degree of dehydration than the number of functioning nephrons. Thus, during changes in hydration, the whole-kidney GFR of birds appears to be less stable than that of mammals but less labile than that of other nonmammalian vertebrates.

8.3.2   Diluting and Concentrating Processes

Representatives of all tetrapod vertebrate classes, as well as many fishes, are capable of producing a ureteral urine of lower osmolality than their plasma. Some fishes—primarily stenohaline marine teleosts—as well as a number of reptiles are not capable of producing a dilute urine. Those animals capable of producing the most dilute urine are those that are subject to the greatest water load. Many of those that cannot produce a dilute ureteral urine are never exposed to a water load, but others, particularly some reptiles that produce a urine of fixed osmolality, are exposed to varying amounts of water in their environment. Some of these animals may be capable of diluting the urine in regions distal to the kidney.

All animals capable of producing a dilute ureteral urine do so by reabsorbing filtered solute, primarily sodium and chloride, without accompanying water somewhere along the nephrons. The site of such reabsorption appears to be essentially the same in all vertebrates, i.e., apparently the early distal segment of teleost, amphibian, and reptilian-type avian nephrons and the thick ascending limb of Henle's loop of mammalian-type avian and mammalian nephrons. Although the site has not been determined for reptilian nephrons, it probably involves the early distal tubule or, possibly, the thin intermediate segment. The mechanism in all cases in which the site is known apparently involves secondarily active, sodium-coupled chloride reabsorption, the energy for which is derived from primary active sodium transport out of the cells at the basolateral membrane via Na-K-ATPase. The coupled, electrically neutral transport step into the cells across the luminal membrane involves a Na-K-2Cl transporter (probably homologs of mammalian NKCC2 in nonmammalian vertebrates).

Changing from a maximally dilute urine to an isosmotic urine conserves large amounts of water, just as changing from an isosmotic urine to a maximally dilute urine permits excretion of large amounts of water. In animals that can produce a dilute urine, dilution always occurs along the tubules as described above. The degree to which the final ureteral urine approaches the osmolality of the plasma then depends on the degree to which the diluted tubule fluid equilibrates with the surrounding interstitium in portions of the nephron distal to the diluting area. The degree of equilibration is apparently controlled by antidiuretic hormone—arginine vasotocin in nonmammalian vertebrates; arginine or lysine vasopressin in mammals—which determines the permeability of the distal nephrons (apparently the collecting ducts) to water. However, the effect of antidiuretic hormone on the permeability of the nephrons to water, although strongly suggested by indirect studies in many vertebrates, has only been demonstrated directly in birds and mammals. In some nonmammalian vertebrates, the hormone clearly acts on structures distal to the kidney, e.g., the urinary bladder of anuran amphibians, to produce the same effect and to alter the osmolality of the urine before it is finally excreted.

As noted above, changes in glomerular filtration rate, usually in the number of nephrons filtering in nonmammalian vertebrates, also controlled by arginine vasotocin, affect the volume flow rate through the collecting ducts. How important changes in volume flow rate are in the equilibration process relative to changes in epithelial water permeability is unknown. Therefore, how important the effect of antidiuretic hormone on the number of filtering nephrons is in the equilibration process relative to its effect on epithelial water permeability is also unknown. These relative effects await direct evaluation in nonmammalian vertebrates.

The production of a urine hyperosmotic to the plasma can conserve additional water, although not as much as changing from a maximally dilute urine to one isosmotic with the plasma. In mammals, the excretion of urea, the major excretory end-product of nitrogen metabolism, with very little water requires the production of a concentrated urine. As noted above, urea also plays a role in the generation of the medullary osmotic gradient required by the mammalian concentrating process.

In birds, urate, the major excretory end-product of nitrogen metabolism, has a very low aqueous solubility. Therefore, its excretion with very little water does not require production of a concentrated urine. Nevertheless, in birds a process remarkably similar to that in mammals has evolved for the production of a moderately concentrated urine. This process does not involve urea in birds. However, in both birds and mammals, as noted earlier, the primary driving force for the production of the medullary osmotic gradient is the sodium chloride reabsorption in the thick ascending limbs of Henle's loops that is also responsible for diluting the urine. Again, in birds, in contrast to mammals, the volume flow rate through the collecting ducts may play some role in equilibration of the fluid in the collecting ducts with the surrounding medullary interstitium.

The urine concentrating ability of birds appears modest compared with that of mammals, but this may be misleading. Direct comparisons of the maximum osmolal U/P ratios of birds and mammals are not justified without considering the fact that the plasma osmolality of birds tends to increase more than that of mammals with dehydration. In addition, inorganic cations included with urate precipitates in avian ureteral urine are excreted without contributing to the osmotic pressure of the urine. Finally, ureteral urine only moderately more concentrated than the plasma may permit efficient osmotically linked water reabsorption in the avian colon and coprodeum.

8.3.3   Sites and Mechanisms of Tubular Transport

Although data on sites and mechanisms of renal tubular transport among the vertebrates are still fragmentary, detailed information being available only for mammals and a few nonmammalian species, some general patterns of importance are emerging. As discussed above, the site and mechanism for dilution of the tubule fluid appear to be the same in all vertebrates—mammals and nonmammals—in which dilution occurs. In contrast to this uniformity among the vertebrates, the quantitative importance of the tubule sites for the reabsorption of filtered sodium and chloride varies. First, reabsorption of filtered sodium and chloride along the renal tubules approaches completion only in mammals and birds. Additional reabsorption in many nonmammalian vertebrates may occur in regions distal to the kidney—colon, cloaca, or bladder. To a significant extent, this is true even for birds. Second, the proximal tubule is the major site for reabsorption of filtered sodium and chloride only in mammals and birds. The distal tubules and collecting ducts assume greater quantitative importance than the proximal tubules in most other nonmammalian vertebrates in which this process has been studied. There is as yet no evidence of an ultrastructural or subcellular biochemical or molecular basis for these quantitative differences.

A somewhat similar pattern in the quantitative importance of distal versus proximal transport sites exists for acid secretion and bicarbonate reabsorption. In mammals, significant acidification of the tubule fluid and the bulk of the bicarbonate reabsorption occur along the proximal tubule. This process involves primarily a sodium-hydrogen countertransport system (NHE3) at the luminal membrane and, therefore, is intimately linked to the sodium reabsorption in this region. It also involves the action of carbonic anhydrase. In contrast, in amphibians, birds, and probably those reptiles that can produce an acidic ureteral urine, no acidification occurs along the proximal tubules. Although some bicarbonate reabsorption must occur in this portion of the tubules, its rate is apparently proportional to the rate of fluid reabsorption. Studies on amphibians suggest that, as in mammals, the process involves sodium-hydrogen exchange and may involve the action of carbonic anhydrase. However, the amount of carbonic anhydrase in the proximal tubule cells appears small and may reflect the limited quantitative importance of hydrogen ion secretion and bicarbonate reabsorption in this tubule region. Instead, the major acidification and the bulk of the bicarbonate reabsorption in amphibians, birds, and some reptiles appear to occur in the distal portions of the nephrons. Of course, significant further acidification and a quantitatively small fraction of filtered bicarbonate reabsorption occur in the distal regions of mammalian nephrons.

Studies on amphibians suggest that the distal acidification and bicarbonate reabsorption may occur in the early distal tubules (the diluting region) or the late distal tubules, the exact site perhaps differing in anurans and urodeles. The data indicate that the acidification and bicarbonate reabsorption processes involve a sodium-hydrogen countertransporter and the action of carbonic anhydrase. Histochemical data on amphibians suggest that the largest concentration of carbonic anhydrase occurs in the distal, rather than the proximal nephrons, again perhaps reflecting the quantitative importance of acid secretion and bicarbonate reabsorption in the distal nephrons.

In chelonian reptiles, important acidification occurs even further distally, in the bladder, and apparently involves a primary electrogenic hydrogen-ion secretory system. This is probably closely related to the vacuolar H-ATPase in intercalated cells of mammalian connecting tubules and cortical collecting ducts. Significant bicarbonate reabsorption is not involved in this region.

Of course, as noted above, data on nonmammalian vertebrates are still fragmentary, and distinct exceptions to these patterns occur in some species. For example, maximum acidification and complete reabsorption of filtered bicarbonate occur along the proximal tubules of elasmobranchs, but there is no involvement of carbonic anhydrase. In alligators, net tubular secretion of bicarbonate, apparently involving the action of carbonic anhydrase, always occurs.

A few distinctive patterns of transport—the net secretion of magnesium and sulfate by marine teleost proximal renal tubules; the net secretion of sodium chloride and water by teleost and elasmobranch proximal tubules; the net secretion or net reabsorption of taurine by proximal tubules of marine teleosts, marine elasmobranchs, and reptiles; and the net secretion of potassium under some circumstances by the distal nephrons of apparently all vertebrates—are being exploited as model systems to determine the mechanisms and controls involved. Studies of the marked net reabsorption of urea by the nephrons of some anuran amphibians have not yet revealed how the apparent active transport of this small polar molecule can occur, but this distinctive process deserves and probably will attract further study.

As is apparent from these summaries of major patterns as well as from the detailed coverage in the preceding chapters, a great deal remains to be learned about renal function in the nonmammalian vertebrates. Future studies may fill these gaps, revealing the adaptive patterns for individual groups and the basic mechanisms common to all groups.

Index

A

Acid–base balance, 197
Acidic amino acids, 186
Acidification, 280
  distal tubules, 121
  proximal tubules, 121
Active reabsorption of sodium, 92
Active secretion of urea, 195
Active urea secretion, 194
Afferent arteriole, 21
  constriction, 56
Aldosterone, 271
Amino acids
  net tubular reabsorption, 184
  net tubular secretion, 184
Ammonia
  produced by the renal tubule cells, 196
  secreted into the tubule lumen, 196
Ammoniagenesis
  enzymes, 198
  steps and control, 198
Ammonia production and secretion
  aquatic and semiaquatic chelonians, 197
  crocodilians, 197
  increase with an acid load, 197
  major end product of nitrogen metabolism, 197
Amphibians, 14–15, 268–270
Angiotensin II, 63, 64
  SNGFR, 64
  whole-kidney GFR, 64
Anion exchange
  brush-border membrane, 211
  peritubular membrane, 213
Antidiuretic hormone, 246

arginine or lysine vasopressin in mammals, 278
hormone—arginine vasotocin in nonmammalian vertebrates, 278
Aquaporin 2 (AQP2) water channels, 254, 256
  luminal membrane, 254
Area available for filtration, A, 37
Arginine vasopressin (AVP), 246
  variation in urine osmolality, 274
Arginine vasotocin (AVT), 246, 270–272
  birds, 62, 256
  constriction of the afferent glomerular arterioles, 61
  number of filtering nephrons, 59
  osmotic water permeability, 256
  permeability to water, 253
  physiological regulator, 59
  regulating the glomerular filtration rate, 62
  SNGFR, 59
  volume flow through the collecting ducts, 256
Atrial natriuretic peptide (ANP), 111
Avian kidney, 16
  medullary cone, 19
Avian renal portal system, 134
Avian renal tubules, 174, 204
  high capacity for reabsorbing glucose, 174

B

Backflux of urate from lumen to peritubular fluid
  avian proximal tubules, 217
  paracellular route, 217
Basement membrane, 24, 26

Printed in the United States
By Bookmasters